"This magnificent book is a must for all mathematics teachers! Its practical value is derived from the fact that the authors are both seasoned and high-quality mathematics teachers who have experienced and have explored every aspect of teaching they discuss in their book. It is comprehensive and challenges the reader to consider the pros and cons of the different strategies described, which go beyond a 'toolbox' of great ideas for teaching!"

Alice F. Artzt
Queens College of the City University of New York

"This resource is a must for all mathematics teachers! New and veteran teachers will find the practical strategies and explicit examples easy to implement in the classroom and helpful in enhancing one's own pedagogy. Authors Bobson Wong and Larisa Bukalov have crafted a fantastic student-focused resource dedicated to ensuring high-quality instruction. Highly recommended!"

Christine DeBono
K-5 Math Instructional Coach, Higley Unified School District, Arizona

"This comprehensive book is an incredible resource for math teachers at any stage of their career. Master Teachers Bobson Wong and Larisa Bukalov do an excellent job describing practical strategies, justifying them with research, and bringing them to life with concrete examples. I highly recommend it."

Michael Driskill
Chief Operating Officer, Math for America

"As someone who has been teaching for 30 years, I find *The Math Teacher's Toolbox* to be a very rich reference of teaching strategies and resources for practitioners, especially mentors and mentees. I plan to use this book in my lessons and my professional development."

Irene Espiritu
Middle School Teacher, Math for America Master Teacher,
New York State Master Teacher

"*The Math Teacher's Toolbox* provides the reader with a summation of research-supported current best practices in mathematics teaching. The layout of this book masterfully helps move the reader from understanding through application of the central ideas most essential to teaching mathematics effectively. Practical ideas for the classroom, as well as discussion about what could possibly go wrong, combine to make this a useful guide for teachers of all experience levels."

Tabetha Finchum
2014 Presidential Awardee for Excellence in Mathematics Teaching

"*The Math Teacher's Toolbox* provides concrete, innovative strategies for adapting often intangible pedagogical theories across a wide range of math content areas and grade levels. Larisa Bukalov and Bobson Wong draw extensively from current research as well as their own years of classroom experience to explore the benefits and possible limitations of each strategy. Having already sought to implement a number of their ideas into my own classroom, I cannot recommend this book enough."

Nasriah Morrison
Math Teacher, Institute for Collaborative Education and Math for America Master Teacher

"This book is truly a 'toolbox' for math instruction. It offers great technology tools and resources for teachers and their students, free online resources for student learning, and practical ideas that every math teacher can use. I will use this book for years to come."

Jendayi Nunn
Mathematics Virtual Instructional Specialist, Atlanta Public Schools

"*The Math Teacher Toolbox* provides a map that can guide new teachers as they begin their journey and help veterans navigate the shifting terrain. The authors summarize current research from many areas of teaching and connect it to structured classroom practices. The experienced writing team organizes complex parts of the profession into a structure that makes it easy for practitioners to put the ideas to use in their classroom."

Carl Oliver
Assistant Principal, City-As-School, New York

"Both new and experienced teachers will have cause to reach into this box of tools and return time and again to dig deeper—and each time you return, you'll find the box just as organized as the last! You'll keep this book nearby throughout your career for its practical, detailed tips, copious references, and teacher-to-teacher tone."

Ralph Pantozzi, Ed.D.
2014 MoMath Rosenthal Prize winner, 2017 Presidential Awardee for Excellence in Mathematics Teaching

The Math Teacher's Toolbox

A winning educational formula of engaging lessons and powerful strategies for math teachers in numerous classroom settings

The *Teacher's Toolbox* series is an innovative, research-based resource providing teachers with instructional strategies for students of all levels and abilities. Each book in the collection focuses on a specific content area. Clear, concise guidance enables teachers to quickly integrate low-prep, high-value lessons and strategies in their middle school and high school classrooms. Every strategy follows a practical, how-to format established by the series editors.

The Math Teacher's Toolbox is a classroom-tested resource offering hundreds of accessible, student-friendly lessons and strategies that can be implemented in a variety of educational settings. Concise chapters fully explain the research basis, necessary technology, standards correlation, and implementation of each lesson and strategy.

Favoring a hands-on approach, this book provides step-by-step instructions that help teachers to apply their new skills and knowledge in their classrooms immediately. Lessons cover topics such as setting up games, conducting group work, using graphs, incorporating technology, assessing student learning, teaching all-ability students, and much more. This book enables math teachers to:

- Understand how each strategy works in the classroom and avoid common mistakes
- Promote culturally responsive classrooms
- Activate and enhance prior knowledge
- Bring fresh and engaging activities into the classroom

Written by respected authors and educators, *The Math Teacher's Toolbox: Hundreds of Practical Ideas to Support Your Students* is an invaluable aid for upper elementary, middle school, and high school math educators as well as those in teacher education programs and staff development professionals.

Books in the *Teacher's Toolbox* series, published by Jossey-Bass:

The ELL Teacher's Toolbox, by Larry Ferlazzo and Katie Hull Sypnieski

The Math Teacher's Toolbox, by Bobson Wong, Larisa Bukalov, Larry Ferlazzo, and Katie Hull Sypnieski

The Science Teacher's Toolbox, by Tara C. Dale, Mandi S. White, Larry Ferlazzo, and Katie Hull Sypnieski

The Social Studies Teacher's Toolbox, by Elisabeth Johnson, Evelyn Ramos LaMarr, Larry Ferlazzo, and Katie Hull Sypnieski

The Math Teacher's Toolbox

Hundreds of Practical Ideas to Support Your Students

BOBSON WONG
LARISA BUKALOV
LARRY FERLAZZO
KATIE HULL SYPNIESKI

The Teacher's Toolbox Series

Copyright © 2020 by John Wiley & Sons, Inc. All rights reserved.

Published by Jossey-Bass
A Wiley Brand
111 River Street, Hoboken NJ 07030—www.josseybass.com

No part of this publication may be reproduced, stored in a retrieval system, or transmitted in any form or by any means, electronic, mechanical, photocopying, recording, scanning, or otherwise, except as permitted under Section 107 or 108 of the 1976 United States Copyright Act, without either the prior written permission of the publisher, or authorization through payment of the appropriate per-copy fee to the Copyright Clearance Center, Inc., 222 Rosewood Drive, Danvers, MA 01923, 978-750-8400, fax 978-646-8600, or on the Web at www.copyright.com. Requests to the publisher for permission should be addressed to the Permissions Department, John Wiley & Sons, Inc., 111 River Street, Hoboken, NJ 07030, 201-748-6011, fax 201-748-6008, or online at www.wiley.com/go/permissions.

Limit of Liability/Disclaimer of Warranty: While the publisher and author have used their best efforts in preparing this book, they make no representations or warranties with respect to the accuracy or completeness of the contents of this book and specifically disclaim any implied warranties of merchantability or fitness for a particular purpose. No warranty may be created or extended by sales representatives or written sales materials. The advice and strategies contained herein may not be suitable for your situation. You should consult with a professional where appropriate. Neither the publisher nor author shall be liable for any loss of profit or any other commercial damages, including but not limited to special, incidental, consequential, or other damages. Readers should be aware that websites offered as citations and/or sources for further information may have changed or disappeared between the time this was written and when it is read.

Jossey-Bass books and products are available through most bookstores. To contact Jossey-Bass directly call our Customer Care Department within the US at 800-956-7739, outside the U.S. at 317-572-3986, or fax 317-572-4002.

Wiley also publishes its books in a variety of electronic formats and by print-on-demand. Some material included with standard print versions of this book may not be included in e-books or in print-on-demand. For more information about Wiley products, visit www.wiley.com.

Library of Congress Cataloging-in-Publication Data:

Names: Wong, Bobson, 1971- author. | Bukalov, Larisa, 1973- author.
Title: The math teacher's toolbox : hundreds of practical ideas to support your students / Bobson Wong, Larisa Bukalov.
Description: Hoboken, NJ: Jossey-Bass, [2020] | Series: Teacher's toolbox | Includes bibliographical references and index.
Identifiers: LCCN 2019051973 (print) | LCCN 2019051974 (ebook) | ISBN 9781119573296 (paperback) | ISBN 9781119573203 (adobe pdf) | ISBN 9781119573241 (epub)
Subjects: LCSH: Mathematics—Study and teaching (Middle school)—Handbooks, manuals, etc. | Mathematics—Study and teaching (Secondary)—Handbooks, manuals, etc.
Classification: LCC QA11.2 .W663 2020 (print) | LCC QA11.2 (ebook) | DDC 510.71/2—dc23
LC record available at https://lccn.loc.gov/2019051973
LC ebook record available at https://lccn.loc.gov/2019051974

Cover Design: Wiley
Cover Image: © malerapaso/Getty Images

Printed in the United States of America

FIRST EDITION

PB Printing   SKY10090436_110824

Contents

List of Tables xix
About the Authors xxi
About the Editors xxiii
Acknowledgments xxv
Letter from the Editors xxvii

Introduction 1
    Our Beliefs about Teaching Math 2
    Structure of This Book 3
    Why Good Math Teaching Matters 4

I   Basic Strategies   5

1. Motivating Students 7
    What Is It? 7
    Why We Like It 8
    Supporting Research 8
    Common Core Connections 9
    Application 10
        Nurturing Student Confidence 10
        Motivating Through Math 11
        Rewards 14
        Motivating Through Popular Culture 15
        Motivating English Language Learners and Students with Learning Differences .. 16
    Student Handouts and Examples 18

CONTENTS

- What Could Go Wrong 18
  - *Using Fear to Motivate* 18
  - *Stereotype Threat* 19
  - *"Why Do We Need to Know This?"* 19
  - *Misreading Students* 20
  - *Limitations to Motivation* 21
- Technology Connections 21
- Figures 22
  - Figure 1.1   Pattern Blocks 22
  - Figure 1.2   Rotational Symmetry 23
  - Figure 1.3   Exponential Growth 24
  - Figure 1.4   Identify a Void 26

2. Culturally Responsive Teaching 27
- What Is It? 27
- Why We Like It 28
- Supporting Research 28
- Common Core Connections 29
- Application 30
  - *Self-Reflection* 30
  - *Building a Collaborative Learning Partnership* 32
- What Could Go Wrong 36
  - *"Color-Blind" Teaching* 36
  - *Good Intentions* 37
  - *Finding the Right Time or Place* 38
- Technology Connections 38

3. Teaching Math as a Language 41
- What Is it? 41
- Why We Like It 41
- Supporting Research 42
- Common Core Connections 42
- Application 42
  - *Eliciting the Need for Mathematical Language* 42
  - *Introducing Symbols and Terms* 43
  - *Translating Between Symbols and Words* 45
  - *Making Connections Between Math and English* 46
  - *Examples of Confusing Mathematical Language* 46
  - *Encouraging Mathematical Precision* 48
  - *Vocabulary Charts and Flash Cards* 49
  - *Visual and Verbal Aids* 51
  - *Word Walls and Anchor Charts* 52

Student Handouts and Examples 53
What Could Go Wrong 53
    Not Treating Math as a Language 53
    Math as a "Bag of Tricks" 54
Technology Connections 55
Figures 57
    Figure 3.1   Concept Attainment 57
    Figure 3.2   Words and Symbols Chart 58
    Figure 3.3   Why the Word "Height" Is Confusing 58
    Figure 3.4   Draw a Picture 59
    Figure 3.5   Functions Anchor Chart 60
    Figure 3.6   Polynomials Anchor Chart 61
    Figure 3.7   Why the Formula $a^2 + b^2 = c^2$ Is Confusing 61

4. Promoting Mathematical Communication 63
What Is It? 63
Why We Like It 63
Supporting Research 64
Common Core Connections 64
Application 64
    Open-Ended Questions 64
    Guiding Students in Conversation 71
    Four-Step Thinking Process 74
    Mathematical Writing 79
    Differentiating for ELLs and Students with Learning Differences 87
What Could Go Wrong 87
    Dealing with Student Mistakes 87
    Dealing with Teacher Mistakes 88
    Problems in Discourse 88
    Finding the Time 89
Student Handouts and Examples 89
Technology Connections 89
Attribution 90
Figures 91
    Figure 4.1   Algebra Tiles Activity 91
    Figure 4.2   Which One Doesn't Belong? 92
    Figure 4.3   Error Analysis 93
    Figure 4.4   Lesson Summary 95

5. Making Mathematical Connections 97
What Is It? 97
Why We Like It 97

Supporting Research 98
Common Core Connections 98
Application 98
    Equivalence 99
    Proportionality 101
    Functions 102
    Variability 104
    Differentiating for ELLs and Students with Learning Differences 107
Student Handouts and Examples 108
What Could Go Wrong 108
Technology Connections 109
Figures 111
    Figure 5.1   Addition and Subtraction of Polynomials 111
    Figure 5.2   Multiplication with the Area Model 112
    Figure 5.3   Division with the Area Model 114
    Figure 5.4   Completing the Square 115
    Figure 5.5   Determining the Center and Radius of a Circle 115
    Figure 5.6   Why $(a + b)^2 \neq a^2 + b^2$ 115
    Figure 5.7   Ratios and Similarity 116
    Figure 5.8   Areas of Similar Polygons 117
    Figure 5.9   Volumes of Similar Solids 118
    Figure 5.10   Arc Length and Sector 119
    Figure 5.11   Proportional Reasoning in Circles 120
    Figure 5.12   Four Views of a Function 120
    Figure 5.13   Rate of Change 121
    Figure 5.14   Characteristics of Polynomial Functions 123
    Figure 5.15   Even and Odd Polynomial Functions 124
    Figure 5.16   Why $f(x) = \sin(x)$ Is Odd and $g(x) = \cos(x)$ Is Even 126
    Figure 5.17   Linear Regression 127
    Figure 5.18   Long-Run Relative Frequency 129
    Figure 5.19   Two-Way Tables 131
    Figure 5.20   Conditional Probability 133

II How to Plan   135

6. How to Plan Units 137
What Is It? 137
Why We Like It 137
Supporting Research 138
Common Core Connections 138
Application 139
    Getting Started 139

  Making Connections Between Big Ideas 139
  Developing a Logical Sequence 140
  Organizing Topics and Problems 141
  Summarizing the Unit Plan 141
  Being Flexible 141
  Developing Students' Social and Emotional Learning 141
  Incorporating Students' Cultures 142
  Differentiating for ELLs and Students with Learning Differences 143
 Student Handouts and Examples 143
 What Could Go Wrong 143
 Technology Connections 145
 Figures 145
  Figure 6.1 Unit Plan: List of Skills 146
  Figure 6.2 Unit Plan: Concept Map 147
  Figure 6.3 Unit Plan: Sequence of Lessons 148
  Figure 6.4 Sample Unit Plan 149

7. How to Plan Lessons 151
 What Is It? 151
 Why We Like It 151
 Supporting Research 152
 Common Core Connections 152
 Application 152
  Defining the Lesson's Scope 152
  Introductory Activity 153
  Presenting New Material Through Guided Questions 154
  Practice 155
  Differentiating for ELLs and Students with Learning Differences 155
  Summary Activity 156
 Student Handouts and Examples 157
 What Could Go Wrong 157
 Technology Connections 159
 Figures 162
  Figure 7.1 Do Now Problem 162
  Figure 7.2 Lesson Plan: Standard Deviation 162
  Figure 7.3 Lesson Plan: Slope-Intercept Form 166
  Figure 7.4 Revised Baseball Field Word Problem 168

8. How to Plan Homework 169
 What Is It? 169
 Why We Like It 169
 Supporting Research 169

Common Core Connections 170
Application 170
    Sources 171
    Homework Format 171
    Homework as Practice 172
    Homework as Discovery 173
    Homework as Transfer 173
    Discussing Homework 174
    Collecting Homework 175
    Grading Homework 176
    Differentiating for ELLs and Students with Learning Differences 177
Student Handouts and Examples 178
What Could Go Wrong 178
    Students Who Don't Do Homework 178
    Mismanaging Class Time 179
    Homework Review Challenges 179
    Choosing the Wrong Problems 180
Technology Connections 180
Figures 183
    Figure 8.1   Homework as Practice 183
    Figure 8.2   Homework as Discovery—Ratios 184
    Figure 8.3   Homework as Discovery—Mean Proportional Theorem 185
    Figure 8.4   Homework as Discovery—Parabolas 186
    Figure 8.5   Homework as Transfer—Similarity 187
    Figure 8.6   Homework as Transfer—Bank Accounts 188

9. How to Plan Tests and Quizzes 189
What Is It? 189
Why We Like It 189
Supporting Research 190
Common Core Connections 190
Application 190
    Types of Questions 190
    Test Format 193
    Quiz Format 196
    Reviewing for Assessments 196
    Creating Scoring Guidelines for Assessments 199
    Grading Assessments 202
    Analyzing Test Results 203
    Returning Tests 204
    Differentiating for ELLs and Students with Learning Differences 207

  Alternate Forms of Assessment 208
Student Handouts and Examples 208
What Could Go Wrong 208
  Poor Scheduling and Preparation 209
  Assessments as Classroom Management 210
  Poorly Chosen Questions 210
  Mistakes on Assessments 211
  Student Cheating 212
  Different Versions of Tests 213
  Grading and Returning Assessments 214
  Test Retakes and Test Corrections 215
Technology Connections 215
  Test Questions, Answers, and Scoring Guidelines 215
  Test Review 216
  Test Analysis 216
Figures 217
  Figure 9.1 Algebra I Test 217
  Figure 9.2 Precalculus Test 220
  Figure 9.3 Quiz 224
  Figure 9.4 Creating Scoring Guidelines 225
  Figure 9.5 Blank Test Corrections Sheet 226
  Figure 9.6 Completed Test Corrections Sheet 228
  Figure 9.7 Test Reflection Form 229

10. How to Develop an Effective Grading Policy 231
What Is It? 231
Why We Like It 232
Supporting Research 232
Common Core Connections 232
Application 232
  Standards-Based Grading 232
  Minimum Grading Policy 234
  Point Accumulation System for Grading 236
  Differentiating for ELLs and Students with Learning Differences 237
  More Than Just a Grade 238
What Could Go Wrong 239
Student Handouts and Examples 240
Technology Connections 240
Figures 241
  Figure 10.1 Grade Calculation Sheet 241
  Figure 10.2 Completed Grade Calculation Sheet 242

III Building Relationships — 243

11. Building a Productive Classroom Environment — 245
- What Is It? — 245
- Why We Like It — 245
- Supporting Research — 245
- Common Core Connections — 246
- Application — 246
  - *Making a Good First Impression* — 246
  - *Learning Names* — 248
  - *Getting to Know Students* — 248
  - *Classroom Organization* — 249
  - *Classroom Rules and Routines* — 250
  - *Course Descriptions* — 252
  - *Soliciting Student Opinion* — 253
  - *Taking Notes* — 254
- What Could Go Wrong — 257
  - *Classroom Tone* — 257
  - *Mishandling the Teacher–Student Relationship* — 258
  - *Taking Notes* — 259
- Student Handouts and Examples — 259
- Technology Connections — 259
  - *Classroom Environment* — 259
  - *Student Surveys* — 260
  - *Note-Taking* — 260
- Figures — 261
  - Figure 11.1   Student Information Sheet — 261
  - Figure 11.2   Course Description — 263
  - Figure 11.3   Brief Handout — 265
  - Figure 11.4   Full-Page Handout — 266
  - Figure 11.5   Annotated Work — 268
  - Figure 11.6   Double-Entry Journal — 269

12. Building Relationships with Parents — 271
- What Is It? — 271
- Why We Like It — 271
- Supporting Research — 272
- Common Core Connections — 272
- Application — 272
  - *Communicating with Parents* — 272

  Addressing Parents' Math Anxiety 273
  Parent–Teacher Conferences 277
  Home Visits 277
  Working with Parents of Culturally Diverse Students 278
  Working with Parents of Students with Learning Differences 279
 What Could Go Wrong 280
 Student Handouts and Examples 281
 Technology Connections 281
 Figures 282
  Figure 12.1 Parent Communication Script 282
  Figure 12.2 Parent Communication Log 283

13. Collaborating with Other Teachers 285
 What Is It? 285
 Why We Like It 285
 Supporting Research 286
 Common Core Connections 286
 Application 286
  Discussing Values 287
  Planning with Other Math Teachers 288
  Interdisciplinary Collaboration 288
  Observing Other Teachers 289
  Co-Teaching 291
  Mentoring 294
  Lesson Study 294
  Professional Learning Community 295
 What Could Go Wrong 297
  Lack of Trust 297
  Reinforcing Negative Stereotypes 297
  Lack of Colleagues 297
  Lack of Time 298
 Technology Connections 298

IV Enhancing Lessons 301

14. Differentiating Instruction 303
 What Is It? 303
 Why We Like It 303
 Supporting Research 304
 Common Core Connections 305

CONTENTS

   Application 305
      Differentiation by Content 305
      Differentiation by Process 313
      Differentiation by Product 315
      Differentiation by Affect 320
   What Could Go Wrong 320
   Student Handouts and Examples 321
   Technology Connections 321
   Figures 323
      Figure 14.1   Tiered Lesson—Literal Equations 323
      Figure 14.2   Tiered Lesson—Midpoint 325
      Figure 14.3   Curriculum Compacting—Coordinate Geometry 328
      Figure 14.4   Tiered Test Questions 331
      Figure 14.5   Review Sheet 331
      Figure 14.6   Fill-In Review Sheet 332
      Figure 14.7   Review Booklet 333

15. Differentiating for Students with Unique Needs 335
   What Is It? 335
   Why We Like It 336
   Supporting Research 336
   Common Core Connections 337
   Application 337
      Strengths and Challenges of Students with Unique Needs 337
      Techniques to Support Students with Unique Needs 340
   What Could Go Wrong 348
   Student Handouts and Examples 349
   Technology Connections 349
   Figures 351
      Figure 15.1   Frayer Model (Blank) 351
      Figure 15.2   Frayer Model—Perpendicular Bisector 352
      Figure 15.3   Concept Map 352

16. Project-Based Learning 353
   What Is It? 353
   Why We Like It 353
   Supporting Research 354
   Common Core Connections 355
   Application 355
      Open-Ended Classwork Problems 355
      Open-Ended Homework Problems 357
      Projects 358

What Could Go Wrong 367
Student Handouts and Examples 368
Technology Connections 368
Figures 369
    Figure 16.1   Discovering Pi 369
    Figure 16.2   Area of a Circle 370
    Figure 16.3   Point Lattice Assignment 371
    Figure 16.4   Paint a Room 374
    Figure 16.5   Project—Bus Redesign Plan 375

17. Cooperative Learning **379**
What Is It? 379
Why We Like It 380
Supporting Research 380
Common Core Connections 381
Application 381
    General Techniques 381
    Differentiating for Students with Unique Needs 384
    Examples 387
What Could Go Wrong 398
Student Handouts and Examples 399
Technology Connections 400
Figures 401
    Figure 17.1   Jigsaw as Practice 401
    Figure 17.2   Jigsaw as Discovery 402
    Figure 17.3   Factoring Station 403
    Figure 17.4   Peer Editing 404

18. Formative Assessment **405**
What Is It? 405
Why We Like It 405
Supporting Research 406
Common Core Connections 406
Application 406
    Asking the Right Questions 407
    Eliciting Student Responses 409
    Responding to Student Answers 412
    Other Methods of Formative Assessment 412
    Differentiating Formative Assessment 413
What Could Go Wrong 414
Technology Connections 415

19. Using Technology 417
- What Is It? 417
- Why We Like It 417
- Supporting Research 418
- Common Core Connections 418
- Application 418
  - *Classroom Organization* 418
  - *Mathematical Content* 422
  - *Using Technology for Culturally Responsive Teaching* ... 425
  - *Using Technology to Differentiate Instruction* 425
- What Could Go Wrong 425
- Student Handouts and Examples 427
- Technology Connections 428
- Figures 429
  - Figure 19.1 Simulation of 1,000 Coin Flips 429
  - Figure 19.2 Transformations of Functions 429
  - Figure 19.3 Centroid of a Triangle 431
  - Figure 19.4 Two Views of a Graph Using Technology 432

20. Ending the School Year 433
- What Is It? 433
- Why We Like It 433
- Supporting Research 433
- Common Core Connections 434
- Application 434
  - *Review* 434
  - *Reflection* 438
  - *Recognition* 439
  - *Maintaining Relationships with Students* 440
  - *Differentiating Year-End Activities* 440
- What Could Go Wrong 441
  - *Year-End Fatigue* 441
  - *"What Can I Do to Pass?"* 441
  - *Running Out of Time* 442
- Technology Connections 443

Appendix A: *The Math Teacher's Toolbox* Technology Links 445

References 461

Index 515

List of Tables

Table 3.1	Vocabulary Chart	49
Table 3.2	Geometry Vocabulary Chart	50
Table 3.3	Visual and Verbal Aids	52
Table 4.1	Class Discussion	67
Table 4.2	Low-Floor, High-Ceiling Problems	69
Table 4.3	Problem-Solving Chart	75
Table 4.4	Scaffolded Lesson Summary	83
Table 4.5	Scoring Guidelines for Lesson Summaries	85
Table 4.6	Scoring Guidelines for Quick Writes	86
Table 5.1	Table for Vehicle Word Problem	105
Table 7.1	Summary Questions	157
Table 9.1	Confidence Level Scoring Guidelines	192
Table 9.2	Comparing Scoring Guidelines	201
Table 9.3	Two Test Questions with Unequal Difficulty	213
Table 9.4	Two Test Questions with Similar Difficulty	214
Table 10.1	Standards-Based Grading	233
Table 10.2	Sample Report Card Grades for a Student with 20% Content Mastery	236
Table 12.1	Parent Communication Outline	274
Table 14.1	Levels of Complexity for Tiered Lessons	306
Table 14.2	Ratio Word Problem Table	314
Table 14.3	Rubric	319
Table 15.1	US and Latin American Prime Factorization Methods	338

Table 16.1	Project Ideas	361
Table 16.2	Basic Project Rubric	365
Table 16.3	Oral Presentation Rubric	366
Table 17.1	Self-Assessment Rubric	385
Table 17.2	Notice and Wonder	389
Table 17.3	Task Cards	397
Table 18.1	Formative Assessment Questions	408
Table 20.1	Using Technology for Multiple-Choice Questions	436

About the Authors

Bobson Wong has taught math at New York City public high schools since 2005. He is a three-time recipient of the Math for America Master Teacher Fellowship, a New York State Master Teacher, and a 2014–2015 recipient of the New York Educator Voice Fellowship. He is a member of the Advisory Council of the National Museum of Mathematics.

He has also worked to improve the quality of high school mathematics standards and assessment in New York. He has served on several committees, including the state's Common Core standards review committee, the state's workgroup to reexamine teacher evaluations, and the United Federation of Teachers' Common Core Standards Task Force. As an educational specialist for the New York State Education Department, he writes and edits questions for high school math Regents exams.

He graduated from the Bronx High School of Science, Princeton University (B.A., history), the University of Wisconsin–Madison (M.A., history), and St. John's University (M.S.Ed., adolescent education, mathematics), where he received his teacher training through the New York City Teaching Fellows program.

He lives in New York City with his wife and children.

Larisa Bukalov has been teaching at Bayside High School since 1998. She has won several awards for excellence in classroom teaching. She is a four-time recipient of the Math for America Master Teacher fellowship, a 2009 recipient of Queens College's Mary Fellicetti Memorial Award for excellence in mentoring and supervising student teachers, and a 2017 recipient of Queens College's Excellence in Mathematics Award for promoting mathematics teaching as a profession. A fourth-generation math teacher, she simultaneously earned degrees from a specialized math high school in Ukraine and a distance learning high school at Moscow State University. After emigrating to the United States, she learned English while earning both her bachelor's degree in math and her master's degree in math education from Queens College, City University of New York.

Over the past 20 years at Bayside, Larisa has taught all levels of math from pre-algebra to calculus, coached the school's math team, and created a math research program in which students wrote papers for the Greater New York City Math Fair, City College Engineering Expo, and the Intel Science and Talent Search.

Larisa has extensive experience providing professional development to pre-service and in-service teachers. She has mentored 16 student teachers. From 2007 to 2009 she provided professional development to early career teachers and math supervisors in New York City on Geometry, Probability, and Problem Solving. As part of her work with Math for America, Larisa has run several professional development sessions for teachers.

She lives in New York City with her husband and children.

About the Editors

Larry Ferlazzo and Katie Hull Sypnieski wrote *The ELL Teacher's Toolbox* and conceived of a series replicating the format of their popular book. They identified authors of all the books in the series and worked closely with them during their writing and publication.

Larry Ferlazzo teaches English, Social Studies, and International Baccalaureate classes to English Language Learners and others at Luther Burbank High School in Sacramento, California.

He's written nine books: *The ELL Teacher's Toolbox* (with co-author Katie Hull Sypnieski); *Navigating the Common Core with English Language Learners* (with coauthor Katie Hull Sypnieski); *The ESL/ELL Teacher's Survival Guide* (with coauthor Katie Hull Sypnieski); *Building a Community of Self-Motivated Learners: Strategies to Help Students Thrive in School and Beyond*; *Classroom Management Q&As: Expert Strategies for Teaching*; *Self-Driven Learning: Teaching Strategies for Student Motivation*; *Helping Students Motivate Themselves: Practical Answers to Classroom Challenges*; *English Language Learners: Teaching Strategies That Work*; and *Building Parent Engagement in Schools* (with coauthor Lorie Hammond).

He has won several awards, including the Leadership for a Changing World Award from the Ford Foundation, and was the Grand Prize Winner of the International Reading Association Award for Technology and Reading.

He writes a popular education blog at http://larryferlazzo.edublogs.org/, a weekly teacher advice column for Education Week Teacher, and posts for *The New York Times* and *The Washington Post*.

He also hosts a weekly radio show on BAM! Education Radio.

He was a community organizer for 19 years prior to becoming a public school teacher.

Larry is married and has three children and two grandchildren.

A basketball team he played for came in last place every year from 2012 to 2017. He retired from league play after that year, and the team then played for the championship. These results might indicate that Larry made a wise career choice in not pursuing a basketball career.

Katie Hull Sypnieski has taught English Language Learners and others at the secondary level for over 20 years. She currently teaches middle school English Language Arts and Social Studies at Fern Bacon Middle School in Sacramento, California.

She leads professional development for educators as a teaching consultant with the Area 3 Writing Project at the University of California, Davis.

She is coauthor (with Larry Ferlazzo) of *The ESL/ELL Teacher's Survival Guide*, *Navigating the Common Core with English Language Learners*, and *The ELL Teacher's Toolbox*. She has written articles for the *Washington Post*, *ASCD Educational Leadership*, and *Edutopia*. She and Larry have developed two video series with Education Week on differentiation and student motivation.

Katie lives in Sacramento with her husband and their three children.

Acknowledgments

Teaching is a collaborative endeavor. We would never have accomplished everything that we've done – including writing this book – without the help of many individuals. We thank all of the people mentioned here. They've made us not just into better *teachers* but also into better *people*.

Several current and former administrators at Bayside High School have provided coaching and professional growth opportunities over the years: Michael Athy, Madeline Belfi-Galvin, Harris Sarney, Susan Sladowski, and Judith Tarlo. Our colleagues in Bayside's Math Department have been a constant source of camaraderie, laughter, and valuable (if sometimes heated) pedagogical discussions. The thousands of students that we've taught at Bayside inspire and challenge us, giving us something to look forward to every day we go to work.

Math for America has created an active, supportive community that trusts and celebrates educators' expertise. Many of our ideas were refined in Math for America's professional development sessions.

Larry Ferlazzo and Katie Hull Sypnieski have been meticulous editors whose frequent questions about our thinking have improved our writing and teaching. Their timely and thought-provoking edits have made this book much better than it would have been without them. David Powell has done an amazing job of formatting our manuscript. His attention to detail is amazing. Pete Gaughan at Wiley and Amy Fandrei at Jossey-Bass patiently guided us through the stressful process of producing this book.

Bayside High School students Safi Ansari, Stefany Flores, Emily Hermida, Hana Ho, Anna Ling, Anyu Loh, Camila Palmada, Hardeep Singh, Navneet Sohal, Jason Sun, Ariana Verbanac, Richard Xing, and Joy Zou contributed their work for this book. Susie Xu and Justin Zhuo helped us proofread the figures and tables in this book.

Finally, our spouses and children deserve special mention for tolerating our many early-morning and late-night conversations about this book and helping us keep things in perspective.

Bobson Wong: Larisa Bukalov has been a mentor, colleague, and friend for 13 years. Prof. Charles Cohen, Sherrill Mirsky, Dr. Carol Nash, Barbara Rockow, and Prof. Robert C.-H. Shell were some of the many educators that I've met over the years that taught me the patience and attention to detail that I needed to write this book. The people that I've met online at #MTBoS and #ITeachMath have influenced and encouraged me. Robert Lebowitz has been a source of mathematical and philosophical conversation for decades; without him, this book would not be possible.

Larisa Bukalov: Bobson Wong has been a colleague and a friend. Thank you for putting my ideas about teaching mathematics in writing and always motivating me to do more. Mary Chiesi hasn't just been my co-teacher but also a mentor and best friend. We developed and practiced many of the strategies in this book together. Dr. Alice Artzt and the faculty at the mathematics education program at Queens College provided countless hours of debates on constructivism, hands-on approach, teaching mathematics as a language, group work, as well as advice on this book's structure and organization. Twenty years after graduation, I know that I can always count on her help and advice. Dr. Nick Metas at Queens College was famous for his attention to detail. A history buff, he shared stories about mathematicians and helped introduce to me the idea of culturally relevant mathematics. My husband, Boris Bukalov, a math and science teacher, checked the math in the book. My late grandfather, Izaya Vayzman, taught me to love mathematics and teaching. Many of the ideas described in this book came from my watching his everyday interactions with students back in the Soviet Union.

Letter from the Editors

We don't teach math, and we don't know much math, either.

We do, however, know pedagogy.

And there's more great pedagogy in Bobson's and Larisa's book than you can shake a stick at.

In fact, there's so much exceptional teaching advice in this book that any teacher – no matter what subject he/she teaches – can learn a great deal of information from this book about effective instructional strategies that can be used in any classroom.

We sure did!

And, because we have so much confidence in Bobson and Larisa, we're sure all the math is great, too.

If you don't believe us, just check out all the math people who have said so many terrific things about *The Math Teacher's Toolbox* – their endorsements can be found in the front of the book.

We're thrilled and honored that Bobson and Larisa's book follows our *The ELL Teacher's Toolbox* in the *Toolbox* series.

It was a pleasure working with them during the 12 months they spent writing it, and we'd wager this won't be the last book you see written by them.

Larry Ferlazzo and Katie Hull Sypnieski
Editors of the *Toolbox* series

Introduction

When people find out what we do for a living, they often admit to us that they hate math or they're not good at it. After repeated negative experiences, many develop *math anxiety*—feelings of fear and tension when doing math (Namkung, Peng, & Lin, 2019, p. 482; Shields, 2007, p. 56). Math anxiety is not simply a set of emotions but a physiological response that affects heart rate and neural activity (Ramirez, Shaw, & Maloney, 2018, p. 145). It can be even more problematic when teachers or parents have it, since they can pass it on to students, which can negatively affect academic achievement (Beilock, Gunderson, Ramirez, & Levine, 2010, p. 1,862; Maloney, Ramirez, Gunderson, Levine, & Beilock, 2015, p. 1,485; Ramirez, Hooper, Kersting, Ferguson, & Yeager, 2018, p. 8).

Unfortunately, math anxiety is common among students, their parents, and teachers. We wrote this book not just to help people overcome math anxiety but also to help them appreciate and use math in the real world. The strategies described in this book reflect techniques and methods that we've used during our combined 35 years of teaching 26,000 lessons to over 5,000 students (including English Language Learners and students with learning differences) from around the world. Many of these strategies rely on *social-emotional learning* (SEL, sometimes called *social and emotional learning*), the process by which people develop the skills necessary to manage their emotions, show empathy for others, and maintain positive relationships with others (Collaborative for Academic, Social, and Emotional Learning [CASEL], n.d.). SEL is a critical part of effective teaching because students' mindsets can affect their cognitive processing. People who experience success in an activity may be motivated and able to learn, while those who experience failure may tend to withdraw, rendering even the most engaging and well-planned lesson useless (Sousa, 2017, p. 61).

Our Beliefs about Teaching Math

What we write in this book reflects four of our core beliefs about math, pedagogy, and students:

1. **Students need to feel safe before they can learn.** Research indicates that when the brain perceives a threat, it instinctively releases adrenaline, which inhibits cognitive functions and other activity viewed as unnecessary (Sousa, 2017, p. 50). Students need to feel safe before they can be receptive to learning. As a result, teaching strategies that make students feel positively about learning can improve student motivation. As we explain in Chapter 11: Building a Productive Classroom Environment, feeling good or safe doesn't guarantee that students will absorb new information, but it is a necessary condition for learning.

2. **Math should make sense to students.** We believe that math should be taught in a way that makes sense to students. In our opinion, part of the reason why math anxiety is so prevalent is that many see it as a collection of disconnected and confusing "tricks." By the time students graduate, they should have the confidence and ability to apply mathematical and critical-thinking skills to real-world situations (Berry & Larson, 2019, p. 40). As we say in Chapter 3: Teaching Math as a Language, using the language-acquisition techniques commonly associated with teaching English Language Learners can help make math more accessible to all students. We also believe that *constructivism*—the idea that students should actively create knowledge by experiencing it and reflecting on it—should be a central part of math instruction.

3. **All students need access to rigorous math.** We believe that *all* students should have access to rigorous math—mathematical learning that includes solving challenging problems and deeper thinking. It abandons outdated notions of the meaning of being good at math—today's powerful calculators have eliminated the need to equate speed and accuracy with mathematical mastery (Devlin, 2019, p. 10; Ruef, 2018). Rigorous math requires both procedural and conceptual understanding (Ben-Hur, 2006, pp. 7–8; Levin, 2018, p. 273; McCormick, 1997, p. 149; Rittle-Johnson, Schneider, & Star, 2015, p. 594). It helps students appreciate the beauty of mathematics and apply it to the world around them. Unfortunately, many barriers (such as low expectations and hidden biases among teachers) can restrict students'

ability to experience math in a positive way (Berry & Larson, 2019, p. 41). When trying to determine what type of work is appropriate for students, we keep two rules in mind:

- **What works for some students often works for others.** Specific strategies (such as using multiple representations or making mathematical connections) designed to help English Language Learners, students with learning differences, or advanced students can frequently benefit *all* students. We discuss this idea more in Chapter 5: Making Mathematical Connections and Chapter 15: Differentiating for Students with Unique Needs.

- **What works for some students often doesn't work for others.** We try to modify instruction to meet the diverse needs of all students (we discuss this more in Chapter 14: Differentiating Instruction). Periodically questioning our attitudes and constantly looking for ways to deepen students' understanding (which we explain more in Chapter 2: Culturally Responsive Teaching) can make us more effective teachers.

4. **Teachers don't have to do everything to succeed.** Nobody (including ourselves!) could possibly implement all of the strategies described here at all times. This book should not be used as a checklist of everything that teachers must do to be effective. Instead, we view it as a collection from which teachers can pick what works for their classrooms. We feel that in teaching, as in life, selecting a few things and doing them well is more productive and sustainable than trying to do everything at once.

Structure of This Book

This book is divided into four parts. Part I: Basic Strategies expands on what we believe to be the central ideas necessary to teach math effectively: motivating students, culturally responsive teaching, teaching math as a language, promoting mathematical communication, and making mathematical connections. Part II: How to Plan discusses strategies for units, lessons, homework, tests and quizzes, and grades. Part III: Building Relationships talks about how to build relationships with students, parents, and co-teachers. Finally, Part IV: Enhancing Lessons contains other important strategies—differentiation, project-based learning, cooperative learning, formative assessment, and technology.

This book is part of a series in *The ELL Teacher's Toolbox* (2018a) by Larry Ferlazzo and Katie Hull Sypnieski. All chapters in the books in this series have the following sections:

- **What Is It?:** brief description of the strategy
- **Why We Like It:** explanation of why we like the strategy

- **Supporting Research:** research that supports the strategy
- **Common Core Standards:** relevant Common Core content and mathematical practice standards
- **Application:** description of ways that the strategy can be implemented
- **Student Handouts and Examples:** list of reproducibles and other figures
- **What Could Go Wrong:** explanation of what could go wrong with each strategy and what can be done in these situations
- **Technology Connections:** links to relevant websites (also available online)
- **Figures:** reproducibles, which are also available online at www.wiley.com/go/mathteacherstoolbox

Why Good Math Teaching Matters

The bottom line is that in today's changing world, we need better thinkers and problem-solvers. We believe that as teachers, we have to do more than just convey mathematical ideas. We also need to be role models for self-confidence, self-reflection, critical thinking, and conceptual understanding. What we say and do affects not only the way in which our students learn math but also their beliefs. We hope that this book can inspire you to do more for your students, your communities, and yourselves.

PART 1

Basic Strategies

CHAPTER 1

Motivating Students

What Is It?

Motivation—*why* people do what they do—affects every aspect of schooling. Without motivation, student learning becomes difficult, if not impossible (Artzt, Armour-Thomas, & Curcio, 2008, p. 48). Motivated students tend to have better performance, higher self-esteem, and improved psychological well-being (Fong, Patall, Vasquez, & Stautberg, 2019, p. 123; Gottfried, Marcoulides, Gottfried, & Oliver, 2013, p. 83; Liu & Hou, 2017, p. 49; Reeve, Deci, & Ryan, 2004, p. 22). Conversely, unmotivated students can become disengaged from academics and, in the worst cases, drop out of school (National Research Council, 2004, p. 24).

According to *self-determination theory*, a theory of motivation developed by researchers Edward L. Deci and Richard M. Ryan, motivation can be *intrinsic* (doing something because it is inherently satisfying) or *extrinsic* (doing something because it leads to some other result) (Ryan & Deci, 2000, p. 55). Many times, motivation is difficult to characterize as purely intrinsic or extrinsic. A student may be drawn by an extrinsic reward but may eventually internalize the values and adapt a more intrinsic motivation (Usher & Kober, 2012b, p. 3).

In addition, motivation is not a fixed quantity (Ryan & Deci, 2000, p. 54). Factors like schools, parents, communities, teachers, and life experiences can positively or negatively affect motivation (Usher & Kober, 2012a, p. 7). Students' motivation can vary from class to class—a student who is highly motivated in one class may be completely disengaged in another (National Research Council, 2004, p. 33).

As a result, educators often need to foster both intrinsic and extrinsic motivation. Students who are intrinsically interested in a topic are more likely to seek challenging tasks, think more creatively, and learn at a conceptual level (National Research Council, 2004, p. 38). However, since many academic tasks may not be inherently interesting, teachers also need to learn how to promote different methods of extrinsic motivation (Ryan & Deci, 2000, p. 55).

To sustain motivation, educators often seek ways to encourage students to internalize values. When students do so, they become more persistent and have a more positive sense of themselves (Ryan & Deci, 2000, pp. 60–61).

Why We Like It

In our experience, keeping motivational strategies in mind can enhance student engagement, academic achievement, and confidence to do math. Boosting their confidence is particularly important since many of our students experience math anxiety (we discuss it more in the Introduction), which can hinder their academic growth.

Supporting Research

Many studies on motivation focus on ways to build inclusive communities that promote learning for *all* students (Kumar, Zusho, & Bondie, 2018, p. 78). Proponents of self-determination theory argue that people are motivated to complete a task if doing so fulfills basic psychological needs, such as autonomy, relatedness, and competence (Ryan & Deci, 2000, p. 64).

However, some researchers have begun to challenge the idea of a universal theory of motivation, arguing that most of the existing work ignores the experiences and members of historically marginalized groups, such as people of color (Usher, 2018, p. 132). These researchers seek a more culturally responsive framework in which motivation is viewed not just as an individual characteristic but as the product of the social and historical context that shapes students' emotions and beliefs (King & McInerney, 2016, p. 2).

Other studies have focused on the effect of emotions on student motivation (Hannula, 2019, p. 310). Students who feel more anxious about math often have decreased motivation and do more poorly in school (Gunderson, Park, Maloney, Beilock, & Levine, 2017, pp. 34–35; Mo, 2019, p. 2; Passolunghi, Cargnelutti, & Pellizzoni, 2018, p. 282). Discouragement from parents, inappropriate or overly difficult work, and lack of support from teachers can further erode students' *self-efficacy*—the realistic expectation that making a good effort will lead to success (Usher, 2009, p. 308). In other words, social-emotional learning is tied to motivation.

Research indicates that as students move through the K–12 school system, their attitudes toward math become less positive (Batchelor, Torbeyns, & Verschaffel, 2019, p. 204; Gottfried et al., 2013, p. 70). As a result, keeping middle and high school students motivated in math class can be particularly challenging.

Despite the different approaches and areas of emphasis in the literature, researchers agree on several ways to improve student motivation:

- Teachers should have meaningful and challenging instruction (Kumar et al., 2018, p. 90).
- Teachers should empathize with students and accept them unconditionally (Wormeli, 2014).
- Administrators should ensure that school personnel are culturally diverse (Usher, 2018, p. 140).

In the Application section of this chapter, we discuss some strategies for improving motivation for all students.

Common Core Connections

Many of the motivational techniques that we describe in this chapter are related to Common Core standards. For example:

- Showing the usefulness of a topic relates to standards such as modeling with mathematics (Mathematical Practice [MP].4), interpreting division of fractions by fractions (6-NS.A.1), using a linear equation to model bivariate data (8-SP.A.3), and interpreting equations as viable in a modeling context (A-CED.A.3) (National Governors Association & Council of Chief State School Officers [NGA & CCSSO], 2010, pp. 7, 42, 56, 65).
- Finding a pattern connects to such standards as describing patterns in bivariate data (8-SP.A.1), writing expressions in equivalent forms (A-SSE.B.3), analyzing functions using different representations (F-IF.C.7), and verifying the properties of dilations (G-SRT.A.1) (NGA & CCSSO, 2010, pp. 56, 64, 69, 77).
- Promoting student autonomy relates to making sense of problems and persevering in solving them (MP.1) and using appropriate tools strategically (MP.5) (NGA & CCSSO, 2010, p. 10). Problem solving and decision making are important elements of self-determination (Heroux, Peters, & Randel, 2014, p. 200).
- Using technology is encouraged in such standards as drawing geometric shapes (7.G.A.2), interpreting scientific notation (8.EE.A.4), finding the solution to the equation $f(x) = g(x)$ (A-REI.D.11), showing key features of the graphs of functions (F-IF.C.7), and describing transformations of functions (F-BF.B.3) (NGA & CCSSO, 2010, pp. 50, 54, 66, 69, 70).

Application

Researchers seeking to merge self-determination theory with culturally responsive teaching (which we describe in more detail in Chapter 2: Culturally Responsive Teaching) have identified five characteristics of effective motivation:

1. **Culture:** Students' learning is affected by the culture that surrounds them. Understanding it can shed light on how inequitable aspects of the dominant culture can negatively affect student learning (Usher, 2018, p. 139).

2. **Meaningfulness:** Students are most likely to see learning as valuable when teachers connect lessons to their lives in meaningful ways (Kumar et al., 2018, p. 83).

3. **Competence:** Competence includes both *cultural competence* (understanding the cultural identities of oneself and others) and *academic competence* (the belief that one can complete a task) (Kumar et al., 2018, p. 83; Usher & Kober, 2012b, p. 2).

4. **Autonomy:** Teaching students how to set goals and make decisions can foster students' sense of *autonomy* (the extent to which they believe they can control their goals and actions) and lead to both individual growth and societal change (Kumar et al., 2018, p. 87).

5. **Relatedness:** When educators develop authentic relationships with students (by learning about their home lives, culture, and values) and use that knowledge to communicate effectively with them, they are more likely to succeed academically (Bonner, 2014, p. 397).

Here are some strategies that apply these characteristics in supporting student motivation.

NURTURING STUDENT CONFIDENCE

As we said in the Supporting Research section in this chapter, building students' self-efficacy and autonomy can help alleviate their math anxiety and improve their motivation, which can in turn improve their academic performance.

One way that we nurture students' self-confidence is to use language that supports their choice whenever possible. Saying, "I recommend that you rephrase this definition in your own words," can often be more effective than simply commanding students to write it down. Explaining why completing a task is necessary can help students understand how they can benefit from doing so (Reeve & Halusic, 2009, p. 150).

Having private conversations with students about areas of concern can also help them feel more in control of their learning ("I've noticed that your work has slipped. Is everything OK? What can we do to improve it?"). Acknowledging and addressing their concerns can often turn complaints into more positive discussions ("Yes, I agree that this wording sounds confusing, but that's what they use on the state tests! Let's make it clearer.") (Reeve & Halusic, 2009, p. 151).

In fact, building meaningful relationships with students—learning about their interests and activities, demonstrating authentic care, and genuinely trying to connect with students of all backgrounds—is the cornerstone of culturally responsive teaching (Heroux et al., 2014, p. 198; Kumar et al., 2018, p. 89). We believe that doing so can help not just students of color but *all* students.

While the techniques we describe above are often successful, they have limitations since many students face constraints that limit their autonomy (Kumar et al., 2018, p. 87). In order for students to become more self-determined, they need more opportunities to set goals for themselves and regulate their learning (Heroux et al., 2014, p. 200). Some students may not complete online assignments because they lack Internet access at home, so simply telling them to "try harder" without figuring out what causes their behavior would accomplish little. Instead, they may need more time or an offline method to complete assignments. We discuss some helpful strategies in Chapter 11: Building a Productive Classroom Environment, Chapter 14: Differentiating Instruction, and Chapter 17: Cooperative Learning.

MOTIVATING THROUGH MATH

Many times, we use the math we teach as a motivational tool. These techniques work best when teachers use them to introduce the concept and elicit the lesson's goals (Posamentier, Smith, & Stepelman, 2010, p. 71).

When implementing these strategies, we also try to be sensitive to their emotional needs. We try to give them meaningful, scaffolded work that presents a moderate challenge. Giving students a reasonable chance to succeed can strengthen their self-efficacy and ease their math anxiety (Margolis, 2014).

Connecting to the Real World

Many times, we try to connect what students are learning to the real world. Practical applications should be brief and accessible enough to advance the lesson, not detract from it (Posamentier et al., 2010, p. 66). For example, students can analyze college loan payments when discussing compound interest or determine which mobile phone plan is most economical when learning about piecewise functions. Real-world applications often provide a strong motivation for mathematical learning (Walkington, Sherman, & Howell, 2014, p. 277).

Personalizing problems based on student interests or cultures can make mathematical tasks more relevant and improve their persistence and learning. Personalizing problems can connect what students already know to abstract mathematical concepts (a phenomenon known as *grounding*) and help them determine if their work is reasonable (Walkington et al., 2014, p. 275). Teachers can get information from students by asking them how they use numbers in their lives or what they are interested in. We talk about specific strategies for learning more about students in Chapter 11: Building a Productive Classroom Environment.

Here are some examples of problems that connect to students' lives and experiences:

- Showing how one unusually low score can negatively affect an average can lead to a meaningful discussion of the harmful effects of outliers and the value of using other statistical measures like the median.

- Students can use pattern blocks or other models of regular polygons to explore why honeycombs are shaped like hexagons. This can introduce a lesson on rotational symmetry, perimeter, and area. Students can relate this topic to chemistry by examining why hexagons appear in molecular structures.

- Precalculus students can discover the need for a polar coordinate system by examining the maps of ancient cities like Paris, Moscow, or Beijing. We first ask students to give directions in a familiar city or town with a rectangular street grid, such as Manhattan. Students are then asked to give directions in a city in which streets radiate outwards from a center. This lesson can lead to a discussion of how math is used in other fields like city planning and navigation.

Students can even create their word problems (we discuss this more in Chapter 4: Promoting Mathematical Communication).

Finding Patterns

Many students learn how to recognize patterns as early as kindergarten. This skill helps them make connections with more complicated ideas in middle and high school (Markworth, 2016, p. 23). Pattern recognition boosts students' feelings of competence since they usually need little prior knowledge (Smith, Hillen, & Catania, 2007, p. 39). Research indicates that the ability to process complex patterns is one of the brain's most important features since it underlies many other cognitive functions, including thought, imagination, invention, and reasoning (Mattson, 2014, p. 13). Pattern recognition is used to make many real-life decisions, including finding the fastest route home, predicting what people are thinking based on their body language, and estimating how much money to spend on a construction project (Barkman, 2018; Miemis, 2010). In our experience, pattern recognition can be incorporated into almost any lesson, often as an introductory exercise (which we discuss in Chapter 7: How to Plan Lessons).

Here are some examples in which we find patterns in lessons:

- Figure 1.1: Pattern Blocks shows an example using triangular numbers that can introduce sequences for middle school or Algebra I students.

- Figure 1.2: Rotational Symmetry has an activity that middle school or Geometry students can use to discover the formula for the minimum angle of rotation that maps a regular polygon back onto itself.

- Figure 1.3: Exponential Growth contains an activity in which students explore exponential growth by looking at the number of layers created when papers are folded in half.

Identifying a Void in Knowledge

Students often want to complete their understanding of a topic, so making them aware of a gap in that knowledge may motivate them to learn more (Posamentier et al., 2010, p. 62). We like this strategy because it allows us to make connections to knowledge that is familiar to students. This helps us promote retrieval practice, which we also discuss in Chapter 4: Promoting Mathematical Communication and Chapter 7: How to Plan Lessons.

Here is an example from Algebra I involving solving systems of linear equations algebraically:

1. Solve for x and y:
   $2x + 5y = 34$
   $-2x - 4y = -28$

2. Solve for x and y:
   $2x + y = 7$
   $3x - y = 3$

3. Solve for x and y:
   $2x + 5y = 34$
   $x + 2y = 14$

In the first two systems, students should be able to add the equations to eliminate one variable and solve for both. However, adding the equations in the third system will not eliminate the variable. Students will then realize that their prior knowledge (in this case, solving by adding equations) will not work for the third system. They could then be motivated to learn what they can do to solve the third system—in this case, multiplying the second equation by −2 so that they can eliminate a variable by adding the equations.

- Figure 1.4: Identify a Void shows an example from Geometry. The first two examples can be solved using the Pythagorean theorem since two side lengths from right triangles are given and students must find the third side. However, the third example can't be solved using the Pythagorean theorem since only one side length is given. This prepares students for seeing the value of trigonometry, which deals with angle measures and the ratios of side lengths in right triangles.

REWARDS

The use of *rewards*—extrinsic incentives like prizes or points—to motivate students is controversial. Many researchers agree that rewards can work when individuals are not initially motivated. However, the effect of incentives on individuals who are already motivated is less clear (Hidi, 2015, p. 87). Some studies conclude that rewards generally undermine intrinsic motivation (Kohn, 1994; Deci, Ryan, & Koestner, 2001, p. 50). Others find that the detrimental effect of rewards on motivation is overstated and that rewards may sometimes have a positive effect (Cameron, Banko, & Pierce, 2001, p. 21; Eisenberger & Shanock, 2003, p. 128).

We believe that rewards, when used with the other strategies that we describe in this chapter, can help keep students motivated to learn. In reality, rewards are necessary for most tasks. After all, almost everyone works for *some* kind of compensation, such as money, awards, or high grades.

When possible, we make rewards meaningful and immediate. Incentives work best when students see an immediate benefit—offering extra computer time now is usually more effective than offering an end-of-year party. Minimizing the gap between the effort and the reward reduces *delay discounting*, in which people assign less value to future rewards (Cheng, 2016).

We use a variety of incentives. Rewards like extra credit, stars, or "good work" tickets are cheap, easy, and accessible to all students (we mention several apps that can keep track of points in the Technology Connections section of this chapter). Students could even be involved in distributing stars or "good work" tickets (Lewis, 2017). To encourage autonomy, teachers can allow students to convert such tickets into more tangible rewards, like extra credit (Chapter 10: How to Develop an Effective Grading Policy describes how we incorporate extra credit into our grades). Teachers may also consider awarding classroom privileges, such as extra computer time (we discuss rules and procedures more in Chapter 11: Building a Productive Classroom Environment).

We sometimes give candy or some other inexpensive prize, such as pens or erasers. However, we find that regularly offering these prizes can be problematic. They obviously require time and money, and they may put students in awkward situations. Some students may be allergic to candy but may not want to tell us publicly.

In our opinion, teachers should *not* use incentives like "no-homework" passes. We feel that such rewards send the message that homework is an unpleasant burden instead of a necessary part of learning (we talk more about the importance of homework in Chapter 8: How to Plan Homework).

Finally, we try to remember that many times, the best reward for a student can be as simple as a smile, high five, or word of encouragement (Wong & Wong, 2004, p. 162). We find that such emotional supports are often more effective or lasting than any incentives.

MOTIVATING THROUGH POPULAR CULTURE

We often use music, art, literature, movies, and other elements from popular culture to motivate students. When we find a pop culture reference to a specific topic that we teach, we can give students a short excerpt at the beginning of a lesson and ask students to discuss it. These excerpts can also be used as a summary at the end of a lesson, in which case students can answer the question that is posed. Teachers can customize pop culture references by finding sources that appeal to students' cultures and interests.

Here are some of our favorite examples of math in pop culture:

- The opening song "Seasons of Love" from the musical *Rent* has the lyrics "Five hundred twenty-five thousand six hundred minutes. How do you measure a year in the life?" Students can use proportional reasoning to calculate the number of hours, minutes, or seconds in a year.

- The satirical book *Flatland* by Edwin Abbott refers to a square living in two-dimensional space that is visited by a sphere. Geometry teachers can use this book or show an excerpt from the movie when talking about cross-sections of three-dimensional objects.

- In the movie *Stand and Deliver*, teacher Jaime Escalante explains to his primarily Hispanic class that the Mayans discovered the concept of zero long before the Greeks and Romans. Teachers can use this example to show that *all* students can do math because mathematicians come from all cultures.

- In the movie *Mean Girls*, Cady Heron and her classmates solve algebra word problems in a math competition and ultimately win the tie-breaker when she determines that $\lim_{x \to 0} \frac{\ln(1-x) - \sin(x)}{1 - \cos^2(x)}$ does not exist. Algebra or calculus teachers can play these excerpts when discussing systems of equations or limits.

- In the Edgar Allan Poe short story "The Pit and the Pendulum," the protagonist is strapped to a frame while a swinging pendulum slowly descends. Algebra students can calculate the time required for a pendulum to descend and determine if the story is realistic. Poe might not be considered "pop culture" by all—especially by students. However, there are appropriately "gruesome" scenes from B-grade movies available online that many would find engaging.
- If fashion dolls were real-life people, would their body measurements be reasonable? Geometry students can use measure dolls and scale the measurements using proportional reasoning.
- Students can examine artwork from various cultures, such as Islamic tiles or Native American quilts. Teachers can ask questions related to pattern recognition, fractions, symmetry, and geometric transformations. Students can even design their own quilts as a project (Paznokas, 2003, p. 255).

One challenge that we find when using pop culture references is that they often depict math in a negative way. The movie *A Beautiful Mind*, which tells the story of Princeton professor John Nash's struggles with schizophrenia, can reinforce stereotypes of mathematicians as being socially awkward or mentally ill. We suggest either incorporating difficult issues into student discussions or avoiding controversial examples entirely if you feel that you can't address them adequately in class. When used appropriately, pop culture references can make math more interesting and fun.

MOTIVATING ENGLISH LANGUAGE LEARNERS AND STUDENTS WITH LEARNING DIFFERENCES

Some motivational strategies, like the ones we discuss in this section that promote self-confidence in math, can work particularly well with both English Language Learners (ELLs) and with students who have learning differences. However, the challenges facing each group can be very different. Most ELLs are just as intellectually capable as most English-proficient students, but face language and cultural challenges. Many students with learning differences may be more English-proficient, but have specific needs related to the way they process new information and concepts. We believe that *always* keeping these distinctions in mind is critical to helping both groups succeed.

English Language Learners

We try to create an environment in which ELLs feel comfortable speaking and writing in another language while learning content. We allow students to ask or answer questions in their own language. In these cases, we ask a different student to translate or use a tool like Google Translate. We encourage ELLs to use glossaries or dictionaries to help them with unfamiliar words. We discuss other strategies to help students build their language skills in Chapter 3: Teaching Math as a Language and Chapter 4: Promoting Mathematical Communication.

In addition, we point out to the class that understanding other languages and cultures is important in today's diverse world. Sometimes we even ask ELLs to teach *us* words and phrases in their native languages. Making an honest effort to learn other languages can not only build a rapport with students (which we discuss more in Chapter 11: Building a Productive Classroom Environment) but can also help remind us of the challenges that they face when learning English.

Students with Learning Differences

We see students with learning differences as *capable* instead of disabled by identifying their strengths as well as ways that we can make them more successful. For example, we may give them specific tasks to perform in class. An autistic student in Larisa's class became the person that students asked for help. His ability to explain problems clearly to his classmates not only made him a valuable member of the class but also helped him improve his social skills.

To help students with learning differences visualize difficult mathematical concepts, we often use *manipulatives* (concrete objects like pattern blocks and algebra tiles that students can use to represent abstract mathematical concepts) and technology. These strategies can often benefit all students, not just those with learning differences. Research indicates that manipulatives and technology have a modest positive effect on student motivation (Bouck & Park, 2018, p. 97; Jensen, 2005; Preston et al., 2015, p. 180). Simply using these strategies does not guarantee that student achievement will improve. They should be supported by high-quality instruction with enough guidance for students (Carbonneau, Marley, & Selig, 2013, p. 396).

Although manipulatives are often associated with elementary school math, we find them especially useful for introducing topics or making connections with prior knowledge. Many of our students find manipulatives less intimidating than other strategies, which can reduce their math anxiety and makes them more receptive to learning. For example, operations with polynomials can be difficult to visualize, so we introduce the topic with algebra tiles, which many students use in elementary school. Algebra tiles help students transition to more abstract tasks, such as multiplying polynomials or completing the square. Chapter 5: Making Mathematical Connections contains other examples of how manipulatives can be used.

We use technology like calculators or online animations to illustrate complicated mathematical ideas. Students who have difficulty factoring polynomials can work backwards by using the calculator to find the zeros. We discuss other benefits, pitfalls, and examples of using technology in the classroom in Chapter 19: Using Technology.

Students can develop their sense of competence and autonomy by receiving work that is properly scaffolded to match their current level of readiness (Usher & Kober, 2012a, p. 4). We do this for all students, and keep it especially in mind for ELLs and students with learning differences. We try to give students work that "meets them where they are" and gradually increases the level of difficulty. For example, a Geometry lesson on applying properties of parallelograms can start with identifying highlighted parts of parallelograms and move to algebraic problems before culminating in formal proofs.

Chapter 14: Differentiating Instruction and Chapter 15: Differentiating for Students with Unique Needs contain other useful strategies.

Student Handouts and Examples

Figure 1.1: Pattern Blocks
Figure 1.2: Rotational Symmetry
Figure 1.3: Exponential Growth

What Could Go Wrong

When *we* try to motivate students, many things can go wrong.

USING FEAR TO MOTIVATE

Research indicates that threatening students generally has a negative effect on their motivation and academic growth (Von der Embse, Schultz, & Draughn, 2015, p. 630). Intimidating students often decreases their intrinsic motivation and causes a fear of failure (Putwain & Remedios, 2014, p. 512). That doesn't mean, however, that we can't communicate frustration or disappointment to students. If we have a good relationship with students, saying something like, "I'm feeling disappointed by what is happening now" can be an effective and appropriate motivational strategy.

In our experience, expressing motivation by explaining how the task will lead to a desired goal ("Doing homework will help you know what types of questions will be on the test") generally works better than browbeating students, especially by linking academic outcomes to unrelated consequences ("If you don't do well on this test, you won't be allowed to go to the prom").

STEREOTYPE THREAT

In a misguided attempt to boost students' confidence, teachers may connect their performance to a prejudice—for example, by saying that girls often struggle with math or praising a Chinese student by saying that Asian students typically excel. This phenomenon, called *stereotype threat*, is a fear of confirming a stereotype about one's gender, ethnicity, or other self-identified group (Steele, 1997, p. 617). People who experience this idea worry so much about being identified with the stereotype that their academic performance suffers (Graham & Morales-Chicas, 2015, p. 25; Laldin, 2016).

To minimize the effect of stereotype threat, we try not to imply that their academic performance is in any way related to their identity. Praising students for being "naturally smart" inadvertently implies that their intelligence is somehow fixed to their identity group. Instead, we try to promote a growth mindset (which we discuss more in Chapter 4: Promoting Mathematical Communication).

In addition, we try to foster a sense of "belonging" among all students in our class. Research indicates that a sense of belonging in school correlates with achievement, self-efficacy, and motivation (Chiu, Chow, McBride, & Mol, 2015, pp. 189-190; Lam, Chen, Zhang, & Liang, 2015, p. 402; Osterman, 2000, p. 327; Reynolds, Lee, Turner, Bromhead, & Subasic, 2017, p. 89). Exposing students to cultural concepts from different ethnic groups can promote self-understanding (Gray, Hope, & Matthews, 2018, p. 101). We discuss this more in Chapter 2: Culturally Responsive Teaching, Chapter 4: Promoting Mathematical Communication, and Chapter 7: How to Plan Lessons. Making *all* students feel welcome in our class and encouraging students from different identity groups to interact in our classrooms (through activities like group work or peer teaching) can also strengthen the ethnic climate in our classroom. A strong ethnic climate can minimize the negative effects of perceiving few classmates of one's own identity group (Graham & Morales-Chicas, 2015, p. 23).

"WHY DO WE NEED TO KNOW THIS?"

During a lesson, students occasionally ask us, "When are we going to use this in life?" or "Why are we learning this?" Since students usually want to know how our lesson is used in their lives, we try to make those connections whenever possible. When discussing compound interest, we talk about student loans. When discussing measures of central tendency, we talk about grades.

However, we can't make real-world connections in every lesson. Many topics—such as formal geometric proofs or rules of exponents—have no immediate connection outside of a math class. These lessons are important as necessary tools for larger mathematical concepts. Trying to find an application in every lesson can backfire, especially if the connections are strained. Constantly connecting math to the real world can actually de-motivate struggling students who may feel even *more* pressure to do well (Hulleman, Godes, Hendricks, & Harackiewicz, 2010, p. 891). Some of our most successful lessons involve straightforward ideas like finding a pattern. Many times, the simplest solution is the most effective one!

In our experience, students often ask us, "Why do we need to know this?" when they get frustrated or bored with our lesson. Since students in these situations are really saying that they don't understand what we're trying to teach, we start by first acknowledging students' concerns ("I can see that you're getting frustrated with this"). We then try to determine the source of those concerns, using some of the questioning and discussion techniques we discuss in Chapter 4: Promoting Mathematical Communication. If necessary, we come up with a clearer or simpler explanation.

MISREADING STUDENTS

If we underestimate our students' potential, we often wind up watering down our instruction by ignoring connections between topics or reducing math to calculator shortcuts and mindless drills. Unfortunately, this can erode student effort, especially for students who face multiple challenges (Mo, 2019, p. 5; Usher & Kober, 2012a, p. 4). In our experience, students can often tell when we have less confidence in their ability to learn and react by being less interested in the lesson, which leads us to slow down even more.

Conversely, giving them tasks that are too difficult can heighten their anxiety. Students may develop perfectionist tendencies, get discouraged when they make a mistake, and even burn out (Mo, 2019, p. 5). Not giving students appropriate support—for example, by teaching without paying attention to their current level of understanding—can make lessons inaccessible to struggling students and discourage them from learning (National Research Council, 2004, p. 47).

We recommend focusing instruction on identifying what students *currently* know, what they *need* to know, and scaffolding instruction with appropriate techniques (such as the ones we discuss throughout this book) so they can learn independently.

In addition, we try to address any emotional concerns that can adversely affect student motivation. Getting to know students better helps us give them emotional comfort when they seem stressed. We also remind them frequently that learning to deal with failure is an important life lesson and share examples from our own lives when we encountered difficulty. We discuss ways to improve the teacher-student relationship in Chapter 11: Building a Productive Classroom Environment.

Sometimes, we mistakenly assume that students have complete control of their environment. Criticizing students for not buying an expensive graphing calculator or not doing homework assumes that they *choose* not to finish a task and ignores economic hardships or personal situations that serve as obstacles. Often, students who we might think are "lazy" have too many responsibilities at home or worry that they won't be valued if they try and fail. Letting students know that we won't humiliate them, empathizing with them, and providing adequate academic and emotional support for tasks can help students move past their fears (Wormeli, 2014).

LIMITATIONS TO MOTIVATION

Occasionally, despite our best efforts, we just can't seem to motivate some students. Getting help from a guidance counselor, psychologist, or some other school professional can be helpful since they often have experience and training to deal with challenging situations. In these cases, we find that putting too much academic pressure on students can often backfire. This can decrease their motivation to stay in school and may make them withdraw even deeper.

Even if we can't motivate them to learn all of the math that we'd like, we can help develop important life skills, such as working with others and coping with adversity. Giving them a friendly greeting and showing genuine concern can build trust and eventually encourage them to open up when ready.

Technology Connections

More information about self-determination theory, including a more detailed explanation and related research, can be found at the Center for Self-Determination Theory's website (http://selfdeterminationtheory.org).

Some sources of real-world math problems include Mathalicious (www.mathalicious.com), which offers real-world middle and high school lessons; Math in Daily Life (www.learner.org); and Real World Math (http://realworldmath.org), which offers free math activities for Google Earth.

Teachers can use virtual manipulatives as motivations for lessons. Didax (http://www.didax.com/virtual-manipulatives-for-math) has a wide variety of virtual manipulatives, including two-color counters, pattern blocks, spinners, and algebra tiles. The National Library of Virtual Manipulatives (http://nlvm.usu.edu) has a wide range of online tools for pre-K to grade 12 math.

Printable algebra tiles can be downloaded from sites like The Math Lab (http://themathlab.com/toolbox/algebra%20stuff/algebra%20tiles.htm). Proof Blocks (www.proofblocks.com) are two-dimensional blocks (with a mathematical fact on each block) that can be physically linked together to form logical chains of reasoning.

In addition, Desmos (www.desmos.com) and Geogebra (www.geogebra.org) have many activities that can be used to motivate students.

Apps like ClassDojo (www.classdojo.com) and LiveSchool (www.whyliveschool.com) allow teachers to track student behavior with a point system. However, if you choose to use one of these tools, we suggest reading articles that share reservations, including "On Using and Not Using ClassDojo: Ideological Differences?" by Larry Cuban (http://larrycuban.wordpress.com/2014/03/15/on-using-and-not-using-classdojo-ideological-differences) and "Classes of Donkey" by David Truss (http://pairadimes.davidtruss.com/classes-of-donkeys).

Harvard math professor Oliver Knill has a list of math-related movie clips on his website (http://www.math.harvard.edu/~knill/mathmovies/index.html) .

The book *Living Proof: Stories of Resilience Along the Mathematical Journey* (Heinrich, Lawrence, Pons, & Taylor, 2019, http://www.maa.org/press/ebooks/living-proof-stories-of-resilience-along-the-mathematical-journey-2), which is available for free online, is a collection of biographical stories from real-life mathematicians about their struggles learning math and how they overcame them.

Figures

1. Complete the table below.

n	1	2	3	4	5	6
Group	□	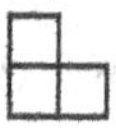 ⌐⌐	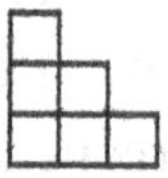 ⌐⌐⌐	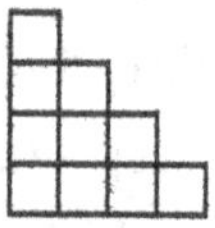 ⌐⌐⌐⌐		
Number of blocks per group						

2. Describe the pattern from the table above in words.

3. Use your pattern to determine the number of blocks in the 20th group. Try to do it without drawing 20 groups of blocks.

Figure 1.1 Pattern Blocks

1. Draw the lines of symmetry that exist for each regular polygon below.

2. Besides line symmetry, regular polygons have another kind of symmetry called rotational symmetry.

   a. For each regular polygon above, fill out the chart below to determine the minimum angle of rotation required to map the figure back onto itself.

Number of sides (*n*)	3	4	5	6
Measure of angle of rotation				

   b. Based on the table above, fill in the blank in the following sentence: In a regular polygon of *n* sides, the measure of the smallest angle of rotation that maps the figure back onto itself is _____ .

Figure 1.2 Rotational Symmetry

1. In this problem, you will examine folding a paper in half.

Number of folds (x)	Number of layers (y)
0	
1	
2	
3	
4	
5	

   a. Fill out the accompanying table to determine the number of layers of paper that are being made with each fold.

   b. Write an equation that relates the number of folds (x) to the number of layers (y).

   c. Graph the equation you wrote in part b using a calculator or other technology. Sketch this graph below.

2. In this problem, you will examine folding three papers in half.

Number of folds (x)	Number of layers (y)
0	
1	
2	
3	
4	
5	

Figure 1.3 Exponential Growth

a. Fill out the accompanying table to determine the number of layers of paper that are being made with each fold.

b. Write a equation that relates the number of folds (x) to the number of layers (y).

c. Graph the equation you wrote in part *b* using a calculator or other technology. Sketch this graph below.

3. Based on your work above, write a formula for each of the following situations. Let x = the total number of folds and let y = the number of layers of paper.

    a. folding 5 pieces of paper in half

    b. folding 2 pieces of paper in half

    c. folding 4 pieces of paper in half

4. How are these equations different from the equations that you have worked with before?

Figure 1.3 (Continued)

Find the value of x.

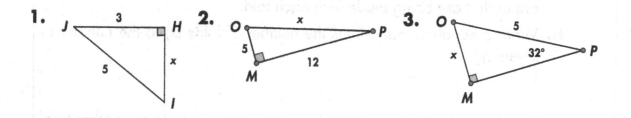

Figure 1.4 Identify a Void

CHAPTER 2

Culturally Responsive Teaching

What Is It?

Culturally responsive teaching is the use of students' prior experiences to expand their learning capacity (Gay, 2018, p. 36; Hammond, 2015, p. 124). According to Gloria Ladson-Billings (1994, p. 18), considered by many to be its founder, culturally responsive teaching empowers students by using aspects of their culture to impart knowledge, skills, and attitudes.

We recognize that many teachers have concerns about the usefulness and practicality of culturally responsive teaching, such as the following:

- Culturally responsive teaching may be unnecessary or hard to implement in an "objective" subject like math.
- Being culturally responsive seems focused on raising students' self-esteem, which won't help them do better on standardized tests.
- White teachers may feel that culturally responsive pedagogy is a personal attack directed at them.
- Some may feel that culturally responsive teaching is driven by a "political" agenda rather than a desire to improve academic outcomes.

In this chapter, we address these and other issues based on our own experience and explain why we ultimately think that being more culturally responsive is worth the effort. We don't claim to have all of the answers—what we describe here reflects our challenges as much as our successes.

Why We Like It

We find that being more culturally responsive has strengthened our relationship with *all* students, not just students of color. As a result, this increases their motivation, which (as we describe in Chapter 1: Motivating Students) also can improve their academic achievement.

In our experience, much of what is considered culturally responsive teaching (such as self-reflection, customizing instruction to match students' prior knowledge, and heightened awareness of teachers' thoughts and actions) is in fact what many teachers call "good teaching" (Ladson-Billings, 1995a, p. 159).

Supporting Research

Data from standardized tests like the National Assessment of Educational Progress shows a large racial gap in math achievement scores—the scores of white students are significantly higher than the scores of historically marginalized groups like black and Hispanic students (Center for Educational Policy Analysis [CEPA], n.d.; National Assessment of Educational Progress, 2017; Reeves & Halikias, 2017). Determining why this gap exists and what can be done to close it, including addressing systemic racism and inequality, has been a focus of a great deal of research (Abdulrahim & Orosco, 2019, p. 1).

Culturally responsive teaching can be traced back to the 1954 *Brown vs. Board of Education* case, in which the US Supreme Court ruled that racial segregation in public schools was unconstitutional. After the Brown decision, more researchers focused on ways to teach diverse students more effectively. In the 1960s, many scholars used the faulty framework of *cultural deprivation* to describe students of color (Schmeichel, 2012, pp. 214–215). These children were often described as coming from "lazy" families living in "inferior" and "deplorable" conditions (Ausubel, 1963, p. 454; Shaw, 1963, p. 92). Students of color who grew up in these cultures supposedly lacked the motivation and learning habits of "mainstream" culture, which was tacitly understood to be white and middle-class (Gordon & Wilkerson, 1966, p. 57).

By the 1970s, scholars began challenging the idea of cultural deprivation. They wrote that describing white, middle-class culture as "superior" to other cultures promoted a self-fulfilling prophecy to new teachers (La Belle, 1971, p. 15). In addition, they argued that teachers could customize their instruction to be more sensitive to students' cultures (Cazden & Leggett, 1976, p. 3).

What has become known as culturally responsive teaching emerged by the 1990s with the work of scholars like Gloria Ladson-Billings, Geneva Gay, and Zaretta Hammond. These researchers criticize the cultural deprivation framework for focusing on what was "wrong" with students of color and failing to acknowledge the inequity that these students face. Instead of blaming students for the achievement gap, they argue, teachers should promote academic success so that students can recognize and critique societal inequity (Ladson-Billings, 1995b, p. 477). Teachers can subconsciously exclude certain groups of students, denying them equal opportunities to succeed (Joseph, Hailu, & Matthews, 2019, p. 135). For example, studies indicate that teachers often perceive black girls as less attentive, more disruptive, and less likely to do well in honors classes (Annamma et al., 2019, p. 22; Francis, 2012, p. 316).

Culturally responsive teaching is supported by neuroscience research. The brain has two main directives—a desire to stay safe (by minimizing social threats) and a desire to be happy (by maximizing chances to connect with others) (Hammond, 2015, p. 47). Positive social relationships help suppress the brain's defense mechanisms and release endorphins, which make the learning experience more successful and improve memory. In addition, learners must connect new information to existing knowledge—including cultural experiences and values—in order for their brains to process and retain it (Sousa, 2017, pp. 50, 94, 165). In other words, in order to store knowledge into long-term memory, students must find meaning in it by connecting it to *their* past experience—not just ours. This explains how students can follow instructions 1 day but forget how to do it the next—if they don't attach a *meaning* to information, it is not stored in long-term memory (Sousa, 2017, p. 57).

Several studies indicate that being more culturally responsive can increase feelings of student belonging and interest in school (Byrd, 2016, p. 4), improve student behavior (Larson, Pas, Bradshaw, Rosenberg, & Day-Vines, 2018, p. 162), raise academic achievement (Aronson & Laughter, 2016, p. 34; Kana'iaupuni, Ledward, & Jensen, 2010, p. 15), and help teachers critique their lessons (Aguirre & del Rosario Zavala, 2013, p. 173).

Unfortunately, much of the research evaluating the effectiveness of culturally responsive teaching is qualitative, often based on case studies rather than controlled, randomized trials. A recent meta-analysis found no strong statistical link between culturally responsive teaching and improved academic outcomes for teachers or students (Bottiani, Larson, Debnam, Bischoff, & Bradshaw, 2017, p. 13). The lack of experimental studies of culturally responsive teaching doesn't mean that educators should dismiss it as unproven. Instead, it highlights the need for more rigorous research.

Common Core Connections

The Common Core Standards emphasize conceptual understanding, which is the goal of culturally responsive teaching. Conceptual understanding can be measured by such standards as asking students to extend the concept of multiplying fractions to division (6.NS.A.1), explain the meaning of a point on the graph of a proportional relationship in context (7.RP.A.2d), understand the connections between proportional relationships and linear equations (8.EE.B.5), justify a method for solving an equation (A-REI.A.1), and understand statistics as a process for making inferences about a population based on a sample (S-IC.A.1) (NGA & CCSSO, 2010, pp. 42, 48, 54, 65, 81).

Application

Culturally responsive teaching is more than a set of engagement strategies. It is a *mindset* in which teachers reflect on and adjust their relationship to their students and the content (Hammond, 2015, p. 52). Culturally responsive teachers think deeply about *what* they teach and ask themselves *why* students should learn what we teach (Ladson-Billings, 2006, p. 34).

To be more culturally responsive, teachers can focus on the following major practice areas:

- **Self-reflection:** becoming more aware of how our beliefs can affect our interpretations of student behavior and academic performance
- **Building a collaborative learning partnership:** building trust with students so that students feel safe enough to take risks
- **Developing information processing skills:** helping students develop their ability to process information and deepen their conceptual understanding (Hammond, 2015, pp. 18–19)

We discuss each of these practice areas below.

SELF-REFLECTION

To be more culturally responsive, teachers should periodically reassess their *beliefs*—the set of assumptions that they have about students, teaching, learning, and mathematics (Artzt, Armour-Thomas, & Curcio, 2008, p. 20). This reflection includes considering how factors like race, culture, upbringing, and social class shape language, communication styles, and worldviews (Kumar, Zusho, & Bondie, 2018, p. 85).

When reevaluating beliefs, Zaretta Hammond (2015, pp. 57–58) suggests thinking about questions like the following:

- How was I taught to interact with authority figures?
- How was I taught to praise or criticize others?
- How was I trained to react to emotional displays, like anger or happiness?

CHAPTER 2: CULTURALLY RESPONSIVE TEACHING

- What did I do as a child that earned praise?
- What do I believe about motivation?
- What do I believe about intelligence?
- What do I believe about how students should learn mathematics?

Since our beliefs play an important role in shaping our instructional practice, we find that self-reflection is a necessary part of good teaching. Most teachers would dispute the notion that they allow stereotypes to affect the way that they teach. However, research indicates that teachers' stated beliefs don't always match their actual classroom practices (Artzt et al., 2008, pp. 74, 21).

Unfortunately, self-reflection of one's instructional practice can be so intimidating and time-consuming that it can delay any meaningful change in teaching (Gonzalez, 2017). One way to make this process more manageable is to keep track of teacher interactions with a subset of students in our classes (such as African American males) to see if any trends emerge (Hammond, 2015, p. 82). Questions that come up when we do so include the following:

- **What help do we give students?** If we tell certain students exactly how to complete a task, we may believe that they are incapable of doing it independently. In contrast, if we provide other students hints or more limited guidance, we may think that they can figure out how to complete the task on their own (Artzt et al., 2008, p. 74).
- **What kind of verbal feedback do we give students?** If we praise students for completing a simple task ("Great job opening your notebook!"), we may send the message that we have low expectations for them (Hammond, 2015, p. 91).
- **Who do we call on in class?** Do we favor boys more than girls? Do we turn to students of certain ethnicities for low-level questions and students of other ethnicities for more challenging questions?

Another opportunity for self-reflection occurs when we react to what we consider to be inappropriate student behavior. Teachers may assume that students are being willfully disrespectful and may not consider other explanations, which can lead to further miscommunication and conflict (Dray & Wisneski, 2011, p. 31). For example, we may phrase a command as an indirect directive ("Would you like to take your seat?") instead of a direct order ("Please sit down") (Hammond, 2015, p. 58). If students misinterpret our statement, we may make matters worse by getting angry ("Why didn't you sit down when I asked you to?"), even though we didn't clearly state our intentions. Insisting that students look us in the eye when we reprimand them can be confusing for students from cultures that view making direct eye contact with adults as a sign of disrespect (Gay, 2018, pp. 29–30).

To help teachers reflect on how they communicate with students, researchers Barbara Dray and Debora Wisneski (2011, pp. 30–31) outline a three-step strategy for self-evaluation:

1. **Describe** what was observed or experienced using objective language ("Maria called out answers 10 times during the lesson without raising her hand").

2. **Interpret** the behavior, keeping in mind that behavior can be interpreted in different ways ("Maria wanted attention" or "Maria understood the lesson"). When interpreting student behavior, teachers should consider alternate explanations, taking into account possible misinterpretations resulting from prejudice or overgeneralizations.

3. **Evaluate** the behavior by attributing positive ("I like that Maria wanted to show that she understood the lesson since she usually struggles in class") or negative ("I feel that she prevented other students from sharing their thoughts") social significance to the behavior, recognizing that such attributions might foster stereotypes. As part of the evaluation process, teachers can think about questions like:

   - Why do I find the student's behavior problematic?
   - How did the student make me feel?
   - What expectations did I have for the student? How does the student's behavior interfere with those expectations?
   - How can I change the environment, my actions, or my expectations to respond differently?
   - What information can I get from the student's family or other school professionals (such as a psychologist, guidance counselor, or other teachers) to better understand the behavior? How can I get this information without judging them?

BUILDING A COLLABORATIVE LEARNING PARTNERSHIP

Building a positive relationship with students can help them reach their full potential while minimizing stress. Neuroscience research indicates that the brain feels safest and relaxed when people feel a connection to others who are trusted to treat them well (Hammond, 2015, p. 73).

Building Trust

To create a learning partnership, teachers need to build trust by first establishing an emotional connection. Building trust is critical since it frees up the brain for learning and higher-order thinking (Hammond, 2015, p. 76).

Simple acts that show genuine care for students can create a stronger relationship with them that can serve as a foundation for the ultimate goal, which is to increase their learning capacity (Gonzalez, 2017). These acts can include sharing hobbies that are similar to students' interests, remembering details from their lives, and being selectively vulnerable (for example, sharing challenges that we had when we were young) (Hammond, 2015, p. 79).

Teachers can learn more about students' cultures and home lives and use this knowledge to communicate more effectively with them. By helping teachers get to know students better, culturally responsive teaching can counter students' feelings of alienation toward school (Kumar et al., 2018, p. 89). Such success is particularly important for students of color who regularly deal with inequity in their lives. Culturally responsive teaching is ultimately designed to help students become more competent in math and use those skills to critique the world around them and fight the inequity that they face (Ladson-Billings, 2006, p. 30).

Using the History of Math

One way to use students' culture in class is to put mathematical ideas into historical context. Showing that the math we teach comes from diverse cultures can help dispel the popular myth, reflected in many textbooks, that it stems primarily from Greek and western European thought. In fact, research indicates that mathematical ideas were often transported across cultures through places like India, China, and the Middle East (Joseph, 2011, p. 12). When students learn more about the history of math, they may see it not as a field in which there is only one correct method but as an area in which different people approach the same problem in different ways. This may in turn encourage students to be more confident in trying another approach.

We like to incorporate history with occasional classroom activities, homework assignments, or group projects such as the following:

- When discussing place value and its relationship to like terms in algebra, students can research the origins of the Arabic numeral system and compare it to other systems, such as Chinese or Roman numerals.
- When discussing the area and circumference of a circle, students can estimate π using methods devised by early Chinese and Indian mathematicians.

- When measuring angles in circles, students can do research to learn how the division of a circumference into 360° relates to the ancient Mesopotamian number system.
- When solving quadratic equations by completing the square, students can use the area model explained by the Islamic mathematician Muhammad al-Khwarizmi, whose book on solving polynomial equations *Hisab al-jabr wa'l muqabala* is the origin of our modern word "algebra" (Joseph, 2011, pp. 463–466, 263, 393, 477–478).

In our experience, trying to incorporate history into every lesson or even every unit is difficult. Including cultural references in lessons will not magically motivate students who feel marginalized or abandoned (Hammond, 2015, p. 3). However, such references, when done occasionally, can help make lessons more interesting to students. For example, a Muslim student in Larisa's class became more interested in math history after she told him that he and al-Khwarizmi were born in the same part of central Asia. This student visited Larisa after graduating from high school to say that what he learned about math history in her class helped inspire him to become a math major in college. We recommend using historical or cultural references as part of a larger process of creating more genuine relationships with students so they will improve academically.

Classroom Environment

Culturally responsive teachers also create a classroom environment in which students feel safe enough to take risks. We create this environment by doing things like the following:

- When students are not sure of an answer, we allow them to express their opinion without fear of being embarrassed ("Even if you don't know how to do it, let's get some ideas out there," "Let's try something and see where it takes us").
- We encourage students to try other methods or use alternate representations to find a more efficient solution ("How could you solve this more easily?").
- When we make a mistake, we allow students to point out our error to us and say to them that we are human ("You have to be really careful here—even *I* can make a mistake!").

In addition, culturally responsive teachers establish routines and talk structures that help build a strong classroom community. This not only appeals to the collectivist values of many students' cultures but also supports deeper learning (Hammond, 2015, p. 151).

We discuss strategies for building an effective learning partnership with students in Chapter 11: Building a Productive Classroom Environment and Chapter 15: Differentiating for Students with Unique Needs.

Developing Information Processing Skills

Culturally responsive teachers plan instruction so that it follows the brain's natural cycle of input, elaboration, and application of knowledge (Hammond, 2015, p. 140). Helping culturally and linguistically diverse students process information more effectively includes techniques like the following:

- Building students' mathematical vocabulary lays a strong foundation for more rigorous thinking (Hammond, 2015, p. 139). We discuss ways to improve students' language skills in Chapter 3: Teaching Math as a Language and Chapter 4: Promoting Mathematical Communication.

- Learning and internalizing *cognitive routines*, the basic mental maneuvers that students use to process information, can strengthen students' higher-order thinking skills. Cognitive routines like graphic organizers and talk activities encourage students to connect new material to what they already know, find relationships and patterns, and determine its place in a larger system (Hammond, 2015, p. 132). We discuss ways to help students make connections in Chapter 5: Making Mathematical Connections and Chapter 17: Cooperative Learning.

- Scaffolding instruction helps students internalize cognitive routines so that they can complete complex tasks on their own. Culturally responsive teachers must be sensitive to the anxiety that students experience when they struggle, which can lead them to resist learning (Hammond, 2015, p. 139). We discuss ways to scaffold instruction while being sensitive to students' social-emotional learning in Chapter 14: Differentiating Instruction and Chapter 18: Formative Assessment.

Using project-based learning that deals with issues relevant to students' lives, such as rising gas prices in urban areas, makes math more meaningful to them (Hammond, 2015, p. 139). Incorporating cultural references in word problems—using a system of equations to calculate the unit cost of empanadas and tacos sold at a restaurant or solving a question from an ancient Chinese mathematical text—can also help students connect to the math we teach. We discuss relevant strategies in Chapter 16: Project-Based Learning.

What Could Go Wrong

Culturally responsive teaching can often be difficult to implement in the classroom. Here are some of the challenges that we've faced.

"COLOR-BLIND" TEACHING

Trying to be "fair" to all students by ignoring racial or cultural differences minimizes the role that race, ethnicity, and culture play in forming students' identities (Gay, 2018. p. 31). In fact, their varied life experiences and prior knowledge can enhance learning for all students. Just as effective teachers customize their instruction to address students' varying amounts of English proficiency, levels of readiness, and learning differences, they should also consider students' racial and ethnic differences in their planning (Ladson-Billings, 1994, p. 33).

Saying "I don't see color when I teach" makes as much sense as saying "I don't think about students' prior knowledge." Experts recommend that parents treat siblings fairly by attending to each child's individual needs as necessary while making them feel that they are valued for their unique talents (Clark, 2012; Pickhardt, 2011). In the same way, teachers can take into account each student's circumstances—including culture—in their instruction.

Furthermore, a "color-blind" approach can erode marginalized students' sense of belonging and harm their academic performance. One study found that students in schools that celebrated cultural diversity (for example, by incorporating it into classroom instruction) and encouraged a sense of school community had higher grades than schools that de-emphasized cultural differences or group identity (for example, by not allowing students to wear religious symbols) (Celeste, Baysu, Phalet, Meeussen, & Kende, 2019, p. 2). Furthermore, claiming to not see students' race or ethnicity can prevent people from recognizing the *implicit biases* (attitudes toward others that unconsciously affect decisions and beliefs) that people have toward others of the same race (Morin, 2015).

We sometimes make the mistake of ignoring students' culture based on a misguided attempt to be fair, as shown in the following examples from our experience:

- In one of Larisa's advanced classes, which consisted mostly of Asians, a black student started writing an Asian-sounding name on tests, homework assignments, and boardwork. When she asked him why he did this, he jokingly replied that he wanted to fit in with the other Asians in the class. Unfortunately, Larisa didn't recognize that he was feeling uncomfortable in the class. Reaching out to him privately, pointing out his strengths to him, finding out what made him uncomfortable, and taking steps to address those concerns would have been helpful to him.

- Bobson frequently got into arguments with a Latina student in his class over minor issues like walking out of class to get a drink of water without permission. Despite several reprimands and phone calls home, her grades spiraled downward and she became increasingly disengaged in class. When he asked her at the end of the school year why she had been so "disrespectful," she replied that her mother taught her to fight for her rights because she felt that authority figures frequently take advantage of people of color. After talking to her further, he discovered that she felt threatened by him and wanted to stand up for herself. Bobson realized that he should have spoken with her much earlier in the year and taken a less confrontational tone.

In short, we find that students' culture is an important part of their identity that should not be ignored. Being more sensitive to students' sense of belonging can not only make them feel more comfortable in our classes but ultimately more productive as well.

GOOD INTENTIONS

Sometimes, we say or do something that students feel is insensitive or even racist. Like most teachers, we consider ourselves fair-minded and dedicated to improving our students' education. As a result, we often feel defensive or upset when we're told that we're being unfair. In response, we say things like, "That was never my intent," or, "I meant well." For example, we have made the mistake of accusing students who go on long vacations during the school year of not caring about their education. Such thinking ignores the context and culture in which students and their families live. The times that families take vacations can be affected by many factors, including when employees allow parents to use vacation days or where cultural or family events fall in the calendar.

In such situations, experts say that our *intentions*, which can be difficult to determine, are less important than the *impact* of our actions (Utt, 2013; van der Valk & Malley, 2019). Our perceptions of what we meant can actually affect our perception of impact. Research indicates people believe that a harmful act is more damaging when it is done on purpose (Ames & Fiske, 2013, p. 1760). However, focusing on our presumably harmless intentions instead of the experiences of the people who have been harmed can reinforce privilege (Tannenbaum, 2013).

We find that judging students and their families is counterproductive and does nothing to improve their learning, which is the ultimate goal of our work. In the example we described earlier, a more productive response to students who leave school for extended periods of time would be to create self-study assignments for them. In addition, building effective relationships with students and parents can increase the odds that students will complete the assignments while they are away.

FINDING THE RIGHT TIME OR PLACE

One of the biggest challenges we face is finding the right time or place to fit culturally responsive teaching into a packed curriculum.

Focusing only on superficial aspects of culture in instruction, such as names or food, can also lead to problems. Simply mentioning ethnic holidays or playing hip-hop music in class will not magically engage students who feel unmotivated and marginalized (Gonzalez, 2017; Hammond, 2015, p. 3; Schmeichel, 2012, p. 225). Of course, making these changes is not a bad practice as long as it is just a small part of what we are doing to become more culturally responsive.

We find that referring to students' ethnicities in our lesson only works best if we have already established a good working relationship with them. Culturally responsive teachers show a connectedness with their students by constantly working to assure students that they are valued as individual learners (Ladson-Billings, 1994, p. 66). Even gestures that might seem minor to some, such as taking the time to learn how to pronounce students' names correctly, greeting them by name every day, or allowing them to express their opinions, can show that we respect them. Such simple acts of humanity don't have to be tied to students' ethnicities. In fact, they should apply to *all* students.

In short, culturally responsive teaching focuses on helping *all* students develop the necessary skills to succeed (Ladson-Billings, 1994, pp. 44, 96). This requires developing the right mindset, in which we see all students as individuals who can learn and look at all aspects of their identity to enable them to grow.

Many of the strategies that we discuss throughout this book, such as using real-life problems relevant to students or using peer learning, can be considered culturally responsive, even if we don't label them as such. We recommend that in order to make these techniques truly successful, teachers should constantly reflect on their beliefs and practices to find ways that students' culture, like other aspects of personality, can be used to strengthen learning.

Technology Connections

Zaretta Hammond's website (http://crtandthebrain.com/resources) has articles related to culturally responsive teaching, including a protocol for checking unconscious bias and an observation guide for school visits. The Cult of Pedagogy website has articles about culturally responsive teaching, including misconceptions about it (http://www.cultofpedagogy.com/culturally-responsive-misconceptions) and tips for making a lesson more culturally responsive (http://www.cultofpedagogy.com/culturally-responsive-teaching-strategies). Edutopia (http://www.edutopia.org/topic/culturally-responsive-teaching) also has articles related to culturally responsive teaching. The Teaching Diverse Learners website (http://www.brown.edu/

academics/education-alliance/teaching-diverse-learners/strategies-0/culturally-responsive-teaching-0), run by the Education Alliance at Brown University, has a summary of culturally responsive teaching and related resources. Other articles appear on the Culturally Responsive Leadership website (http://culturallyresponsiveleadership.com).

The New York State Education Department (http://www.nysed.gov/common/nysed/files/programs/crs/culturally-responsive-sustaining-education-framework.pdf) has published guidelines for teachers, schools, and districts to make their teaching more culturally responsive. The Region X Equity Assistance Center has a list of evidence-based strategies for culturally responsive teaching (http://educationnorthwest.org/sites/default/files/resources/culturally-responsive-teaching.pdf).

The Radical Math website (www.radicalmath.org) has lesson plans, articles, data sets, and other information for educators who want to integrate social justice into their curricula. Many of the resources on this site are relevant for teachers who want to make their teaching more culturally responsive. The "We the People" National Alliance, which is dedicated to sharing experiences related to promoting math literacy as a constitutional right for all students, has resources and other information on its website (http://iris.siue.edu/math-literacy). The *Journal of Urban Mathematics Education* (https://journals.tdl.org/jume/index.php/JUME) is an open-access peer-reviewed journal that has research related to culturally responsive teaching.

The Story of Mathematics (www.storyofmathematics.com) has a history of mathematics from different parts of the world as well as a list of important mathematicians.

CHAPTER 3

Teaching Math as a Language

What Is It?

A language is a systematic way of communicating ideas using symbols, sounds, and words that follow a set of accepted rules ("Language" [Merriam-Webster, n.d.-a]; "Language" [Oxford], n.d.-b). Many researchers and mathematicians have supported the idea that mathematics can be thought of as a language (Bruun, Diaz, & Dykes, 2015, p. 531; Devlin, 2012; Jones, Hopper, & Franz, 2008, p. 307; Orlin, 2018). They note that people use math to perform many tasks that require language, such as communicating, solving problems, and creating mechanical tools (Adams, 2003, p. 786). Furthermore, in both mathematics and language, symbols and rules must be applied uniformly and consistently in order to communicate (Jones et al., 2008, pp. 307–308).

Why We Like It

Many math teachers may feel uncomfortable thinking about themselves as literacy instructors. However, the two of us, like most math teachers we know, often teach literacy without realizing it. For example, we show students how to read and interpret symbols and text, translate them into more familiar language, and write coherent prose. Like Curcio and Artzt (2007, p. 257), we believe that "all teachers are teachers of reading and writing—and mathematics teachers are no exception."

Teaching math as a language allows us to adopt successful techniques used for English Language Learners (ELLs), such as cooperative learning, using students' prior knowledge, and repeating ideas (Jones et al., 2008, p. 308). Students must be able to engage in mathematical discourse before they can learn new material (Kotsopoulos, 2007, p. 302). This approach can create a low-risk environment in which students will feel comfortable taking risks and may reduce math anxiety.

Supporting Research

Researchers have argued that teachers of content areas like math and history should include literacy instruction specific to their subjects (Curcio & Artzt, 2007; Gabriel & Wenz, 2017; Shanahan, 2012). *Disciplinary literacy* involves teaching students not just the essential content but also the tools for reading and writing in that field, including the specialized vocabulary, language types, text structures, and symbols (Annenberg Foundation, n.d.; Shanahan, 2012). Math teachers are the best people to help students develop these specialized skills since they have experience reading mathematics texts (Adams, Pegg, & Case, 2015, p. 500).

Common Core Connections

The Common Core Standards for Mathematical Practice emphasize many language-related skills, such as modeling with mathematics (which can be seen as articulating thoughts or translating from English to mathematics), using appropriate tools strategically (which can be seen as choosing the correct vocabulary), and attending to precision (which can be seen as using the correct grammar and syntax) (NGA & CCSSO, 2010, pp. 7–8).

Application

In this section, we focus on strategies for teaching symbols and vocabulary—the basic building blocks of language. We discuss how to apply all of the skills for reading, writing, speaking, and listening in Chapter 4: Promoting Mathematical Communication.

ELICITING THE NEED FOR MATHEMATICAL LANGUAGE

When we introduce a new symbol or term, we try to show students its relevance to their daily learning in math. Many times, demonstrating the need can be as simple as showing that a term or symbol lets us represent commonly used concepts more succinctly. For example, using the symbol = is shorter than writing "is equal to," and using the word *ratio* is less cumbersome than saying "comparison of two numbers using division."

When possible, we also show students how the symbol or term is relevant outside of a math class. For example, when we discuss the difference between the mean and median, we discuss why governments report typical income levels using the median instead of the mean. When we talk about exponential growth, we give real-world examples that use compound interest.

Some symbols and terms represent concepts that are so central to a unit that they require a longer discussion when we introduce them. One technique that we have found useful in these situations is a *concept attainment strategy*, an inductive learning method developed by Jerome Bruner and his colleagues (Bruner, Goodnow, & Austin, 1956). In this method, students develop a definition for an idea after comparing several examples and nonexamples (Ferlazzo & Sypnieski, 2018a, pp. 123, 128; Gay, 2008, p. 219; Silver, Strong, & Perini, 2009).

We often use concept attainment as a motivation at the beginning of a lesson. One example appears in Figure 3.1: Concept Attainment, taken from our lesson on prisms. In this example, students consider pairs of solids to develop the definition of a prism. Each pair consists of a solid that *is* a prism and one that is *not* a prism.

We reveal the solids one pair at a time on our classroom projector. If we have physical models, we bring them in to help students visualize the solids. Students discuss them—in partners, small groups, and eventually with the entire class—refining their definition after each set. After seeing the first pair, students may conclude that a prism is a cube. After the second, they may change the definition to say that the sides may be rectangles, but all pairs of opposite sides need to be parallel. After the third, they may note that one pair of opposite sides doesn't have to be rectangles. After the fourth, they may say that all pairs of opposite sides have to be polygons.

Students may then work with partners to write a final definition of a prism based on all of their analyses, which we then discuss as a class. Though the exact wording may differ slightly, the common definition should be that a prism is a solid with one pair of parallel polygons and parallelograms as the other sides.

INTRODUCING SYMBOLS AND TERMS

We find that students often struggle to pronounce mathematical symbols. In our experience, if students can't pronounce a symbol, then they often struggle to understand its meaning. For example, they may read the summation notation $\sum_{k=1}^{n} k^2$ as "that thing that looks like a big E" instead of "the sum from k equals 1 to n of k-squared." Learning how to pronounce symbols and terms correctly helps students acquire language skills and improve understanding (Alkire, 2002; Kidd, 1992, p. 51). Research indicates that students learning a new language who struggle with pronunciation tend to feel more anxious about learning the language (Baran-Lucarz, 2011, p. 509; Szyszka, 2011, p. 293) and are less successful on tests (Awan, Azher, Anwar, & Naz, 2010, p. 34; Tang, Zhang, Li, & Zhao, 2013, p. 78).

As we introduce a new symbol or term, we state the pronunciation clearly ourselves, followed by a *choral response* technique, in which students as a group repeat what we say. This technique has the advantage of being quick and simple, so we use it often when students hear a term for the first time.

Another useful technique is *paired dictation*, in which students are divided into pairs with whiteboards, markers, and erasers. One student reads a mathematical statement while the other writes it down on the board. The reader can then check for accuracy and offer constructive feedback, or all the writers in the room can hold up their boards simultaneously while the class looks around the room, compares answers, and engages in a brief whole-class discussion (Ferlazzo & Sypnieski, 2018a, p. 244). The amount of paired dictation depends on the amount of vocabulary that has been introduced, how much time is available, and how comfortable students generally are with speaking mathematical language. Paired dictation doesn't have to be an elaborate full-period activity. It could be as simple as a Do Now activity at the beginning of a lesson or a summary activity at the end of a lesson to reinforce pronunciation.

Linking mathematical symbols and terms with their pronunciation helps students both read and write them correctly. For symbols, we display typed versions of them on our classroom projector so students can see how they would appear in a textbook or on a test. Since our students usually write math by hand, we spend a great deal of time showing students how mathematical symbols are written, often giving tips to help students write them clearly. For example, we say that the Greek letter μ should not be written like the English lower-case "u" but has a longer tail on the left that extends below the baseline of the written text. Teachers with students required to take computer-based tests may also want them to practice typing symbols as well.

Teaching students how to write and pronounce mathematics precisely is particularly important because inaccurate handwriting or speaking can easily lead to misinterpretation. For example:

- Students who misread the power x^2 (which should be read as "x-squared") as "x two" may confuse it with the subscript x_2 ("x-sub-two") or the product $x \cdot 2$ ("x times 2"). This can be especially confusing in mathematical statements that use both superscripts and subscripts, such as $d = \sqrt{(x_2 - x_1)^2 + (y_2 - y_1)^2}$.

- Writing trigonometric expressions with parentheses around the angles—i.e., sin (37°) instead of the more common sin 37°—and pronouncing them with the word "of"—i.e., "sine of 37 degrees" instead of "sine 37"—helps students see that trigonometric expressions represent functions whose inputs are angle measures.

- Pronouncing $\log_3 (9)$ as "log base 3 of 9" instead of "log 3 of 9" helps students remember that 3 is the base in the logarithmic expression. Linking the correct pronunciation and writing can also help reduce errors like confusing the expression $\log_3 9$ with $\log 3^9$ (read as "log of 3 to the ninth") (Thompson & Rubenstein, 2000, p. 570).

We often ask students to rephrase symbols in their own words. Paraphrasing helps students clarify their thinking (Curcio & Artzt, 2007, p. 263) and improve their performance on tests (Burstein et al., 2012, p. 23; Moran, Swanson, Gerber, & Fung, 2014, p. 102). In addition, asking students to paraphrase helps us identify misconceptions. One of Bobson's students explained the rule $x^a \cdot x^b = x^{a+b}$ as "When you multiply xs by xs, you leave it as x and add the exponents." Bobson pointed out that she seemed to understand a key requirement of the rule—that the bases must be the same. However, her description needed to be more specific in order to avoid confusion. He then asked students to compare her version with the more formal statement, "When you multiply powers with the same base, we add the exponents and leave the bases unchanged." Students pointed out that using the phrase "powers with the same base" is important because the rule cannot be applied if the bases are not identical.

Learning mathematical symbols can present many opportunities for ELLs to become more engaged with the class. We have noticed that ELLs often find reading symbols to be less stressful than reading English words. In addition, students who know logographic writing systems like Chinese (where a character represents a word or phrase) understand that a misplaced stroke can dramatically change a character's meaning, so they can share their experience with their classmates.

In short, we recognize that teaching students to read, write, and pronounce mathematical language takes valuable time. However, we believe that doing so can help students overcome their math anxiety and build confidence. Additionally, students may realize that they can improve their mathematical skills in the same way that they can become more fluent in any other language.

TRANSLATING BETWEEN SYMBOLS AND WORDS

To help students translate between symbols and terms, we have them create charts that list phrases commonly associated with each symbol. Figure 3.2: Words and Symbols Chart shows an activity in which students identify all of the phrases associated with the inequality symbols $<$, $\leq$, $>$, and $\geq$. These charts can be simple enough to be used as a Do Now activity at the beginning of a lesson. (We talk more about Do Now activities in Chapter 7: How to Plan Lessons.)

While these charts can help students translate phrases word-for-word, we find that they have limitations. Mathematical symbols and terms, like their counterparts in any language, must be read and interpreted in context. For example, the phrase "x is less than 5" is written as "$x < 5$" but "x less than 5" is written as "$5 - x$," which would confuse students who simply memorize that "less than" is always written as "$<$." We point out to students that the sentence $x < 5$ requires a verb (in this case, "is"), but the expression x less than 5 is a phrase that has no verb.

MAKING CONNECTIONS BETWEEN MATH AND ENGLISH

Another technique that we emphasize when teaching mathematical terms and symbols is to note their word origins. Students who understand the roots of words can make connections between familiar English words and related mathematical words. Common examples include the word *quadrilateral* (meaning "a shape with *four sides*"), which consists of the roots *quadri-* (meaning "four") and *lateral* (meaning "side"). This technique can be especially helpful with words that at first glance have no obvious relationship to English words, such as *perpendicular*, which comes from the root *pend* (meaning "to hang") and is related to more familiar words like *pendant* and *pendulum* (Thompson & Rubenstein, 2000, pp. 572–573).

We often encourage students to look up the origins of mathematical terms. This can be done easily through homework or classwork. In addition, students can make connections between the etymologies of mathematical terms and words in other languages that they know.

When possible, we try to highlight words that have similar meanings in both everyday English and in math. For example, when we discuss factoring, we remind students that in the same way that a *factor* is a contributor and a *product* is a result (something that is produced), a *factor* in math is a quantity that is multiplied with other factors to produce a *product* (Thompson & Rubenstein, 2000, p. 573). A *line of best fit* is a *line* that *best fits* data shown in a scatterplot. When we *evaluate* in math, we *find the value* of an algebraic expression by substituting appropriate numbers for variables, just as we *evaluate* something by determining its value.

EXAMPLES OF CONFUSING MATHEMATICAL LANGUAGE

We find that many of our students get frustrated with mathematical language because it relies heavily on symbols. Here are some common sources of confusion and our suggestions for how to address them.

Many symbols (such as $\leq$, ∞, and Δ) have no direct relationship to their sounds in English (Kenney, Hancewicz, Heuer, Metsisto, & Tuttle, 2005, p. 51). When students have no frame of reference for a symbol, we pronounce it several times and ask students to practice it repeatedly. We also show students how to write the symbol step-by-step and ask them to practice that as well. When possible, we offer tips on how to remember the symbols. For example, we compare the "greater than" symbol > to a mouth that opens toward the larger quantity. When symbols are similar to others that students may know, we also make those connections. For example, we note that the symbol for congruence $\cong$ is a combination of the equality symbol $=$ and the similarity symbol $\sim$, thus illustrating the idea that congruent shapes have the same size (i.e., are equal) and shape (i.e., are similar).

Some symbols and terms look alike but have different meanings. Symbols can be especially confusing because they can also have different pronunciations. For example, parentheses can refer to multiplication (read as "times") or function inputs (read as "of"), which can be especially confusing when both uses occur together in a mathematical sentence, such as $f(3) = 5(3)$ (which is read "f of 3 equals 5 times 3"). A *base* can refer to a side of a triangle or a number raised to a power. In these situations, we point out the other meanings that these symbols and terms have.

Other symbols look different but have the same meaning (such as $\cdot$ and $\times$ for multiplication) (Kenney et al., 2005, p. 4). In these cases, we point out to students that these symbols are like synonyms in English. We also note that mathematicians prefer some symbols over others to avoid confusion. For example, we discourage students from using the $\times$ symbol for multiplication in Algebra because it can easily be confused for the variable x. Instead, we suggest using the $\cdot$ symbol or parentheses.

Subtle word differences (such as the number *ten*, the *tens* place, and the *tenths* place) can be easily overlooked (Kenney et al., 2005, p. 51). We ask students to pay particular attention to the spelling and pronunciation of words. We also carefully enunciate potentially confusing words and highlight differences in spelling to emphasize their meaning. To highlight these differences, we often give students problems in which they must apply their understanding of these differences. For example, we may ask students to round to the tens place in some examples but to the tenths place in others.

Some words can confuse students because they look alike in English and math but have different meanings. For example, the *height* of a triangle is the distance from a vertex to the line containing the opposite side, but students may mistakenly think it is a distance that is always measured from the horizontal and conclude that the height changes if a triangle is rotated. In Figure 3.3: Why the Word "Height" Is Confusing, students may mistakenly think that the height of $\triangle ABC$ has changed from 3 to 7 after the triangle is rotated, but in both cases the height is defined to be the distance from B to $\overline{AC}$, which is 3. In these cases, we point out the different meanings of the word in English and math. When possible, we use precise mathematical language that is less likely to be confused in this context, such as *altitude* instead of height. We also remind students that math is a language with words that are borrowed from English but often have other meanings. In addition, we include examples in our lessons that require students to apply the mathematical definitions, so we may ask students to identify the heights of triangles that are rotated at various angles.

Some terms are so particularly confusing that we discourage students from using them. For example, *cancel* can refer to different operations, such as adding two opposites to get zero or dividing the numerator and denominator of a fraction by the same nonzero factor (Gay, 2008, p. 222). We hear students use the term *break down* to refer to factoring a number into primes (e.g., $72 = 2^3 \cdot 3^2$) or writing a number in expanded form (e.g., $125 = 1 \cdot 10^2 + 2 \cdot 10^1 + 5 \cdot 10^0$). *Reduce* is often used to refer to writing an equivalent fraction but implies that the new fraction is somehow smaller in value. Problematic terms like these hide their mathematical meanings and often leave students frustrated (Gay, 2008, p. 222). Instead, we refer to more specific mathematical terms, such as "add the opposite," "factor," or "simplify the fraction."

ENCOURAGING MATHEMATICAL PRECISION

We often ask students to verbally explain how to solve a problem. As they speak, we write on the board exactly what they say to show how easily their words can be misinterpreted. If students tell us to "draw a table" without specifying how many rows or columns, we sketch a coffee table. If students tell us to "find x" instead of "find the value of x," we circle x on the board and point to the variable. We encourage precision by reminding students in this light-hearted way that they need to use exact mathematical language. In addition, using this strategy helps us think about what language *we* may use that might confuse students.

Another activity to encourage mathematical precision is similar to the *picture dictation* strategy described by Ferlazzo and Sypnieski (2018a, p. 246), in which the teacher draws an image and, without showing it to students, dictates what it looks like while students attempt to draw it.

In our version, we draw a picture using mathematical shapes, as shown in Figure 3.4: Draw a Picture. The complexity of the picture we use depends on our students' mathematical knowledge. Students work in pairs, placing a barrier between them (such as a backpack) so each student can't see what the other is doing. One student (the "speaker"), without showing the picture we created, attempts to describe the picture verbally to the other student (the "artist"), who draws it on graph paper. We may allow the artists to ask clarifying questions or repeat what speakers say, but we don't allow the speakers to see what the artists are doing. We put the image on a grid to enable students to describe distances and directions more precisely. We end the activity by asking students to hold up their papers simultaneously and compare, which often results in very amusing drawings!

Before introducing new terms and symbols, we think about how they might be used in everyday language or in other math classes. Looking up words in dictionaries helps us see these connections. We also find that talking to teachers of students in other grades or referring to K–12 math resources helps us see how symbols and terms have been used in lower grades and how they will be used in the upper grades. Using a common vocabulary throughout the K–12 curriculum reduces the need for reteaching and helps students become more familiar with math terminology. Using consistent vocabulary helps students see that what they are presently learning is related to their prior knowledge (Karp, Bush, & Dougherty, 2016, p. 62).

VOCABULARY CHARTS AND FLASH CARDS

To help students learn mathematical symbols, we often work with them to create a chart of vocabulary terms, as shown in Table 3.1: Vocabulary Chart. We include a column for any symbols associated with each term and how the symbols are read. (If the term is particularly difficult for students to pronounce, we encourage them to write a phonetic pronunciation.) We have another column for the formal mathematical definition and a numerical example.

At the start of a unit, we give students a vocabulary chart with the terms listed in alphabetical order in the first column. Students can then fill out the remaining columns for each term as we introduce them in class.

We encourage ELLs to include the term and the definition in their home language. Many ELL students already have printed or electronic dictionaries (often on their mobile phones), which we allow them to use in class to help them translate terms. Printed dictionaries can be more effective since the process of spelling out the word and physically looking for it engages various cognitions that improve memory retention (Kipfer, 2013). We find that this translation strategy works best when students already know how to express the concepts in another language so that they can transfer that knowledge to English. Encouraging students to transfer their knowledge from a familiar language to English can be an effective instructional tool (Cummins, 2005).

When we teach geometry, we use a slightly modified version that includes diagrams and proof steps, as shown in Table 3.2: Geometry Vocabulary Chart. Our

Table 3.1 Vocabulary Chart

Term	Symbol and pronunciation	Definition	Example
Absolute value	$|x|$ "Absolute value of x"	Distance from 0 to the number on a number line	$|-5| = 5$
Counting numbers	$\mathbb{N}$ "set of counting numbers" OR "set of natural numbers"	Set of numbers used for counting (starts with 1)	1, 2, 3, . . .
Factorial	$x!$ "x factorial" (fac-TOR-ee-uhl)	Product of all counting numbers from 1 to the number	$5! = 5(4)(3)(2)(1) = 120$
Whole numbers	{0, 1, 2, 3, . . .}	Set of counting numbers and 0 (starts with 0)	{0, 1, 2, . . .}

Table 3.2 Geometry Vocabulary Chart

Term	Definition	Symbol and pronunciation	Example with diagram	Conclusion and reason
Congruent	Two figures are congruent if one can be mapped onto another using a set of rigid motions. "same shape and size"	≅ "is congruent to"	A•———•B C•———•D $\overline{AB} \cong \overline{CD}$	$AB = CD$ (If two segments are congruent, then they have the same measure.)

Parallel	Never intersecting and always equidistant "same distance apart"	∥ "is parallel to"	A• 1/2 •B
			  3/4
			C• 5/6 •D
			  7/8
			$\overline{AB} \parallel \overline{CD}$
				$\angle 1 \cong \angle 5, \angle 2 \cong \angle 6, \angle 3 \cong \angle 7, \angle 4 \cong \angle 8$
				(If two parallel lines are cut by a transversal, then the corresponding angles formed are congruent.)
				$\angle 3 \cong \angle 6, \angle 4 \cong \angle 5$
				(If two parallel lines are cut by a transversal, then the alternate interior angles formed are congruent.)

Perpendicular	Intersecting to form right angles "makes a corner"	⊥ "is perpendicular to"	A• •C B•—— $\overline{AB} \perp \overline{BC}$	$\angle B$ is a right angle. (Perpendicular lines intersect to form a right angle.) $m\angle B = 90$. (A right angle has a measure of 90.)

geometry vocabulary charts include a column for a diagram since every geometry term can be illustrated with a picture. These charts also include a column for logical conclusions that can be drawn from each term, which are useful when writing geometric proofs.

One advantage of vocabulary charts is that students can test their knowledge of different representations of the terms by folding their papers to reveal only certain columns. For example, they can test themselves to see if they know the terms associated with the pictures or if they can identify the terms given their definitions.

Another similar strategy that we use for teaching vocabulary is to have students create flash cards. The flash cards typically have the same information as a vocabulary chart. However, it is written on a 3 × 5 or 5 × 7 index card. The front of the card has the mathematical symbol or term and the back of the card contains the definition, as well as any symbolic representations and pictures. Students can either create cards by hand or type them (the Technology Connections section of this chapter shares resources for creating online flash cards).

When using flash cards, students can order and group terms differently based on their preferences. They can group cards by level of familiarity (so that they study cards that they are less familiar with) or concept (so students can create their own graphic organizers by rearranging their cards).

We usually give students the option to choose which format—vocabulary charts or flash cards—works best for them.

VISUAL AND VERBAL AIDS

When possible, we encourage students to use visual aids to help them remember the meaning of terms and symbols. Research indicates that drawing by hand improves student understanding (Kenney et al., 2005, p. 71). Drawing can lead to better recall because people use several skills simultaneously—they must create the image in their mind and translate it to their hands as they sketch an image (Fernandes, Wammes, & Meade, 2018, pp. 304–305; Wammes, Meade, & Fernandes, 2016, p. 1773).

Some examples are shown in Table 3.3: Visual and Verbal Aids. Teachers can use these pictures to help students, particularly those who struggle with written or verbal techniques, to master mathematical vocabulary. After seeing examples, students can be encouraged to create their own examples throughout the year (Thompson & Rubenstein, 2000, p. 572). We tell students to write these examples in their notebooks as well as on any vocabulary charts, flash cards, or other notes that they create.

Using informal language can also help students remember definitions. Allowing students to invent and justify their own terminology can help foster student creativity and thinking (Rubenstein & Thompson, 2002, p. 109). We often label these tips "kidspeak" or "mathslang" in class to show that they are equivalent to

Table 3.3 Visual and Verbal Aids

Term or symbol	Visual and verbal aid
Whole numbers	{O, 1, 2, 3, 4, . . .} (numbers that start with a "hole")
Union	Union — shaded Venn diagram of A ∪ B
Intersection	∩nd — Venn diagram with A ∩ B shaded. "Bayside High School is at the *intersection* of 32nd Avenue *and* Corporal Kennedy Street."
Horizontal	(image of sun rising over a horizontal line) orizontal

slang—acceptable for casual conversation but inappropriate for formal communication, where more precise language is often required. For example, the informal definition of similar polygons as shapes that have "the same shape but different size" fails to mention that their corresponding sides are in proportion—a fact that is often used to solve problems involving similar polygons.

WORD WALLS AND ANCHOR CHARTS

When possible, we try to use a *word wall*, a collection of high-frequency terms placed on part of a classroom wall so that students can easily see and refer to them (Jasmine & Schiesl, 2009, p. 302). Word walls can help students recall, spell, and pronounce difficult words more accurately (Cleaver, 2018). Although often associated with ELLs, we find that word walls can help all students master math vocabulary.

When we use word walls, we typically put each term on a large card or piece of construction paper and attach it to the wall using magnets (if we have a magnetic board in the room) or tape. We also include a picture if possible. We don't include the definition on word walls because we prefer to use the word wall as a prompt for students. Including the definitions would also clutter the limited space on our walls. As we introduce a term, we add it to the word wall and leave it up for the rest of the unit.

In addition, we often put *anchor charts*, which summarize important content or procedures, on the walls for students to use in class. Anchor charts support literacy by making the thought processes visible (Mulvahill, 2018). We like to use anchor charts for procedures that we use frequently, such as multistep rules like the order of operations or steps for performing calculations on a graphing calculator. While we make up many anchor charts ourselves, we frequently encourage students to make their own.

An example of an anchor chart for functions appears in Figure 3.5: Functions Anchor Chart. This anchor chart includes the definition of a function, different representations of functions (tables and graphs), and examples of functions (both real-life and mathematical). Another example of an anchor chart appears in Figure 3.6: Polynomials Anchor Chart.

Students can be involved in creating word walls and anchor charts. Teachers can set up a basic outline for a word wall or anchor chart and allow students to create the words and pictures (Gonzalez, 2018). As students create the material, they apply the vocabulary and improve their mathematical language skills.

Student Handouts and Examples

Figure 3.1: Concept Attainment
Figure 3.2: Words and Symbols Chart
Figure 3.4: Draw a Picture
Table 3.1: Vocabulary Chart
Table 3.2: Geometry Vocabulary Chart
Table 3.3: Visual and Verbal Aids
Figure 3.5: Functions Anchor Chart
Figure 3.6: Polynomials Anchor Chart

What Could Go Wrong

Here are some of the things that can go wrong when teaching mathematical symbols and vocabulary.

NOT TREATING MATH AS A LANGUAGE

One of the most serious mistakes we've made is not emphasizing reading and writing in our classes. As Larisa's grandfather, also a math teacher, used to say, "A definition must make sense to you before you can use it to solve problems." Often, we feel so much pressure to move through the curriculum quickly that we gloss over the development and practice of language skills. This usually results in lost time later in the year and increased student (not to mention *teacher*) frustration. Not creating a solid foundation for mathematical language is as counterproductive as teaching students how to read English without teaching them the alphabet first.

One of the biggest obstacles we face is student resistance to the idea that math is a language. Our students often complain, "This is a math class, not an English class!" We try to address these concerns by not separating our teaching of language acquisition from our teaching of mathematical skills. Instead, we integrate the strategies described in this chapter into our lessons throughout the year so that language learning becomes a regular part of our class.

Another mistake we've made is to not reinforce formal language after we introduce it to students. When first learning a concept, students can use informal language and connect new ideas to what they already know (Rubenstein & Thompson, 2002, p. 108). However, we find that when we don't encourage students to use formal language, they often struggle to articulate their thoughts clearly. They may resort to saying "you know what I mean," which usually leads to confusion and frustration. Giving students multiple opportunities to practice language in class and in homework helps them become more comfortable with the material. We discuss some useful strategies in Chapter 4: Promoting Mathematical Communication.

Sometimes, we get so carried away with word walls and anchor charts that we wind up cluttering our room. This surplus of print can overwhelm students as they struggle to find the appropriate word or procedure on the wall. Students may also rely exclusively on these tools, mistakenly thinking that the resources will somehow remain available indefinitely. We try to limit our word walls and anchor charts to a predefined area of our classroom and retire words or charts as students become more familiar with them. Putting too many charts on the walls can overwhelm students, who will often ignore them (Lenihan, 2015; Routman, 2018).

MATH AS A "BAG OF TRICKS"

One of the most serious mistakes we've made is to reduce math to a "bag of tricks." Mnemonic devices and shortcuts can make mathematical tasks easier, but they should not be the only takeaway from a lesson. Math, like any other language, is much more than a set of rules. Just as some students misspell *neighbor* as "*nieghbor*" because they remember only part of the "*i* before *e* except after *c*" rule, students who only remember mathematical snippets (often because they only remember a mnemonic device or formula) without their context are more likely to make mistakes.

For example, many of our students remember the Pythagorean theorem as the formula $a^2 + b^2 = c^2$ but forget that c represents the hypotenuse and that the theorem only applies to right triangles. Not remembering what the variables represent can lead to confusion and incorrect thinking. In the triangles shown in Figure 3.7: Why the Formula $a^2 + b^2 = c^2$ Is Confusing, students may correctly think that $3^2 + 4^2 = x^2$ but incorrectly think that $3^2 + 4^2 = y^2$ and $3^2 + 4^2 = z^2$.

A better way to help students remember the theorem is to use descriptive words to represent the variables, such as "In a right triangle, $leg^2 + leg^2 = hypotenuse^2$." To help with vocabulary, a memory aid like "In a right triangle, the *longest* side has the *longest* word (hypotenuse), and is opposite the *largest* angle (the right angle)" also helps.

We find that in most cases, students use mathematical facts incorrectly because they don't fully understand them in context. Like students who remember songs learned in a foreign language class but cannot use the language to express a thought, students who only know isolated mathematical facts cannot communicate mathematically (Kenney et al., 2005, p. 4).

Technology Connections

To help us find precise definitions of words, we use online dictionaries like Merriam-Webster (http://m-w.com) and the Oxford Dictionaries (www.oxforddictionaries.com). An online search for "math online dictionaries" can yield many useful sites, such as Mathwords (http://mathwords.com). The Common Core Standards website contains a math glossary (http://www.corestandards.org/Math/Content/mathematics-glossary). In addition, many states have mathematical glossaries in English. Examples include the New York State Education Department (http://www.p12.nysed.gov/assessment/nysaa/2013-14/glossaries/glossarymathrev.pdf), the Texas Education Agency (http://www.texasgateway.org/resource/interactive-math-glossary), and the California Department of Education (www.cde.ca.gov/ci/ma/cf/documents/mathfwglossary.pdf). Check with your district or state to see what resources are available.

The Using English for Academic Purposes website (http://www.uefap.net/speaking/speaking-symbols/speaking-symbols-mathematical) has a list of common mathematical symbols and their pronunciations.

Many school districts and states provide glossaries and other print resources for ELLs to use on standardized tests. For example, New York State provides bilingual mathematical glossaries in over 30 languages (http://steinhardt.nyu.edu/metrocenter/resources/glossaries). One limitation of bilingual glossaries is that since they provide only literal translations, they are only useful for students who already know mathematical terms in other languages. Some publishers have online glossaries with terms in different languages. For example, Holt, Rinehart, and Winston has glossaries for middle school (http://my.hrw.com/math06_07/nsmedia/tools/glossary/msm/glossary.html) and high school (http://my.hrw.com/math06_07/nsmedia/tools/glossary/aga/glossary.html).

Many websites have pre-printed math word wall cards or flash cards. For example, the Virginia Department of Education (http://www.doe.virginia.gov/instruction/mathematics/resources/vocab_cards) has a list of freely available word wall cards for grades K–12 in both Microsoft Word and PDF formats. Websites like Quizlet (www.quizlet.com), Easy Notecards (www.easynotecards.com), Classmint (www.classmint.com), and Cram (www.cram.com) allow you to use premade flash cards or create your own.

For online vocabulary quizzes, teachers can use sites like Kahoot! (www.kahoot.com) or Quizlet (www.quizlet.com). Google Classroom (http://classroom.google.com) also has the ability to create an online quiz with multiple-choice or drop-down questions using Google Forms.

CHAPTER 3: TEACHING MATH AS A LANGUAGE   57

Figures

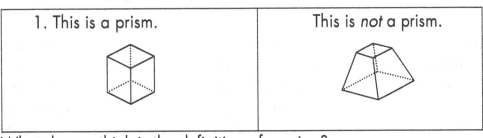

What do you think is the definition of a prism?

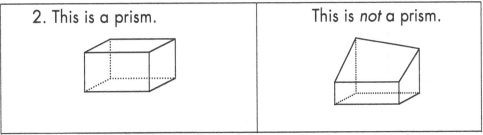

Refine your definition of a prism based on the examples above.

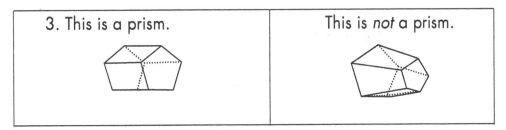

Refine your definition of a prism based on the examples above.

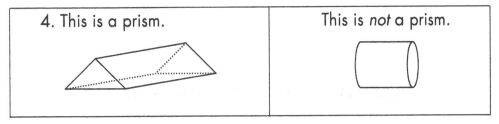

Refine your definition of a prism based on the examples above.

Figure 3.1 Concept Attainment

58 THE MATH TEACHER'S TOOLBOX

Match each phrase listed below with the inequality symbol used to represent it. Write each phrase under its symbol in the accompanying chart.

less than	under	at most
a maximum of	does not exceed	not greater than
not more than	greater than	over
more than	not under	not less than
exceeds	a minimum of	at least

<	≤	>	≥

Figure 3.2 Words and Symbols Chart

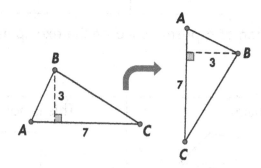

Figure 3.3 Why the Word "Height" Is Confusing

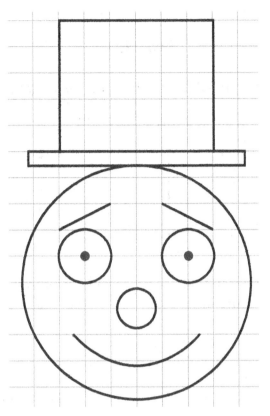

Figure 3.4 Draw a Picture

Function

A function is a relation in which each input has exactly 1 output

Is this a function?

x	y
1	1
2	4
-2	4

YES since every x-value (input) gets its own y-value (output)

x	y
1	1
4	2
4	-2

NO since x = 4 (input) has y-values of 2 and -2 (2 outputs)

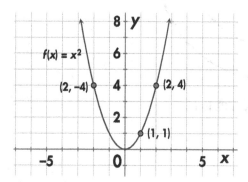

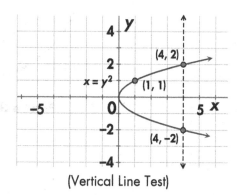

(Vertical Line Test)

ID number	Student
205-273-669	José
629-027-338	Sara
225-188-139	George

YES since every ID number (input) is associated with only 1 student (output)

Birthday	Student
Jan. 16	José
Feb. 4	Sara
Feb. 4	George

NO since a birthday (input) can be associated with more than 1 student (output)

Figure 3.5 Functions Anchor Chart

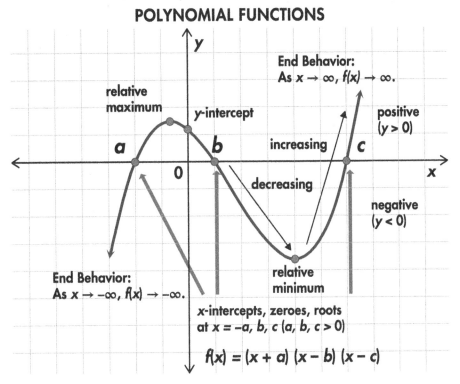

Figure 3.6 Polynomials Anchor Chart

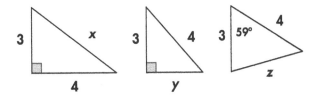

Figure 3.7 Why the Formula $a^2 + b^2 = c^2$ Is Confusing

CHAPTER 4

Promoting Mathematical Communication

What Is It?

Math can be viewed as a language with its own set of symbols, vocabulary, and rules for communication. In Chapter 3: Teaching Math as a Language, we discussed strategies for teaching math vocabulary and symbols. In order to be able to use these words and symbols, students must first understand what they mean and then learn how to combine them into mathematical statements and contextual problems.

This chapter focuses on the strategies that we use to promote *mathematical communication*, which we define to be reading, writing, speaking, and listening in mathematics. We find that these strategies work best when we incorporate them into our classroom routines, which we discuss more in Chapter 11: Building a Productive Classroom Environment.

Why We Like It

Many of our students fear math because they struggle to communicate mathematically. We often hear complaints like "I get it, but I can't explain it," "I hate word problems because I never know where to start," and "I read the textbook, but it didn't make any sense to me." Many of the ideas in this chapter are language acquisition strategies used for English Language Learners (ELLs). We believe that good teaching for ELLs is good teaching for everybody. These techniques can help both ELL and English-proficient students acquire greater fluency in mathematical language, which, in turn, can build self-confidence and assist them in overcoming any math fears.

Supporting Research

Research indicates that students who learn how to communicate mathematically are more prepared to solve problems (Curcio & Artzt, 2007, p. 259), to succeed in a math class (Bruun, Diaz, & Dykes, 2015, p. 531), to work independently (Adams, Pegg, & Case, 2015, p. 499), and to see the connections between math and the world around them (Kenney, Hancewicz, Heuer, Metsisto, & Tuttle, 2005, pp. 10–11).

In order to build mathematical understanding, students must learn how to share their strategies and ideas with others. Effective math lessons should train students to engage in discourse so that they can see the underlying mathematical meaning behind the language (Kenney et al., 2005, p. 74; Kotsopoulos, 2007, pp. 304–305; Lepak, 2014, p. 218).

In addition, students who improve their mathematical communication skills are more likely to experience less math anxiety, to understand that mathematical knowledge can be increased with the right training, and to develop a more positive mathematical identity (Allen & Schnell, 2016, p. 400).

Common Core Connections

The Common Core's Standards for Mathematical Practice list several important communication skills that are required for mathematical proficiency, including constructing viable arguments and critiquing the reasoning of others (MP.3), looking for and making use of structure (MP.7), and looking for and expressing regularity in repeated reasoning (MP.8) (NGA & CCSSO, 2010, pp. 6–8).

In addition, the Common Core's Standards for Mathematical Content include many reading and writing skills, such as describing situations in which opposite quantities combine to make zero (7-NS.A.1a), explaining each step in solving a simple equation (A-REI.A.1), and interpreting the parameters of a linear or exponential function in context (F-LE.B.5) (NGA & CCSSO, 2010, pp. 48, 65, 71).

Application

We use several strategies to help students improve their ability to communicate mathematically (many of them were inspired by techniques that Larisa learned when she was learning English). Here are some of our favorites.

OPEN-ENDED QUESTIONS

In order to talk about and discover important concepts, students need to learn mathematical discourse. To encourage discourse, we try to show students that we value understanding more than simply getting the correct answer (Stein, 2007,

p. 286). We find that students who can only solve problems with mechanical procedures often struggle to connect different ideas in solving more complex questions. For example, some students who can quickly solve a linear equation struggle to solve a verbal problem like the following, which requires that they interpret the context, write an appropriate linear equation, and solve it:

> Taxicabs in a large city charge an initial fee of $2.50 and an additional $1 surcharge for road improvements. Taxis also charge an additional 50 cents per 1/5 mile. Julio hires a taxi to go to the airport. His total fare, which includes a 20% tip, was exactly $42. To the nearest tenth of a mile, how long was Julio's trip?

One of the best ways that teachers can help students communicate is to use *open-ended questions*, which have unlimited answers. Open-ended questions allow students with different levels of understanding to engage in the mathematics. When all students can contribute—even in a small way—they can build their mathematical confidence, learn how to participate in the mathematical community, look for different ways to solve problems, and rely on each other instead of the teacher for the answer (Hodge & Walther, 2017, p. 432).

Some of the techniques that we use to foster mathematical dialogue among students include the following:

- Acknowledge the legitimacy and importance of students' questions—often, if one student asks a question, many other students are wondering the same thing.
- Respond to student questions with open-ended ones like:
  - "What should we do next?"
  - "Why do we have to do this?"
  - "What made you think of that?"
  - "What more can we say about this?"
- Give enough time for students to process questions. We often prompt students with cues like "Let's take a moment to think about this." Sometimes, we've had to wait as long as 30 seconds or a minute before someone responds. Increasing *wait time*—the amount of time that teachers pause after asking a question—to as much as 10–15 seconds encourages students to process information and give longer and more thoughtful responses (Goodwin, 2014; McCarthy, 2018). We find that increasing wait time makes the class feel less rushed and puts less pressure on students. This practice is helpful to all students, especially to ELLs.

- Try another technique. If nobody responds to our question after sufficient wait time, we rephrase our question or ask students to talk to a partner using the Think-Pair-Share technique we describe later in this chapter. When asking open-ended questions, we try to wait several seconds before allowing students to answer.

- Call on a variety of students using wooden ice-cream sticks or a random student selector. We discuss ways to call on students in Chapter 18: Formative Assessment.

- Allow students to use imprecise mathematical language—what we call "kidspeak" or "mathslang"—so that students can speak freely during the exchange. We may occasionally clarify the language to facilitate the student conversation.

- Confirm that students' questions have been answered before moving on and thank students for asking them.

Although we incorporate open-ended questions in our lessons, we sometimes find that important discussions can occur when a student asks a question that we didn't anticipate. In these situations, we stop our lesson to address possible misconceptions before they lead to further confusion. For example, Larisa's class once had a discussion about simplifying the solutions to quadratic equations. Her class had solved three equations, with solutions $2 \pm \sqrt{4}$, $2 \pm \sqrt{10}$, and $2 \pm \sqrt{18}$ that simplified to $\{4, 0\}$, $2 \pm \sqrt{10}$, and $2 \pm 3\sqrt{2}$, respectively. The conversation, which appears on the left in Table 4.1 (accompanied by our comments on the right), started when Andre, one of Larisa's students, asked a question about the solutions.

After this discussion, Larisa asked students to think silently for a few minutes and then answer Andre's original question in writing in their own words. Students then shared their summaries in pairs. After her class had a brief whole-group discussion, she shared her version of the answer with more precise mathematical language (if the radicand has a perfect square factor, the radical can be simplified).

In reality, many of these kinds of discussions turn out to be far messier than our example above. However, we find that with enough practice in class using the techniques we outline, our students learn to ask the clarifying questions themselves so that we don't have to.

Icebreakers for Mathematical Communication

We find that many students come to our class with negative emotions about math. We try to address this challenge with activities that ask students to write about some of their mathematical successes or failures. (See Chapter 11: Building a Productive

Table 4.1 Class Discussion

Class discussion	Teacher thoughts and action
Andre: "I'm confused. In some of these questions, you got a square root and kept going. In others, you stopped. How do you know when to stop?"	During her lesson, Larisa noticed that several students had gotten different solutions for these three equations.
Teacher: "This is an excellent question. How do you know when to stop? Let's take a moment to think about how to explain that."	Larisa paused her lesson to address Andre's question. After asking this question, she waited for several seconds until someone answered.
Katelyn: "Because sometimes you simplify and sometimes you don't."	Looking around the room, Larisa noticed that several students still looked confused. She guessed that although many of her students were familiar with the word *simplify*, having heard it in class many times, they may not have understood what it means.
Teacher: "Hm. What do you think she means by that?" *(pause)* Teacher: "OK, turn to the person next to you and discuss. When do you simplify radicals?"	Larisa asked a clarifying question and waited, but she got no response. She then used the Think-Pair Share technique (which we describe in Chapter 17: Cooperative Learning) to facilitate student dialogue.
Teacher: "Lena, you made a really interesting comment earlier. Please share it with everyone."	Walking around the room, Larisa heard Lena say something similar to what many groups said, so she asked Lena to share.
Lena: "If the radical can't be simplified more, then you're done. Otherwise, you keep going."	The language that students use here isn't mathematically precise or accurate.
Chelsea: "How do you know if it can be simplified more?"	
Jason: "If the number under the square root is a perfect square, then you square root it and keep simplifying."	

Table 4.1 (*Continued*)

Class discussion	Teacher thoughts and action
Andre: "But 18 isn't a perfect square. Why'd you keep simplifying?" Gabriel: "Because you can break 18 into 9 and 2, and 9 is a perfect square, so you square root it to get 3 and bring it outside the radical."	Lena's comment that "you keep going" is vague. However, to keep the conversation moving, Larisa refrained from commenting since many students now wanted to talk and she didn't want to interrupt the flow of conversation. After the Think-Pair-Share, many more students wanted to speak, so Larisa called on volunteers to share their thoughts.
Teacher: "So you can *factor* 18 into 9 and 2. But you can also factor 18 into 6 and 3. Why do we pick 9 and 2 here?" Hannah: "Because 9 is the biggest perfect square that goes into 18."	Larisa gently corrected Gabriel by using the more precise word *factor* since the colloquial term *break down* often confuses students. She asked a clarifying question to elicit the idea that in order to simplify radicals, we find the *largest* perfect square factor of the radicand.
Teacher: "So you find the largest perfect square that is a factor of the radicand (the number under the square root symbol). You simplify the perfect square radicals and multiply them by the remaining radical. *(pause)* Andre, does this answer your original question?" Andre: "Yes, now I get it." Teacher: "Thank you all. This was excellent work!"	Larisa summarized the conversation using correct mathematical language, pointing to the appropriate part of the work on the board as she spoke. She then confirmed that Andre felt his question was answered, thanked him for asking, and praised the class for their discussion.

Classroom Environment for more about this topic.) We also like to give students accessible open-ended mathematical tasks that introduce them to techniques that we use throughout the school year.

We can't do *all* of these activities every year, but we try to incorporate as many as we can reasonably fit into our lessons.

Table 4.2 Low-Floor, High-Ceiling Problems

Problem	Discussion
ALGEBRA: Why will a four-function calculator and a graphing calculator give different answers when evaluating the expression $6 - 1 + 4 \cdot 2$?	This leads to a discussion about order of operations (graphing calculators, unlike less-sophisticated four-function calculators, follow the order of operations) and the misconceptions that can arise from using a mnemonic like PEMDAS (which implies that addition is evaluated before subtraction).
GEOMETRY: Why do camera stands have only three legs and not four or five?	This leads to a discussion of planes, which are uniquely defined by three points. (This also explains why a four-legged table can be less stable than a tripod.)
STATISTICS: Between 60% and 80% of students who take the Advanced Placement Chinese, BC Calculus, and Physics C exams typically get a 4 or 5 (the highest possible scores). In contrast, only 25–40% of students who take the AP History or English exam typically get a 4 or 5. Why?	This leads to a discussion of variables that affect data, even though they are not included—in this case, the prerequisites for each course.

Low-floor, High-ceiling Problems

One way to encourage mathematical discourse is to give students a *low-floor, high-ceiling problem*, a problem that has multiple entry points so all students can work on it but can be extended to higher levels. Table 4.2 provides some examples of low-floor, high-ceiling problems that we use and some of the points that we make during the subsequent discussion.

Manipulatives

Manipulatives can promote multiple representations and help students develop a deeper mathematical understanding (Safi & Desai, 2017, p. 488). We find that using manipulatives helps engage students in a mathematically meaningful task that is less intimidating and more accessible for students. They can also improve student motivation (which we discussed in Chapter 1: Motivating Students).

Figure 4.1: Algebra Tiles Activity shows an activity involving algebra tiles, which would be appropriate for middle school, Algebra I, and Algebra II students. We discuss more specific uses of manipulatives in Chapter 15: Differentiating for Students with Unique Needs.

Which One Doesn't Belong?

Another activity uses problems based on Christopher Danielson's book *Which One Doesn't Belong?* (Danielson, 2016). Each problem consists of a square divided into four parts, each of which contains a number, shape, expression, or other mathematical representation. Students compare the similarities and differences between each representation and talk with their classmates to determine which representation doesn't belong (Buchheister, Jackson, & Taylor, 2019, p. 204).

We like this strategy because we can create an infinite number of variations depending on the topic and desired level of difficulty. These problems engage all students since they usually need little prior knowledge. In addition, each problem has more than one correct answer depending on what criteria students use to compare, so we often have lively mathematical debates!

Two examples of Which One Doesn't Belong? appear in Figure 4.2. The first contains four different equations. Students can notice the following:

- $x^2 + 7x + 12 = 0$ is the only equation whose solutions are negative.
- $x^2 + 4x + 6 = 0$ is the only equation whose solutions are imaginary.
- $(x - 3)(x + 2) = 0$ is the only equation in factored form.
- $2^x - 32 = 0$ is the only equation that is exponential.

The second example in Figure 4.2 is appropriate for Geometry and can be used to elicit the following:

- The square is the only shape in the diagram with right angles.
- The rhombus is the only shape shown that has four congruent sides and no right angles.
- The parallelogram shown here is the only shape with noncongruent sides (it is the only shape shown that is not a regular polygon).
- The equilateral triangle is the only shape shown that doesn't have four sides.

When we use this strategy, we emphasize that all of these answers can be considered correct and that how students justify their thinking matters just as much as what answer they reach. We want students to learn that if they can clearly show their thought process, they can earn partial credit even if their final answer is wrong. However, an incorrect answer without any work helps neither the student nor the teacher.

GUIDING STUDENTS IN CONVERSATION

We find that many students often feel intimidated by talking about math. To overcome these concerns, we try to create many small opportunities for students to talk to each other about the math. Here are some of the ways that we set classroom expectations for mathematical dialogue.

Think-Pair-Share

We use the *Think-Pair-Share* (which is similar to the *turn-and-talk* technique) strategy frequently because it can be done easily and quickly at almost any point in a lesson to engage all students in a mathematical conversation. We like to use this technique when we sense that students are confused about something or need to choose from many possible options. Think-Pair-Share can also be used to allow students to check their understanding or share questions that they may have about a task (Ghousseini, Lord, & Cardon, 2017, p. 426). We discuss Think-Pair-Share more in Chapter 17: Cooperative Learning.

Explaining Classwork and Homework

Another way that we encourage students to communicate mathematically is to have them explain their work to the rest of the class. At the beginning of a lesson, we often model the first few problems. We then ask students to write their work on the board and talk through their solutions. Teachers who have a document camera may simply use it to display student work. We use a similar method for homework review at the beginning of class. (We discuss homework review in more detail in Chapter 8: How to Plan Homework.)

We use several techniques to encourage students who may not feel comfortable speaking publicly. One way is by offering extra credit for explaining problems. We allow students to write a detailed explanation on the board that they can read to the class. Students may also bring a friend or interpreter to stand in front with them as they explain their work. Often, students will feel more comfortable speaking if a friend is standing with them. Classmates can also jump in and help to explain the problem. If necessary, we guide the presenter through the solution by asking questions like "What did you do first?" or "Why did you choose to do that?" Finally, we always thank students for coming up to the front of the class. With these strategies, our students often gain more confidence throughout the year.

In our experience, students tend to feel more empowered to manage a class discussion when they stand at the board. At the beginning of the year, we model possible comments and open-ended questions from students, such as "If anyone got a different answer, please explain what you did," "How did you know to do that?" or "How did other people approach this question?" If everyone got the same answer and used the same method, then we move on. If not, we encourage students to share what they did.

To help students share their thoughts when they get a different answer, we encourage them to use prompts like:

- "I got a different answer. Here's what I did..."
- "I disagree with your answer because..."
- "I agree with your answer, but I used a different method..."
- "I don't understand why you did..." or "I don't understand how you found/used/knew..."

Teachers can post these sentence starters in the classroom to help students remember them. We practice these sentence starters at the beginning of the year by using the Think-Pair-Share strategy described earlier in this chapter. Students can then use the prompts by agreeing or disagreeing with their partner's answer or asking for clarification. Some students can then summarize their discussion with the class. If students feel uncomfortable discussing their own work, they could explain what the other student did. Another way to encourage students to speak up is to ask them to identify the step in the work on the board where they got confused or where their answer differs.

We try not to intervene during these mathematical conversations, even when students ask questions and look to us for the answers. Sometimes, especially at the beginning of the year, we allow long pauses to let students think and give them space and time to speak up. In general, we interject to move a conversation along, to ask clarifying questions, or to suggest an alternative if the entire class agrees on an incorrect solution.

Modeling Student Dialogue by Thinking Aloud

Many times, we find that we need to model dialogue for students. Modeling a conversation can help students focus on content instead of determining whether an answer is correct (Ghousseini et al., 2017, p. 424). Furthermore, helping students visualize more challenging situations can help empower students and improve both their self-confidence and competence in math (Usher, 2009, p. 309). The *think-aloud* strategy (often used in literacy instruction) can teach students how to improve their comprehension and share their thoughts (Trocki, Taylor, Starling, Sztajn, & Heck, 2014, p. 278).

Think-alouds work best when they clearly model a broader mathematical reasoning process. Teachers can use them to help students retrieve prior knowledge and check if their thinking leads to a solution (Ghousseini et al., 2017, p. 425; Trocki et al., 2014, pp. 279–280).

CHAPTER 4: PROMOTING MATHEMATICAL COMMUNICATION

For example, here is a sixth-grade word problem followed by a think-aloud for solving it.

To create a bag of cranberry granola trail mix, a supermarket adds 3 ounces of granola and 2 ounces of cranberries to each pound of base mix. How many ounces of granola and cranberries will the supermarket need to make 50 pounds of the trail mix?

"So let's find the key phrases from the problem. To make 1 pound of mix, you need 3 ounces of granola and 2 ounces of cranberries."

"What do we need to find? Let's look at the last sentence. We need to make 50 pounds of the mix."

"I'm having a hard time visualizing how this works. Let me write out a few examples to understand what's going on. We know what we need to make 1 pound. What do we need to make 2 pounds? Three pounds?"

"This is a lot of writing! I'd like to organize this information better so that I see what's going on. Let me create a table of values. If I write out enough rows in the table, maybe I can figure out a pattern and find a shortcut."

Here is an Algebra II problem along with its think-aloud strategy:

If $\sin(\theta) = \frac{\sqrt{3}}{2}$ and $\frac{\pi}{2} < \theta < \pi$, then find the exact value of $\tan(\theta)$.

"So let's see what we're given. We know that $\sin(\theta) = \frac{\sqrt{3}}{2}$ and $\frac{\pi}{2} < \theta < \pi$."

"What do we need to find? I look at the end of the problem. We need to find the exact value of $\tan(\theta)$. The word *exact* means we can't approximate with a calculator."

"I remember that whenever we solved these trigonometry problems before, we sketched a graph of the angle and found the reference angle. We also labeled the sides of the angle based on the given information. So I'll sketch a graph here."

"I also remember that we'll need to know the appropriate sides for the trigonometric ratios sine and tangent. I can use the Pythagorean theorem to help me find the missing value."

Modeling a "bad" think-aloud (in other words, making intentional mistakes) can help students see what *not* to do, as shown by the following example:

What is $(7.1 \times 10^5) + (8 \times 10^3)$ in scientific notation?

"So I remember that in scientific notation we have powers of 10 and other numbers, so we multiply the 10s and we multiply the other numbers."

"We first figure out what operation we're doing. We're *adding* here, so we add the powers of 10. Then we add 7 and 8.1 together to get the coefficient."

"Then we write our final answer in scientific notation: first the coefficient, then the × symbol, then the power of 10."

During a "bad" think-aloud, students could take notes on the errors made (not converting the numbers to the same power of 10) and what should have been done instead. Students could then work individually or in pairs to use a better think-aloud to solve the problem correctly, such as the following:

What is $(7.1 \times 10^5) + (8 \times 10^3)$ in scientific notation?

"So I remember that in scientific notation we have powers of 10 and other numbers. We first figure out what operation we are doing to the numbers (adding)."

"Then we change the numbers so they have the same power of 10. We keep the powers of 10 the same and add the other numbers to get the coefficient."

"Then we convert our final answer to scientific notation. We convert the coefficient to scientific notation, combine the powers of 10 using exponent rules, and write our final answer: first the coefficient, then the × symbol, and then the power of 10."

ELLs may have difficulty following the think-aloud strategy. Providing extra support, such as explaining vocabulary terms and using alternate representations (such as manipulatives, pictures, or physical objects) can often be helpful (Trocki et al., 2014, p. 280). In addition, the simple act of writing down what you are saying can benefit everyone! Teachers may also want to consider creating online presentations with closed captioning added so that students, especially ELLs, can simultaneously see and hear what's being said (see the Technology Connections section of this chapter for more information).

FOUR-STEP THINKING PROCESS

We find that many students are intimidated by word problems because they often don't know where to start or what to do next. To give them additional support, we created a chart that divides the solution process into four steps: Given, Find, Solve, and Check, as shown in Table 4.3: Problem-Solving Chart. We based this chart on the four-step problem-solving process outlined by Stanford mathematician George Pólya in his classic text *How to Solve It* (Pólya, 1945). Although we use this chart for word problems, it could also be used to solve any type of mathematical problem.

Table 4.3 Problem-Solving Chart

1. Given...	
1a. What information am I given? Highlight key phrases. Rewrite text with math symbols or words. Create a table of values or draw a diagram.	**1b. What unknown information do I need to find?** Look at the last sentence. *(Include units and rounding.)* Look for words like: find, calculate, express, determine, simplify, create, prove, justify, explain, which, what, how many.
2. Find...	
2a. What does the given information tell me? • How is this *similar* to a problem I can solve? • How is this *different* from a problem I can solve?	**2b. How do I find the unknown information?** • Write an equation or inequality. • Use formulas, definitions, or theorems. (Check conditions first!)
3. Solve...	
Carry out the plan you outlined in step 2 above.	
4. Check...	
• Did I find what I was supposed to find? • Does my answer make sense in the context of the problem?	

When we start doing word problems at the beginning of the year, we give students a handout with a blank copy of our chart (one for each problem). Students can use it as a note-taking tool or an outline of the steps needed to solve problems. We find that the chart is especially useful for students who need additional scaffolding for word problems. As students become more familiar with the four-step method, they write their solutions without using the chart, although we provide blank copies of the chart for them as a reference. In our experience, allowing students to decide when they feel comfortable not using the chart promotes *metacognition*, the process of thinking about thinking. People with good metacognitive skills are better at solving problems since they constantly monitor their thinking to see if they are progressing towards a solution (Fadel, Trilling, & Bialik, 2015). Once they see flaws in their thinking, they are more likely to adjust their strategy (Darling-Hammond, Austin, Shulman, & Schwartz n.d., "Thinking," p. 160).

We don't require students to use this chart for every problem. We use it several times at the beginning of the year when we first use word problems as a model until they become more comfortable with and eventually internalize the steps. Researchers argue that *automaticity*—the ability to do a task without making a conscious effort to perform the required tasks—can be achieved through consistent repetition and consistent practice (Segalowitz, 2008, p. 402).

Step One: Given (Understand the Text)

In this first step, students determine what information is given in the problem. We encourage students to read the text out loud when possible. Reading out loud promotes the development of language skills in young children (Duursma, Augustyn, & Zuckerman, 2008, p. 554; Hinds, 2015; Matthiessen, 2018) and helps children internalize language and structures (Sharpe, 2009). We find that students often understand the text better if they read it out loud. If the text is especially difficult, we read the text to the class. After reading it out loud, we may ask students to read it to themselves or to a partner. When reading aloud is impossible (for example, when students are taking a test), we tell them to mimic the read-aloud process as much as possible by mouthing the words silently to themselves.

We ask students to read through the text once without doing anything else so that they get an overall sense of the problem. Then, starting with the second reading, we ask students to highlight or underline key parts. We often have highlighters in our classrooms to facilitate this process. Some researchers question the effectiveness of highlighting as a learning strategy (Dunlosky, Rawson, Marsh, Nathan, & Willingham, 2013, p. 45; Marsh & Butler, 2013, p. 6). However, other research indicates that highlighting can facilitate long-term retention, especially if students are trained to highlight effectively (Yue, Storm, Kornell, & Bjork, 2015, p. 76). We find that highlighting helps to promote engagement and gives students another reason to actually read the text and not just go through the motions of reading.

To determine which parts of the text are important, we ask students to read through it at least four times using the following steps:

1. Read the text silently to themselves.
2. Read the text out loud, while underlining or highlighting relevant words or phrases. If we are doing a problem in class, we ask a volunteer to read it for the

class. If students are taking an assessment, then they obviously read the text to themselves.

3. Read through the text again to identify *what is given* in the problem. They should ignore extraneous contextual information and focus on phrases that describe "who is doing what to whom" in the problem. Students should include relevant subjects, verbs, numbers, and units—highlighting "Jenna drove 288 miles" has more meaning than simply highlighting "288." Students should also rewrite key phrases using their own words, math symbols, or a different representation such as a table or a labeled diagram (Pólya, 1945, p. 7). ELLs can also rewrite key phrases in a more familiar language, using a glossary if necessary.

4. Read through the text again to identify *what we need to find*. We encourage students to look at the end of the problem. We point out to students that the most important sentences in mathematical paragraphs often occur at the end, unlike traditional paragraphs, which usually start with its most important sentence and is followed by supporting evidence (Kenney et al., 2005, p. 12). Often in mathematical text, the end of the problem states what specific question must be answered. We tell students to look for keywords like *calculate*, *express*, *create*, *determine*, or *prove* or questions like *which?*, *what?*, or *how many?* When students identify the unknown quantity with a variable, we ask them to state what they need to find as precisely as possible. Writing a sentence like, "Let x = coats," doesn't specify whether we need to find the *number* of coats, the *cost* of one coat, or the *total* cost of the coats.

Step Two: Find (Create a Plan to Find the Unknown)

Once students identify what information is given, we ask them to create a plan to find the unknown. We remind students that to solve a mathematical problem, students need to recall prior mathematical knowledge (Pólya, 1945, p. 9). To prompt students, we encourage students to ask themselves two questions:

- How is this problem *similar* to another problem *that I know how to solve*?
- How does this problem *differ* from a problem *that I know how to solve*?

When we organize our classwork by increasing difficulty, we can phrase this question in terms of levels, such as "How is this Level 3 problem similar to (or different from) a Level 2 problem?" We find that when students identify similar problems that they know how to solve, they make the problem more accessible and reduce their anxiety about solving it.

We encourage students to discuss ideas with a partner, when possible, to improve their planning skills. Asking students to communicate with a partner as they create a plan helps promote metacognition as they identify their mistakes and correct misconceptions (Raymond, Gunter, & Conrady, 2018, pp. 277–278).

Step Three: Solve

In this step, students carry out the plan that they outlined in the previous step. This typically involves solving an equation or inequality by applying relevant formulas and theorems.

We find that students often blindly move forward with an inefficient or incorrect method until they get stuck, but by the time they reach this point, they have no time left to try another method. To prevent this from happening, we remind students to monitor their progress as they solve the problem, asking themselves questions like:

- Do I know that this step is correct?
- Does my work resemble the work for a similar or related problem that I can solve?
- Am I getting closer to what I need to find?

We tell students that if the answer to any of these questions is no, then their work may not lead to the correct answer. In these cases, we encourage them to do any or all of the following:

- Look over their work to see if they can find a mistake.
- Redo their work for one or more steps.
- See if another mathematical definition, theorem, or formula could be used.
- Try a different method.

Step Four: Check

After students answer the question, they should check to see that it is reasonable. This can mean substituting a solution into the original equation or checking to see that the final answer makes sense in the context of the original problem. If a word problem asks to find the height of a flagpole in feet, an answer of 360 is unreasonable—unless the calculations were done in inches, in which case converting 360 in. to 30 ft makes more sense.

Students can also reflect on their solution or compare it to the solutions of their peers. Doing so allows students to find a more efficient solution method or refine their thinking for future problems (Pólya, 1945, p. 36; Raymond et al., 2018, p. 280). For example, students who use the quadratic formula to solve the equation $x^2 + 7x + 12 = 0$ could also try solving the equation by graphing or factoring. If they see that these methods are simpler and faster than their original method, they may remember to try graphing or factoring first when solving other quadratic equations in the future.

As we said earlier, we don't necessarily require this four-step process for *every* word problem used in class. We use it as a model and practice it in class as an explicit problem-solving method that students can eventually use automatically.

MATHEMATICAL WRITING

Research indicates that writing in a math class can improve student understanding (Curcio & Artzt, 2007, p. 262) and help teachers see how students think (Kenney et al., 2005, p. 31).

However, many students can be intimidated by the amount of effort required to write (Williams & Wynne, 2000, p. 132). In addition, teachers may worry that student writing can take time away from class instruction and require a great deal of time to grade.

We find that giving our students a variety of small writing assignments throughout the year improves their learning and our teaching while keeping the tasks manageable. We recognize that every class and every teacher is unique. We recommend trying different strategies at various intervals to determine what works best.

Here are some of our favorite writing activities.

Quick Writes

One easy way to introduce writing in a classroom is to use a *quick write*, in which students are given a few minutes to write a response to a prompt. Quick writes can be given at the end of class as an exit ticket to help students summarize the lesson. They can also be given during a lesson. For example, students can record in their notebooks what they were thinking or what difficulties they encountered while solving a problem (Curcio & Artzt, 2007, p. 263).

As we mentioned earlier, we often include prompts that allow students to discuss their attitudes or feelings about math. Such questions are often more accessible to students since no prior mathematical knowledge is required (Bixby, 2018, p. 145). They can also help us to get to know students better and build a relationship with them. Here are some examples of these prompts:

- Apps like PhotoMath allow students to take a picture of an equation and immediately see the solution. Should students be allowed to use these apps on homework? Why or why not?
- How do you feel about math?

- Describe a time in any previous math class where you experienced success. Why do you think you were successful?
- Describe a time in any previous math class where you faced a challenge. How did you deal with it? What, if anything, do you wish you had done differently?

We read through students' expressions of their feelings about math to get to know them better and adjust our instruction as necessary. We can be conscious that students who say they struggle may need additional encouragement in class. We may also rearrange our classroom seating so that students who feel less comfortable with math sit with students who are more confident.

We find that for these types of prompts, students can often write substantial responses without much additional guidance. However, some students may need more support. In these cases, teachers can show model responses or provide additional prompts, such as "For example . . ." or "This made me feel . . . because . . . ."

For example, several of Bobson's students responded to the prompt "How do you feel about math?" at the beginning of the year by saying that they disliked math. They had trouble memorizing formulas and felt that much of the math that they learned in school had no connection to their lives. In response, Bobson made sure that his lessons emphasized solving problems by finding patterns and reasoning instead of simply memorizing formulas. He also highlighted lessons that had practical applications (such as calculating compound interest) and discussed real-world problems relevant to their community (like analyzing the distribution of local resources).

We also give assignments that ask students to recall prior knowledge. Researchers have found that *retrieval practice*—the act of recalling something learned in the past and thinking about it right now—increases the likelihood that students will remember the information later and apply it in new situations (Lemov, 2017; Smith & Weinstein, 2016). Retrieval practice is especially important in math, where many concepts build on previously learned ones. Since these prompts should elicit writing, they should not ask how to solve a specific problem but focus instead on explaining procedures, stating mathematical facts, or describing methods. Here are some examples of retrieval practice prompts:

- How do you multiply powers that have the same base?
- List all of the quadrilaterals (and their characteristics) that you remember.
- State the conditions that must be true in order to use the quadratic formula.

Over time, we give students more advanced prompts in which students reflect on material discussed in a lesson. These prompts can ask students to synthesize information, evaluate different solution methods, or compare and contrast two concepts. Here are some examples of more advanced prompts:

CHAPTER 4: PROMOTING MATHEMATICAL COMMUNICATION

- Describe the similarities and differences between a rotation and a reflection.
- Explain the problems that could arise when using the shortcut PEMDAS to remember the order of operations.
- What types of quadratic equations are best solved by graphing? Which are best solved by using the quadratic formula? Explain.

Student-generated Word Problems

Typically, students are given a word problem and asked to write an equation that must be solved to answer it. Another way to get students to write is to allow students to write their own word problem. To help prompt students, teachers can ask students to write a problem that is solved using an equation or diagram that we provide (Curcio & Artzt, 2007, p. 264).

Students can also ask their classmates to solve their problems. When students write for an authentic audience (as opposed to writing only for their teacher), they often become more enthusiastic about writing and take both themselves and their work more seriously (Schulten, 2018).

ELLs can benefit from writing word problems as they become more familiar with both math and English (Dong, 2016, p. 537). We encourage students to create problems that reflect their cultures or contexts familiar to them. Here is an example of a student-generated word problem:

> On Chinese New Year, Joy's extended family gathers together to eat. This year, Joy will cook with her mother and father. When working together, the three of them can cook all of the food in 2 hours. Working alone, her mother can finish cooking in 5 hours. Working alone, her father can finish in 6 hours. How long would it take Joy to cook by herself?

Error Analysis

Asking students to analyze their mistakes can help them prevent future ones. In order for students to identify the mistakes they made, they must engage in higher-order thinking since they must compare their work to what they already know (Marzano, 2000, p. 76).

Figure 4.3: Error Analysis contains a worksheet from a lesson that we often teach in the first few weeks of school. This sheet has four samples of work for a four-point question. (We vary the question on the sheet to match the course.) Students grade each sample out of four points. As a hint, we tell students that one of the samples should get four points, one should get three, one should get two, and one should get one point.

Usually, students agree on the following:

- Anna should get four points because she did all the work correctly.
- Bob should get three points because he made a "small" mistake—he said that $3x - 5x = 2x$ instead of $-2x$.
- Carlos should get two points since he made a "big" mistake—he added $5x$ twice to only one side of the equation instead of combining like terms or adding $5x$ to both sides.
- Davina made "many" mistakes to earn only one point—she subtracted $3x$ twice from only one side of the equation and wrote that $3 - 19 = 16$ instead of -16.

We then help students articulate the difference between the mistakes in the sample work by defining *conceptual errors* (errors in thinking) and *computational errors* (errors in calculation) and explaining the difference between them. (We discuss this difference more in Chapter 9: How to Plan Tests and Quizzes.) After sharing the scoring guidelines for the problem, we ask students to regrade the sample work and explain the error contained in each sample using the language from the scoring guidelines. Students should then be able to say that Bob made a computational error, Carlos made a more serious conceptual error, and Davina made both.

Lesson Summaries

Another way to encourage writing on a regular basis is to ask students to write a short summary of each lesson. We ask students to include important mathematical ideas from each lesson as well as one example with all work correctly shown. Lesson summaries can be written either at the end of the notes for a lesson or in a separate notebook. One advantage of having lesson summaries in a separate notebook is that students can then use them as a review book for the course.

We provide several scaffolds to help students write lesson summaries. As we teach our lessons, we also highlight important definitions, theorems, and other mathematical facts on the board so that students know that they should include these facts in their summaries. When time permits, we allow students to write a brief summary at the end of class. To give students more structure when writing summaries, teachers can provide prompts like "What were the big ideas from today's lesson?" or "How does today's lesson relate to previous lessons?" In addition, we publish our lesson presentation files with our boardwork online in a format that students can easily access, such as PDF. Doing so not only allows students to refine or write their summaries at home but it also enables students to look through the lesson at their own pace. Publishing notes online can also help ELLs process the lesson more easily.

CHAPTER 4: PROMOTING MATHEMATICAL COMMUNICATION

Students can find lesson summaries overwhelming since a lesson summary could be done almost every day. Teachers who want to review the summaries may find that grading them can take too much time. One option is to ask students to write weekly instead of daily summaries, in which they could explain three or four big ideas from the week. Students could also work in groups, so each student in a group would be responsible for writing one summary.

We find that lesson summaries are an effective strategy for getting students to synthesize mathematical content. We prefer that students focus on summarizing the material discussed in the lesson. However, we encourage them to include any comments or questions that may come up as they write their summaries.

Figure 4.4: Lesson Summary shows a student example of a lesson summary. Table 4.4: Scaffolded Lesson Summary shows an example of a scaffolded lesson summary.

Table 4.4 Scaffolded Lesson Summary

DATE : _____

AIM # _____ : _____

BIG IDEAS (such as definitions, theorems, formulas):

MODEL EXAMPLES:

Problem: **Solution:**

Problem: **Solution:**

ADDITIONAL COMMENTS (such as questions, concerns, suggestions):

Correspondence Journal

Students can also maintain a correspondence journal, in which students can write about their anxiety, lack of confidence, frustration, pride, and satisfaction (Cole & Feng, 2015, p. 12). Teachers can then periodically collect them and add comments. Correspondence journals can create a personal dialogue between teachers and students, which is especially useful when teachers don't have enough time for each student to talk through ideas in class. They can also provide a private space for students to reflect thoughtfully and express themselves freely without worrying about what their peers might think (Cole & Feng, 2015, p. 13). Obviously, the time involved for maintaining correspondence journals with 150 students can be quite challenging.

Although all of the writing strategies that we describe can be done on separate sheets of paper, we usually ask students to keep their mathematical writing in a separate notebook. Journal notebooks can thus be a convenient way to keep all of the different writing exercises that we describe in this chapter.

Maintaining a journal notebook has several advantages. Students can more easily monitor their progress. In addition, students can personalize their journal notebooks with decorations, which helps them take more pride in their work and motivates them to continue writing. Using a separate notebook also facilitates a written dialogue between students and teachers (Cole & Feng, 2015, p. 11).

Since many of our students aren't used to expressing their thoughts by hand, we also give students the option of keeping an electronic journal. We find that many students find graphing easier on a computer (since they can simply paste graphs into a document) but often have difficulty typing complicated mathematical expressions. One possible compromise is to allow students to print out typed copies of their journals and write in mathematical symbols by hand. The Technology Connections section of this chapter lists resources that students can use to maintain an electronic journal.

Real-world Connections

Another writing technique that teachers can use is to ask students to write about how a math lesson connects to the real world. Mathematical modeling can increase student motivation by showing students how math can be used to answer important questions (Garfunkel & Montgomery, 2016, p. 8). For example, students can briefly write about how they could use a lesson in real life. Teachers could also ask students to do some research for homework about how concepts used in the lesson are used in the real world.

We don't try to incorporate mathematical modeling into every lesson since many don't have an immediate application. However, we take advantage of it whenever a lesson has a direct connection to the real world. We discuss how we use mathematical modeling as a motivational tool, along with how we motivate students when we *don't* use modeling, in Chapter 1: Motivating Students.

As we stated earlier, we are not suggesting that you use each of these writing ideas every day, every week, or even every month. We encourage you to use these ideas as frequently or as infrequently as possible, depending on what makes sense for you and your classes.

Grading Student Writing

We find that grading writing assignments is much easier when we use scoring guidelines. Scoring guidelines clearly communicate our expectations to students and remove much of the uncertainty around grading.

Writing assignments that contain primarily mathematical content have scoring guidelines that focus on organization and correct use of mathematical language. Table 4.5 shows the guidelines that we use to grade lesson summaries. The guidelines emphasize mathematical accuracy while also encouraging students to express their comments or concerns.

Table 4.5 Scoring Guidelines for Lesson Summaries

Criteria	Points
FORMAT: Summaries are written in a separate graph paper notebook in chronological order	_____ out of 2 points
COMPLETENESS: Appropriate length (1 paragraph) for each summary	_____ out of 2 points
MAIN IDEAS: Summarizes all important mathematical ideas (such as formulas, theorems, definitions, diagrams)	_____ out of 2 points
LANGUAGE: Uses proper mathematical text and symbols	_____ out of 2 points
EXAMPLES: Includes at least one relevant example from each main idea *with all correct work*	_____ out of 2 points
ADDITIONAL COMMENTS: Includes comments, concerns, or suggestions for improvement	_____ out of 2 points (extra credit)
TOTAL:	_____ **out of 10 points**

In contrast, when we give other types of writing assignments that allow students to share their thoughts, we use scoring guidelines that focus more on effort than mathematical correctness. Research indicates that praising effort and reflecting on mistakes can motivate students (Stein, 2007, p. 287) and encourage them to think more deeply (Bixby, 2018, p. 146).

These actions support the development of a growth mindset (the idea that mathematical knowledge can be increased) instead of a fixed mindset (the idea that people are born with an unchangeable amount of mathematical knowledge) (Allen & Schnell, 2016, p. 400). Offering positive feedback that praises students' effort ("You gave a lot of thought to this response") can strengthen a growth mindset by suggesting that hard work will help them improve. Students with a growth mindset are more likely to learn from mistakes and keep trying. They see the mistakes as opportunities to increase their knowledge. In contrast, students who are praised for their intelligence ("You're really smart") often lose confidence when they begin to struggle. They become less motivated when they encounter setbacks since they don't believe they can improve their knowledge (Dweck, 2007).

Table 4.6 is an example of scoring guidelines for student quick writes that focuses on students' reflection of the material.

In addition to using scoring guidelines to grade, we *try* to write personalized comments for each student. In general, we try not to focus exclusively on grammatical or mathematical errors when we evaluate student writing. When students are graded mostly on their mistakes, they can feel more frustrated, which can discourage them from writing more and ultimately prevent them from getting enough practice to improve (Cole & Feng, 2015, p. 7). We find that writing comments that elicit more thought, such as "Explain what you mean by this," encourages students to continue writing.

Table 4.6 Scoring Guidelines for Quick Writes

Criteria	Points
PROMPT: Answers the question posed in the prompt	____ out of 2 points
EVIDENCE: Provides sufficient evidence (examples, details, etc.) to support statements	____ out of 3 points
CLARITY: Writing is clear, concise, and logically organized	____ out of 3 points
MATHEMATICS: Uses appropriate mathematical language correctly	____ out of 2 points
TOTAL:	____ **out of 10 points**

DIFFERENTIATING FOR ELLS AND STUDENTS WITH LEARNING DIFFERENCES

ELLs and students with learning differences may face particular challenges when communicating mathematically. Here are some strategies that we find effective in working with them, as well as with all students.

We find that applying many of the strategies that we describe in Chapter 3: Teaching Math as a Language—such as using vocabulary lists, flash cards, and word walls—can help ELLs develop academic language. ELLs can also benefit from writing collaboratively on an assignment (Read, 2010, p. 5). Students can be paired in different ways. Working with English-proficient students can expose ELLs to more fluent English. Pairing ELL students with others with similar ability in English and another language may make students more willing to participate. Teachers may also pair ELLs with speakers of the same language, speakers of different languages, or speakers proficient in English, depending on the resources and students that teachers have (De Jong & Commins, n.d.).

One strategy that can be particularly helpful for ELLs is to show them videos of a mathematical explanation in their home language or in English with captions (see the Technology Connections section for these resources). They can watch these videos on a device in class or at home. Ideally, teachers would tell them in advance which topics will be discussed.

ELLs can also benefit from additional support to organize their writing. For example, teachers can provide additional prompts for correspondence journals or other reflective writing exercises, such as:

- I didn't understand _____ because _____ .
- One question that I wish I had asked in class today but didn't was _____ .
- One thing I wish I had said in class today but didn't was _____ .
- Today's lesson reminded me of something I learned earlier in math: _____ .

To help students write lesson summaries, teachers can create scaffolded writing guides like the one in Table 4.4: Scaffolded Lesson Summary. This scaffolded writing guide gives students more structure for summarizing lessons. Students can fill in the date, aim, main ideas for the lesson, examples, and comments.

What Could Go Wrong

DEALING WITH STUDENT MISTAKES

If a teacher criticizes a student for making an error, a lot can go wrong.

Often, students get discouraged by their mistakes. We recommend treating mistakes as learning opportunities in which all students can collaborate (Stein, 2007, p. 287). In other words, we try to treat student ideas as "works in progress" that can be refined (Staples & Colonis, 2007, p. 259). For example, we may say, "These ideas can't both be true. Which one is right?" Depersonalizing student mistakes—referring to a mistake as "this way of thinking" instead of "what Maria did wrong"—also helps students feel less ashamed. Sometimes we encourage students to rethink mistakes by saying, "This is the right answer to a different question. What question is that?"

Many times, students are reluctant to share their work with others because they are afraid of making mistakes or looking foolish in front of their peers. We deal with this by frequently inviting students to share incorrect work with the class as a way of illustrating common errors and having classwide discussions about them.

Sometimes, students laugh at their classmates' mistakes. In these situations, we immediately say that laughing at someone shows disrespect and violates our classroom rule about respecting ourselves and each other. (We say more about rules in Chapter 11: Building a Productive Classroom Environment.) We sometimes deflect the conversation away from the student by saying that we made a similar mistake when we were in school. We then talk about why the mistake might be more common than students realize and discuss the misconceptions that can arise.

DEALING WITH TEACHER MISTAKES

When we model mathematical reasoning through a think-aloud or similar strategy, we often make the mistake of not making *enough* mistakes. We may simply present the correct solution method and not include common errors that students may make. As a result, students get the impression that they, too, should get problems right on the first try. They can then get frustrated when they can't reproduce our apparent perfection.

When we think through a problem, we sometimes try methods that don't lead to the correct solution, but we model the reflective process of constantly looking at our work to see if it leads to a reasonable solution. For example, if we are solving a word problem where we need to find the height of a building but our answer appears to be negative, we then reevaluate our work.

Other times, the mistakes we make are actually unintentional. In these situations, we freely acknowledge them and work with students to find our error. This not only helps model metacognitive thinking but also makes us more human!

PROBLEMS IN DISCOURSE

One mistake that we've made when trying to encourage students to talk is to give insufficient wait time after asking a question. As we said earlier in this chapter, waiting up to 15 seconds gives students time to think and respond (Goodwin, 2014; McCarthy, 2018).

Sometimes, we find that students simply don't talk. They may feel awkward about speaking in class, they might not be fluent enough in English to articulate their question, they may not understand the question, or they may just be tired. In these cases, we try to rephrase the question or approach it from a different perspective. We find that providing more background information (such as giving additional direct instruction) often encourages more students to speak up.

At other times, we elicit student answers too quickly. In these situations, we often wind up calling on the same people. This can erode the motivation of other students who may give up trying to answer questions if they think that we won't call on them or allow them to think. When possible, we resist the temptation to answer every question ourselves.

In the end, being flexible in class—addressing important student questions as efficiently as possible while making sure that we have enough time to meet the lesson's objectives—works best.

FINDING THE TIME

We often struggle to find enough opportunities for students to write about math. For example, students may not have much experience writing in a math class, or we may be unsure where we can fit writing into a packed curriculum. Incorporating short writing opportunities in our lessons or homework can make them more manageable for both students and teachers.

Student Handouts and Examples

Figure 4.1: Algebra Tiles Activity

Figure 4.2: Which One Doesn't Belong?

Table 4.3: Problem-Solving Chart

Figure 4.3: Error Analysis

Figure 4.4: Lesson Summary

Table 4.4: Scaffolded Lesson Summary

Technology Connections

The Which One Doesn't Belong? website (http://wodb.ca) has many free examples of the activity. The *Which One Doesn't Belong?* book website by Christopher Danielson (http://www.stenhouse.com/content/which-one-doesnt-belong) also has a collection of free resources related to the book.

Several online tools allow students to maintain journals. The free Google Docs (http://docs.google.com) word processor, which is integrated with the Google Classroom (http://classroom.google.com) content management system, allows teachers to grade assignments and offer feedback online. To insert math symbols into digital documents, students, and teachers can use software like EquatIO (http://equat.io) or MathType (http://store.wiris.com).

To pick students randomly for group or whole-class discussion, teachers can use any of the widely available online random student name selectors available. Random student selectors are also available for mobile devices (search your phone's app store for examples) and presentation software like SMART Notebook and Microsoft PowerPoint. A random student name generator is also available for Google Classroom as a free extension to the Chrome web browser (http://chrome.google.com/webstore/detail/random-student-generator/kieflbdkopabcodmbpibhafnjalkpkod).

Khan Academy (http://khanacademy.com) has math videos in a variety of languages that students can use to help them learn mathematical vocabulary. Several publishers have online glossaries with terms in different languages. For example, Holt, Rinehart, and Winston has glossaries for middle school (https://my.hrw.com/math06_07/nsmedia/tools/glossary/msm/glossary.html) and high school (https://my.hrw.com/math06_07/nsmedia/tools/glossary/aga/glossary.html). Many school districts and states provide glossaries and other print resources for ELLs to use on standardized tests. For example, New York State provides bilingual mathematical glossaries in over 30 languages (http://steinhardt.nyu.edu/metrocenter/resources/glossaries).

Both Microsoft PowerPoint (http://products.office.com/en-us/powerpoint) and the free Google Slides (http://slides.google.com) have the ability to add closed captioning to presentations, which can help students (particularly ELLs). Exact instructions can be found by searching online for "enable closed captioning in [software name]."

Attribution

Christopher Danielson's book *Which One Doesn't Belong?* (Danielson, 2016) inspired us to create our own eponymous examples.

George Pólya's classic book *How to Solve It* (Pólya, 1945) has been an invaluable resource in helping us teach students to think and communicate mathematically.

Figures

Algebra II

AIM #1.01: How do we use algebra tiles to represent polynomials?

Do Now

Find the area of each rectangle below.

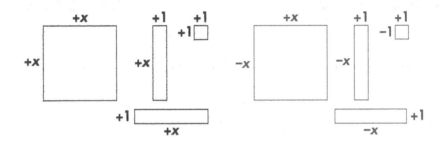

Level 1

Write the polynomial in standard form that is represented by each set of algebra tiles.

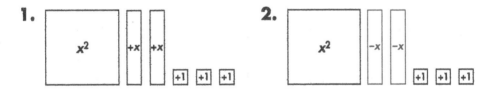

Level 2

Represent each polynomial below using the least number of algebra tiles possible.

3. $2x^2 - 4x + 1$ **4.** $-x - 4$

Level 3

For each problem below, draw the algebra tiles that you use. Label each tile $+1$, -1, $+x$, $-x$, x^2, or $-x^2$.

5. Represent 0 using 2 tiles. (Why do you think this is called a "zero pair"?)

6. Represent $+2$ using 8 tiles.

7. Represent 0 using 6 tiles in at least two different ways.

Figure 4.1 Algebra Tiles Activity

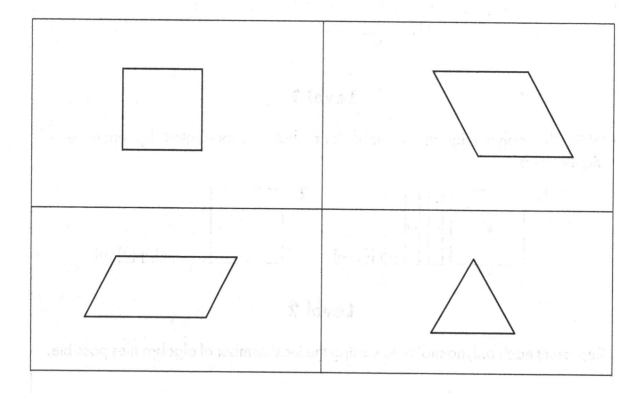

Figure 4.2 Which One Doesn't Belong?

Algebra I

AIM #___: How do we grade constructed-response questions?

Four students were asked the following question on an exam:

Solve algebraically for x : $3(x + 1) - 5x = 12 - (6x - 7)$

1. The four students' work is shown below. The question is worth 4 points. Give each student 1, 2, 3, or 4 points. (HINT: One student should get 1 point, one should get 2 points, one should get 3 points, and one should get 4 points.) Write a short (1–2 sentence) explanation of each grade.

2. One of these students made what is called a "conceptual error." One student made what is called a "calculation error." One student made *both* a conceptual and a computational error. One student made *no* errors. Determine which student made which errors.

3. Based on your work in the previous questions, explain the difference between a conceptual error and a computational error.

4. Now read the scoring guidelines for the question and re-grade each response using the guidelines. Write a short (1–2 sentence) explanation of each grade using language from the rubric to help you.

Figure 4.3 Error Analysis

THE MATH TEACHER'S TOOLBOX

Anna
Solve algebraically for
x: 3(x + 1) − 5x = 12 − (6x − 7).

$3x + 3 - 5x = 12 - 6x + 7$
$3 - 2x = 19 - 6x$
$+2x +2x$
$\overline{}$
$3 = 19 - 4x$
$+4x +4x$
$\overline{}$
$4x + 3 = 19$
$-3 -3$
$\overline{\frac{4x}{4} = \frac{16}{4}} \quad \boxed{x = 4}$

1st Grade:
___ out of 4

Error made:
☐ Conceptual
☐ Computational

2nd Grade:
___ out of 4

Explanation of Anna's error:

Bob
Solve algebraically for
x: 3(x + 1) − 5x = 12 − (6x − 7).

$3x + 3 - 5x = 12 - 6x + 7$
$2x + 3 = 19 - 6x$
$+6x +6x$
$\overline{}$
$8x + 3 = 19$
$-3 -3$
$\overline{8x = 16}$
$\div 8 \div 8$
Answer: $x = 2$

1st Grade:
___ out of 4

Error made:
☐ Conceptual
☐ Computational

2nd Grade:
___ out of 4

Explanation of Bob's error:

Carlos
Solve algebraically for
x: 3(x + 1) − 5x = 12 − (6x − 7).

$3x + 3 - 5x = 12 - 6x + 7$
$+5x +5x$
$\overline{}$
$8x + 3 = 12 - 6x + 7$
$8x + 3 = 19 - 6x$
$+6x +6x$
$\overline{}$
$14x + 3 = 19$
$-3 -3$
$\overline{\frac{14x}{14} = \frac{16}{14}} = \boxed{\frac{8}{7}}$

1st Grade:
___ out of 4

Error made:
☐ Conceptual
☐ Computational

2nd Grade:
___ out of 4

Explanation of Carlos's error:

Davina
Solve algebraically for
x: 3(x + 1) − 5x = 12 − (6x − 7).

$3x + 3 - 5x = 12 - 6x + 7$
$-3x -3x$
$\overline{}$
$3 - 8x = 12 - 6x + 7$
$+8x +8x$
$\overline{}$
$3 = 12 + 2x + 7$
$3 = 19 + 2x$
$16 = 2x$
$x = 8$

1st Grade:
___ out of 4

Error made:
☐ Conceptual
☐ Computational

2nd Grade:
___ out of 4

Explanation of Davina's error:

---

5. What parts of the scoring guidelines surprised you? Why?
6. Based on this activity, explain what you will do differently on a question like this.

Figure 4.3 (Continued)

CHAPTER 4: PROMOTING MATHEMATICAL COMMUNICATION

Aim 1.29: When are rational expressions undefined?

A rational expression is a quotient of two polynomials with a nonzero divisor OR A rational expression is a fraction of the form $\frac{p}{q}$ where p and q are polynomials and q is not a zero. You just need to equal denominator to zero and write undefined when variable (x) = ____

8) $\frac{9}{3x-4}$

$3x - 4 = 0$
$+4 +4$
$\frac{3x}{3} = \frac{4}{3}$
$x = \frac{4}{3}$

Undefined when $x = \frac{4}{3}$

16) $\frac{x-2}{x^3 - 4x^2 + 4x}$

Undefined when $x^3 - 4x^2 + 4x = 0$
or
when $x(x^2 - 4x + 4) = 0$
$x(x-2)(x-2) = 0$
$x = 0$ or $x - 2 = 0$

Undefined when $x = \{2, 0\}$

Figure 4.4 Lesson Summary

CHAPTER 5

Making Mathematical Connections

What Is It?

As we said in the Introduction, mathematics is not a collection of isolated topics. Algebra, geometry, statistics, trigonometry, number theory, and other areas of mathematics are all related to each other. Functions can be represented algebraically or graphically. Probability can be calculated algebraically or geometrically. Fractions are used in number theory, algebra, trigonometry, geometry, and statistics. Making connections is central to mathematical thinking.

In this chapter, we focus on *mathematical connections*, the ways in which the content that we teach relates to other mathematical concepts. As students acquire knowledge in different areas, they can get better at comparing information and seeing the underlying structure that connects them (Willingham, 2002). Making these mathematical connections helps students *transfer learning*—extend what was learned in one situation to another context (Darling-Hammond, Austin, Shulman, & Schwartz, n.d.-a, "Lessons," p. 190).

This chapter doesn't focus on showing how math is used in other subjects, such as science or economics. Instead, it focuses on the ways that what we teach can be related to other mathematical ideas.

Why We Like It

We find that creating opportunities for students to see mathematical connections has many benefits. When students see math as a logical and consistent system of ideas, they become better at solving problems since they can use multiple viewpoints (Darling-Hammond et al., n.d.-a, "Lessons," p. 190). In addition, when they relate new material to previously learned ideas, they can see the "big picture" and may gain a greater appreciation of math.

As we said in Chapter 1: Motivating Students, real-world applications, while important (which is why we discuss them throughout this book), can often be difficult to find for many topics. In these situations, we find that making mathematical connections are easier since they usually relate more readily to students' prior knowledge.

Supporting Research

Many studies indicate that students who represent mathematical ideas in different ways (Beckmann & Izsák, 2015, p. 35; Blanton, Brizuela, Gardiner, Sawrey, & Newman-Owens, 2015, p. 542; Earnest, 2017, p. 217; Hackenberg & Lee, 2015, p. 230; Roy et al., 2016, p. 495) and connect procedures to larger concepts (Woodbury, 2000, p. 230) can improve their reasoning skills and deepen their understanding.

In addition, neuroscience research indicates that experts, unlike novices, organize knowledge around core concepts in their field (National Research Council, 2000, p. 36). To save information to long-term storage, learners must make sense of it by placing it in the context of past experiences (Sousa, 2017, p. 55).

Common Core Connections

Using appropriate tools strategically (MP.5), looking for and making use of structure (MP.7), and looking for and expressing regularity in repeated reasoning (MP.8) are included in the Common Core's Standards for Mathematical Practice (NGA & CCSSO, 2010, pp. 7–8). These standards emphasize examining and comparing different mathematical representations.

In addition, the Common Core standards were designed to cover several major themes, such as statistics, geometry, and ratios, coherently across grades. Many of the mathematical connections in the Common Core are explained in the progressions documents written by the authors of the standards (Common Core Standards Writing Team, 2013a, p. 4). In the Application section that follows, we explore some of these connections in more detail.

Application

Since the point of mathematical connections is to strengthen understanding and improve problem-solving skills, we want to make these links visible to students in our teaching. Many mathematical connections fit into four major themes that span across K–12 Common Core math:

- **Equivalence:** *Equivalent* quantities are mathematically equal but represented in different ways.

CHAPTER 5: MAKING MATHEMATICAL CONNECTIONS

- **Proportionality:** Pairs of numbers are in a *proportional* relationship if they are in equivalent ratios (Common Core Standards Writing Team, 2011, p. 14).
- **Functions:** *Functions* are mathematical relationships in which each input value determines exactly one output value (Common Core Standards Writing Team, 2013b, p. 1).
- **Variability:** *Variability* is the extent to which values taken from a sample differ from the rest of the population. Measuring this variability helps mathematicians to predict future behavior.

In this section, we discuss some of the ways that we make mathematical connections among these four "big ideas."

EQUIVALENCE

From elementary to high school, students must demonstrate that expressions are equivalent by composing and decomposing them (Common Core Standards Writing Team, 2013a, p. 10). Often referred to as *transforming, rewriting,* or *seeing structure in* expressions, this skill is critical for making connections between topics. Here are some of our favorite examples of using equivalence.

Arithmetic from Numbers to Polynomials

The algebra that we teach in high school is an extension of the arithmetic that students learn in elementary school. We often introduce algebraic topics by making connections with their prior experiences in elementary math.

When we teach adding and subtracting polynomials, we show that it follows many of the same rules as adding and subtracting whole numbers, as shown in Figure 5.1: Addition and Subtraction of Polynomials. These operations with polynomials are based on the same concepts as the corresponding operations with numbers. We use the questions as an activity at the beginning of a lesson to activate prior knowledge on place value and arithmetic operations and relate it to polynomial operations.

Multiplication and Division with the Area Model

To multiply polynomials, we use the *area model* (sometimes called the *box method* or *tabular method*). The area model is based on the idea that multiplication can be represented using rectangles, whose area is the product of their lengths and widths.

We introduce the area model with the introductory activity shown in Figure 5.2: Multiplication with the Area Model. We start by using the area model to multiply whole numbers. In the example 32 · 21, students represent the product as a rectangle whose length is 32 (written in expanded form as 30 + 2) and width is 21 (written as 20 + 1). Students find the area of each rectangle by multiplying its length and width. The total area, which represents the product, equals the area of each of the four smaller rectangles, so $32 \cdot 21 = (30 + 2)(20 + 1) = 600 + 40 + 30 + 2 = 672$. If necessary, we do additional problems with multiplying whole numbers. We then proceed to multiplying polynomials, showing students that calculating $32 \cdot 21$ is similar to simplifying $(3x + 2)(2x + 1)$.

The area model can also be used to divide whole numbers by reversing the process of multiplication, as shown in Figure 5.3: Division with the Area Model. Since many students are familiar with long division, we ask students to describe the steps in an example while showing the corresponding steps using the area model.

We find that the area model helps students solve quadratic equations without having to memorize a formula. Figure 5.4: Completing the Square shows the solution to the equation $x^2 + 6x - 2 = 0$. Students draw an incomplete square to represent the trinomial $x^2 + 6x +$ ___ as $(x +$ ___$)^2$ and add a constant (in this case, 9) to both sides of the equation so that they can rewrite the expression $x^2 + 6x +$ ___ as the perfect square $(x + 3)^2$. The area model thus provides a clear visual explanation of why this method is called *completing the square* (Wong, 2018).

We also use the area model to help students determine the characteristics of a circle given its equation in general form, as shown in Figure 5.5: Determining the Center and Radius of a Circle. Students rewrite the equation in standard form by completing the square twice, once for each variable (Wong, 2018).

In short, we find that using the area model has several advantages:

- Since it covers topics from elementary school to precalculus, teachers across grades can use the same language to explain and demonstrate concepts. This makes math more coherent and avoids learning separate procedures for multiplying whole numbers, expanding binomials, dividing whole numbers, multiplying complex numbers, and completing the square.
- Students can see that operations with numbers and operations with polynomials have similar structure.
- Students can improve their conceptual understanding and create visual representations for various operations.
- Students will be able to develop vocabulary to explain why common misconceptions like $(a + b)^2 = a^2 + b^2$ are incorrect by providing an appropriate area model, as shown in Figure 5.6: Why $(a + b)^2 \neq a^2 + b^2$.

PROPORTIONALITY

Ratios and proportional relationships are critical for mathematical understanding. They extend elementary schoolwork in units and division into high school algebra (average rate of change), geometry (slope and similar figures), and trigonometry (trigonometric ratios) (Common Core Standards Writing Team, 2011, p. 1).

Here are some examples of the mathematical connections that we make that deal with proportional reasoning.

Similarity

When we discuss similarity in high school geometry, we connect it to students' prior knowledge of ratios, proportions, and dilations from middle school. Figure 5.7: Ratios and Similarity shows questions that we use to summarize our lesson on corresponding parts of similar triangles. The questions show that dilations (which middle school students discuss with scale diagrams) of similar triangles have a scale factor that is the ratio of the corresponding sides.

We extend this proportional reasoning in geometry to relate the areas of similar polygons, as shown in Figure 5.8: Areas of Similar Polygons. By comparing the areas of similar polygons, students can see that the ratio of the areas of similar polygons is the square of the ratio of the corresponding side lengths. Similarly, we extend proportionality into three dimensions, as shown in Figure 5.9: Volumes of Similar Solids. Here, students see that the ratio of the volumes of similar solids is the cube of the ratio of the corresponding side lengths.

Circles

Another example of proportional reasoning occurs when we discuss parts of circles. Figure 5.10: Arc Length and Sector shows the connection between the central angle measure, the arc length, and the area of a sector. In this problem, students notice that the ratio of the arc length to the circumference (in context, the ratio of the amount of pizza crust eaten to the total amount of pizza crust) and the ratio of the sector area to the total area (in context, the ratio of the pizza eaten to the pizza's area) are proportional to the central angle measure.

Proportional reasoning in circles is helpful to introduce radian measure. When we use this example, we point out that the division of a full circumference into 360° makes many calculations in advanced math more complicated. In contrast, using radians relates angle measure to circumference since a full rotation has a radian measure of 2π, which is also the length of the circumference when the radius is 1. This allows students to simplify calculations with angle measure, arc length, and sector area, as shown in Figure 5.11: Proportional Reasoning in Circles. In this example, students can see that in a unit circle, a central angle measure in radians is equal to its arc length and proportional to its sector area.

We also point out that using radians has advantages in advanced math classes. For example, $\lim_{\theta \to 0} \frac{\sin(\theta)}{\theta} = 1$ when measured in radians. When measured in degrees, this limit equals a much messier number, $\frac{\pi}{180}$. We obviously don't expect middle school or geometry teachers to explain or even show students this formula! Instead, we tell students that mathematicians have devised different ways of measuring angles to simplify calculations.

FUNCTIONS

Functions—bivariate relationships in which each input gets mapped to exactly one output—are one of the most important topics in high school math. In fact, New York State's EngageNY curriculum calls grade 9–12 math a "story of functions" (EngageNY, 2013, p. 2).

Students lay the groundwork for understanding functions in elementary school, when they learn how to recognize and describe patterns. In middle school, students start to describe numerical relationships between two quantitative variables more formally using multiple representations of a function (Common Core Standards Writing Team, 2013b, p. 4). To match the different ways that real-world functions can be modeled, students learn to express functions in four ways: verbally, numerically (with a table of values), graphically, or algebraically (with an equation).

We use Figure 5.12: Four Views of a Function when we introduce functions. Given a verbal representation of a function, students can work in groups of three so that each person completes a representation of a function. By seeing the representations side-by-side, students can begin to see the relationships between the functions' characteristics more easily.

Here are some of the ways that we use multiple representations of functions to make mathematical connections.

Rate of Change

One common application of multiple representations of a function is calculating and interpreting the rate of change. The rate of change of a linear function ties equivalent ratios on the coordinate plane to the slope of the line (Common Core Standards Writing Team, 2013b, p. 4). Figure 5.13: Rate of Change is an example of a problem that can be used to relate slope, rate of change, and proportionality in functions. Students use multiple representations of a linear function to calculate the rate of change and relate it to the slope of the line. The rate of change is later extended to the *average* rate of change for a nonlinear function and ultimately the *derivative*, the instantaneous rate of change.

Characteristics of Functions

We also use multiple representations of a function to relate its intercepts to its zeros, as shown in Figure 5.14: Characteristics of Polynomial Functions. We incorporate these questions into a lesson in our polynomials unit. Students can work in groups of two or three so that each student can complete different tasks. For example, one or two students can complete the table of values and write the equation, while the other can draw the graph. Students should see that the x-intercepts of the graph are also the zeros of the polynomial function and the solutions of the equations formed when each factor is set equal to 0. They can also relate the term *intercept* with the table of values. The x-intercept, which is at the point where the graph *intercepts* the x-axis, is the value in the table where $y = 0$. Similarly, the y-intercept, which is at the point where the graph *intercepts* the y-axis, is the value in the table where $x = 0$.

Even and Odd Functions

Other applications of multiple representations occur when we discuss other types of functions. To introduce even and odd functions in Algebra II, we use the introductory activity in Figure 5.15: Even and Odd Polynomial Functions. We typically divide the class into pairs or groups using a technique like the jigsaw method (which we describe more in Chapter 17: Cooperative Learning). One group of students (or one person in each pair) completes the table and graph for the even functions while the other group (or the other person in each pair) completes the table and graph for the odd functions. Students will then come together to complete the summary, stating the characteristics of even and odd polynomial functions based on their equations, tables, or graphs. They should see that even polynomial functions have even exponents, follow the formula $f(x) = f(-x)$, and have line symmetry over the y-axis, while odd polynomial functions have odd exponents, follow the formula $-f(x) = f(-x)$, and have point symmetry over the origin. Since many students have trouble articulating the pattern in the tables, we provide the formal mathematical definitions.

We use a similar technique in Algebra II for analyzing trigonometric functions later in the year, as shown in Figure 5.16: Why $f(x) = \sin(x)$ Is Odd and $g(x) = \cos(x)$ Is Even. As with the activity that we described earlier with even and odd polynomial functions, we divide the class into pairs or groups, with each person or group completing the questions for one function and then coming together to answer the summary questions. Students can see that even though the sine and cosine functions have no exponents, they still follow many of the rules of even and odd functions. They can also generalize that $\sin(-x) = -\sin(x)$ and $\cos(-x) = \cos(x)$.

Although we find multiple representations of functions useful, we don't require that students create a table, write an equation, and draw a graph *every* time they encounter a function. Doing so would be unnecessarily tedious and time-consuming. Furthermore, using just *one* representation is often enough to answer a question. Our goal is to help students become more fluent with all three representations so that they can choose the one that works best for them.

VARIABILITY

Many students (and teachers) struggle to find a connection between statistics (which seems "fuzzy" because it deals with the variability of data) and other areas of math, which appear to be more clear-cut. However, we find that in many ways, teaching statistics enables students to make connections to algebra, geometry, proportionality, and functions, all of which are major topics in the Common Core standards. Here are some of our favorite examples.

Linear Regression

When we discuss *linear regression*—modeling a relationship between two quantitative variables with a linear function, we use many of the tools and strategies that we applied when analyzing proportional relationships. However, we modify our interpretation of our mathematical model to reflect the variability that results when we sample from a population.

One example of these modifications is shown in Figure 5.17: Linear Regression. We use this example after teaching the mechanics of creating a line of best fit from a data set and calculating the correlation coefficient r. This activity could also be modified to introduce linear regression by omitting part a of #2. Students start by writing the equation of a line based on the given graph. They can interpret the slope in #1 with a sentence like "for every increase of 1 hr in rental time, the rental cost increases by $6." However, since there is more variability in #2, students cannot say that there is a direct variation between the number of miles run and the running time. In fact, none of the points are on the regression line. Since the regression line represents an *average* increase, students must interpret its slope with a sentence that reflects the data's variability, such as "for every increase of 1 mile that Isabella runs, her running time increases by 6.64 min *on average*."

Using Tables to Represent Probability

Students use a wide variety of tools to represent and calculate probability, which can often confuse them. In Grade 7, students develop probability models to represent verbal descriptions of chance events and find the probability of compound events using tables, tree diagrams, and sample spaces. In high school, students transition to formulas for the probability of compound events, conditional probability, and independence (NGA & CCSSO, 2010, pp. 50–51, 82). Many students find this topic disjointed since they see few connections between probability and other concepts in mathematics.

In our experience, using tables when solving probability problems helps make it more coherent and accessible for students. Middle school students create tables to represent various relationships between two quantities in problems like the one below:

> On a school field trip, 756 people traveled by bus or van. Each bus held 55 people, while each van held 12 people. There were 4 more buses than vans. How many people in all rode by bus? How many people in all rode by van?

Students can solve this problem by creating a table similar to Table 5.1.

They can then write the equations $12v + 55b = 756$ and $v + 4 = b$ and solve the system to answer the question.

When we introduce probability to students, we like to use tables to tie probability to ratios and functions, as shown in Figure 5.18: Long-Run Relative Frequency. Students can work in pairs or small groups to track the relative frequency of tails that appear when a fair coin is flipped 50 times. By creating a table of values and graphing the results, students represent relative frequencies as a function of the number of tosses in different ways. In addition, students can see that the empirical probability of an event approaches its theoretical probability after many repeated trials. They can conclude that probability represents a *long-term* relative frequency of a chance outcome.

When we teach probability formulas, we use *two-way tables* (also called *contingency tables*) that show the observed frequencies for two sets of qualitative data. Figure 5.19: Two-Way Tables shows questions that we incorporate into our Algebra lesson on calculating the probability of the intersection and union of events. Using two-way tables of race and ethnicity data from the US Census Bureau, students can work independently or in pairs to find different compound probabilities. The table

Table 5.1 Table for Vehicle Word Problem

Vehicle type	Number of vehicles	Number of people per vehicle	Total number of people
Van	v	12	$12v$
Bus	b	55	$55b$

provides a visual explanation of the formulas for the probability of the intersection and union:

- The probability of the *intersection* of two events is connected to finding the joint frequency at the *intersection* of the appropriate row and column in the table. For example, the probability that a randomly selected American is white and Hispanic is $\frac{38}{326} \approx 0.12$.
- The probability of the *union* of two events is connected to finding the *sum* of all joint frequencies in the appropriate row and column. For example, the probability that a randomly selected American is white or Hispanic is $\frac{(38+1+3+17+197)}{326} \approx 0.79$.

After doing many problems, students notice a pattern in the solution, which we write more formally with the formula $P(A \text{ or } B) = P(A) + P(B) - P(A \text{ and } B)$.

We also like to use two-way tables to calculate *conditional probability*, the probability that an event happens given that another event has already occurred. Figure 5.20: Conditional Probability shows questions that we give to our classes at the beginning of our lesson on conditional probability. From the two-way table of race and ethnicity data from Figure 5.19: Two-Way Tables, students calculate different conditional probabilities. Since many students have trouble understanding how to interpret the word "given" in this context, we explain to them that they can use the appropriate marginal frequency (the row or column total) as the denominator in their conditional probability calculation. For example, by reading the table properly, students can see that out of 235 million white Americans, 38 million are Hispanic, so the probability that a randomly chosen American is Hispanic given that they are white is $\frac{38}{325} \approx 0.16$. Students can thus calculate conditional probability without using the formula $P(A \text{ given } B) = \frac{P(A \text{ and } B)}{P(B)}$.

Using two-way tables can be challenging for students, especially English Language Learners (ELLs) and students with learning differences, since they require a great deal of language skills. Here are some of the techniques that we use to make them less intimidating for *all* students:

- We spend a great deal of time in class emphasizing the meaning of the numbers in two-way tables, translating between numbers and words so that students become more familiar with the language.
- Students can create an anchor chart that highlights vocabulary associated with two-way tables as well as the interpretation of numbers in tables (40 = "40 million Americans who are black and non-Hispanic"). Anchor charts summarize important concepts or procedures in a chart that is

typically posted on a classroom wall for students to use in class. We discuss how students can make anchor charts in Chapter 3: Teaching Math as a Language.

- We develop the probability formulas in class so that students could use them if they prefer.
- We learn more about students' lives and interests so that we can analyze data that are meaningful to them when possible (we discuss this more in Chapter 1: Motivating Students, Chapter 2: Culturally Responsive Teaching, and Chapter 11: Building a Productive Classroom Environment).

We discuss other strategies for improving students' language skills in Chapter 3: Teaching Math as a Language.

We find that solving questions with two-way tables is worth the effort because it applies many familiar skills learned from other units. This helps make probability and statistics, a unit that confuses many students, more achievable. Like the area model, which we discussed earlier in this chapter, two-way tables can also provide a visual representation that helps students avoid remembering a formula that many find hard to memorize. In addition, students who take more advanced statistics courses like AP Statistics use two-way tables to determine if an association exists between two qualitative variables. In the end, we believe that two-way tables can help build a coherent bridge between the probability that students begin to learn in middle school and the advanced statistics that they may see in high school and beyond.

DIFFERENTIATING FOR ELLS AND STUDENTS WITH LEARNING DIFFERENCES

We encourage ELLs to rewrite concepts and problems in a more familiar language. Rewriting key phrases helps them to make mathematical connections more easily by extracting essential ideas and focusing on the math. ELLs can also use graphic organizers (such as Frayer models and concept maps) to express mathematical ideas visually while learning vocabulary.

Students with learning differences can benefit from using manipulatives, which make abstract ideas more concrete and often represent ideas in multiple ways. We discuss manipulatives more in Chapter 15: Differentiating for Students with Unique Needs. Project-based learning activities (which we talk about in Chapter 16: Project-Based Learning) can also enable students with learning differences to understand the relationships between different mathematical concepts.

We discuss these and other relevant strategies for ELLs and students with learning differences extensively in Chapter 14: Differentiating Instruction and Chapter 15: Differentiating for Students with Unique Needs.

Student Handouts and Examples

Figure 5.1: Addition and Subtraction of Polynomials
Figure 5.2: Multiplication with the Area Model
Figure 5.3: Division with the Area Model
Figure 5.7: Ratios and Similarity
Figure 5.8: Areas of Similar Polygons
Figure 5.9: Volumes of Similar Solids
Figure 5.10: Arc Length and Sector
Figure 5.11: Proportional Reasoning in Circles
Figure 5.12: Four Views of a Function
Figure 5.13: Rate of Change
Figure 5.14: Characteristics of Polynomial Functions
Figure 5.15: Even and Odd Polynomial Functions
Figure 5.16: Why $f(x) = \sin(x)$ Is Odd and $g(x) = \cos(x)$ Is Even
Figure 5.17: Linear Regression
Figure 5.18: Long-Run Relative Frequency
Figure 5.19: Two-Way Tables
Figure 5.20: Conditional Probability

What Could Go Wrong

Some students don't want to make mathematical connections. Instead, they may insist on using a formula or "trick." They may lack the confidence to think that they can achieve a deeper understanding, or they may want to get the answer quickly so they can move on to the next problem. In these situations, we try to resist the temptation to just show them a formula and move on. We don't *avoid* shortcuts, but we try to provide alternate explanations so students can make more sense of mathematics and not misapply formulas or short cuts. For example, students who learn "flip and multiply" for dividing fractions may invert the wrong fraction or even use it to *multiply* fractions (Gojak, 2013). Research indicates that students who believe that a piece of information is relevant to them are more likely to store it in their long-term memory (Sousa, 2017, p. 55). We believe that depriving students of mathematical discovery can erode their motivation and lead to greater inequity in academic outcomes.

Instead, we use a variety of strategies. For struggling students, we try to build their confidence by providing more direct instruction, giving more scaffolding, or reteaching material from prior courses. For advanced students, we may ask them to work on more challenging questions, explain the connections between mathematical ideas, or prove that their preferred shortcuts work in all cases. In the end, though, we don't ban students from using a shortcut unless it is mathematically incorrect or can lead to confusion. We focus on giving them other ways to remember how to solve a problem in case they forget the shortcut.

Other issues arise when we rely only on mathematical connections to motivate students. What might be simple or obvious to us might not be so clear to our students. Being enthusiastic about what we teach is obviously important, but just having a high level of energy won't engage students who didn't get enough sleep or who are dealing with personal issues that prevent them from focusing. We find that getting to know students better (which we discuss in Chapter 11: Building a Productive Classroom Environment) helps us be more aware of how to meet their social and emotional needs and customize our instruction, which can then improve their academic performance.

Technology Connections

Achieve the Core has an interactive coherence map at http://achievethecore.org/coherence-map that shows the connections between Common Core K–12 math standards. The University of Arizona's Institute for Mathematics and Education houses an online collection of *progressions* (http://ime.math.arizona.edu/progressions), which are narrative documents for major Common Core themes, such as expressions and equations, statistics and probability, geometry, and functions. Each progression describes the development of a topic across grade levels. The Underground Mathematics website (http://undergroundmathematics.org) has resources and a concept map that explain how many big mathematical ideas, such as averages, symmetry, and transformations, are related.

Bobson wrote a blog article on using the area model with numbers (http://bobsonwong.com/blog/20-think-inside-the-box-1) and polynomials (http://bobsonwong.com/blog/21-think-inside-the-box-2). Mathematician James Tanton's website (www.jamestanton.com) contains several mini-lessons explaining how to use the area model and how it shows mathematical structure. Tanton's Exploding Dots website (www.explodingdots.org) shows another representation that connects elementary school arithmetic to the polynomial algebra of high school. Mathies (http://mathies.ca/apps.php#Aa1) has online algebra tiles and other web-based manipulatives.

YouCubed has a short video at http://www.youcubed.org/resources/tour-mathematical-connections that explains how number sense, geometry, and algebra are connected under the big idea of ratio and proportion. MathCounts (http://www.mathcounts.org/resources/video-library/mathcounts-minis/mini-66-similarity-and-proportional-reasoning) has a video with examples on how to solve problems using similarity and proportional reasoning.

The Connecting Representations protocol (http://curriculum.newvisions.org/math/course/getting-started/instructional-routine-connecting-representations) from New Visions for Public Schools is an instructional routine to help students connect mathematical ideas and reflect on them.

Wikipedia (https://en.wikipedia.org/wiki/Multiple_representations_(mathematics_education)) has an overview of multiple representations of a function. New Visions for Public Schools's online curriculum (http://curriculum.newvisions.org/math/course/algebra-i/modeling-with-functions) lists several big ideas for functions, including the rate of change and multiple representations of a function.

Explanations of how to use two-way tables to represent categorical data and calculate probability appear on http://MathBitsNotebook.com (http://mathbitsnotebook.com/Algebra1/StatisticsReg/ST2TwoWayTable.html), MathBootCamps (http://www.mathbootcamps.com/conditional-probability-notation-calculation), and Math and Stats (http://www.mathandstatistics.com/learn-stats/probability-and-percentage/using-contingency-tables-for-probability-and-dependence).

The online book Nix the Tricks (http://nixthetricks.com) by Tina Cardone and the online community known as the MTBoS has alternatives to many of the shortcuts used in math education and explains why the tricks are bad for understanding math.

Figures

1. Calculate each of the following.

   a.
   $$\begin{array}{r} 2\ 0\ 4 \\ +1\ 5\ 0 \\ \hline \end{array}$$

   b.
   $$\begin{array}{r} 4\ 5\ 6 \\ -2\ 4\ 3 \\ \hline \end{array}$$

   c.
   $$\begin{array}{r} 2(10^2)\ +\ \ldots\ \ \ 4(1) \\ +1(10^2)\ +\ 5(10^1)\ \ldots \\ \hline \end{array}$$

   d.
   $$\begin{array}{r} 4(10^2) + 5(10^1) + 6(1) \\ -2(10^2) + 4(10^1) + 3(1) \\ \hline \end{array}$$

   e.
   $$\begin{array}{r} 2(x^2)\ +\ \ldots\ \ \ 4(1) \\ +\ 1(x^2)\ +\ 5(x^1)\ \ \ \ldots \\ \hline \end{array}$$

   f.
   $$\begin{array}{r} 4(x^2) + 5(x^2) + 6(1) \\ -2(x^2) + 4(x^1) + 3(1) \\ \hline \end{array}$$

2. What similarities and differences do you notice in the examples above?

3. To solve parts *a* and *c* above, Sheldon writes the work below. Explain the error in his reasoning.

   1a. $204 + 150$

   $$\begin{array}{r} 2\ 4 \\ +\ 1\ 5 \\ \hline 3\ 9 \end{array}$$

   1c. $2(10^2) + 4(1) + 1(10^2) + 5(10^1)$

   $$\begin{array}{r} 2(10^2)\ +\ \ \ 4(1) \\ +\ 1(10^2)\ +\ 5(10^1) \\ \hline 3(10^2)\ +\ 9(10^1) \end{array}$$

Figure 5.1 Addition and Subtraction of Polynomials

1. In rectangle ABCD, AB = 5, BF = 2, and FC = 3.

    a. Find the areas of ABFE, EFCD, and ABCD.

    b. Write an equation that relates the three areas that you found in part a.

    c. What property is illustrated by the equation you wrote in part b?

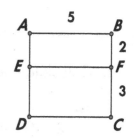

2. The rectangle at right is divided into squares.

    a. Using different colors, shade squares in the rectangle above to illustrate 32 · 21 = 30(20) + 2(20) + 30(1) + 2(1).

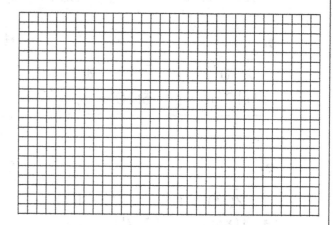

    b. Represent 32 · 21 by filling in the boxes below and simplifying the result. (HINT: Use your work from part a.)

3. The accompanying rectangle can be used to calculate $(3x+2)(2x+1)$.

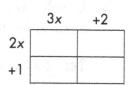

    a. Fill in the boxes in the rectangle above, using your work in #2 as an example. Express your product in standard form.

    b. How does your work in #3a compare to your work in #2b?

Figure 5.2 Multiplication with the Area Model

4. Multiply the following by filling in the boxes. Express the product in standard form.

a. $(2x^2 + 3x + 1)(2x + 1)$

	$2x^2$	$+3x$	$+1$
$2x$			
$+1$			

b. $(3x^2 - 2x + 4)(5x + 3)$

	$3x^2$	$-2x$	$+4$
$5x$			
$+3$			

Figure 5.2 (Continued)

The work below shows two different methods of calculating 672 ÷ 32. Write a short description of each step.

Long division	Area model	Description of step
32)672	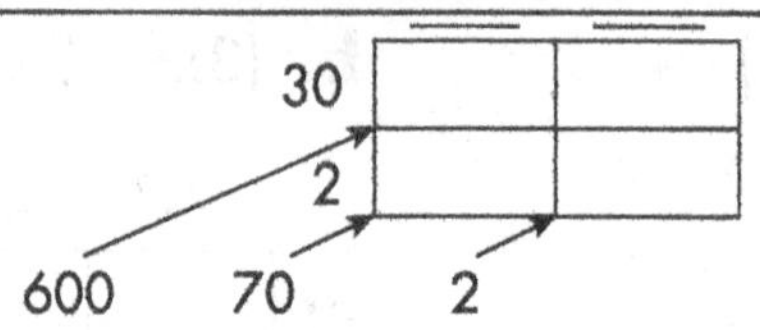30 2 600 70 2	
2 32)672	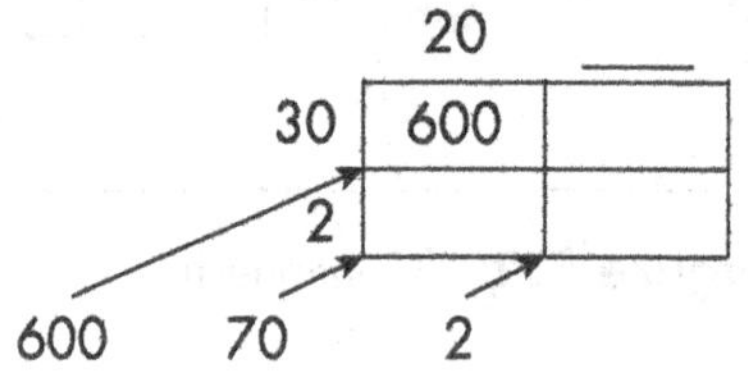20 30 600 2 600 70 2	
2 32)672 −64	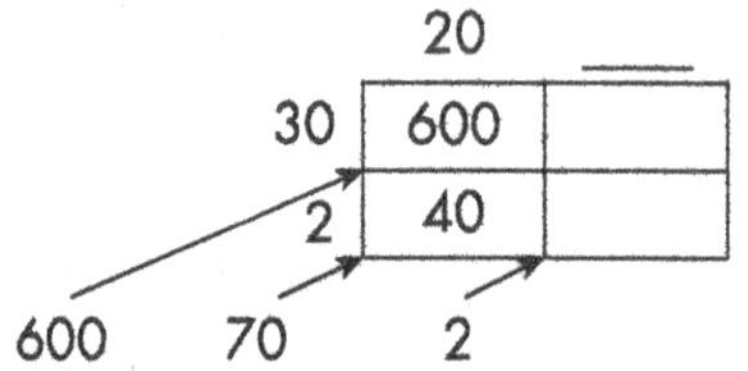20 30 600 2 40 600 70 2	
2 32)672 −64 32	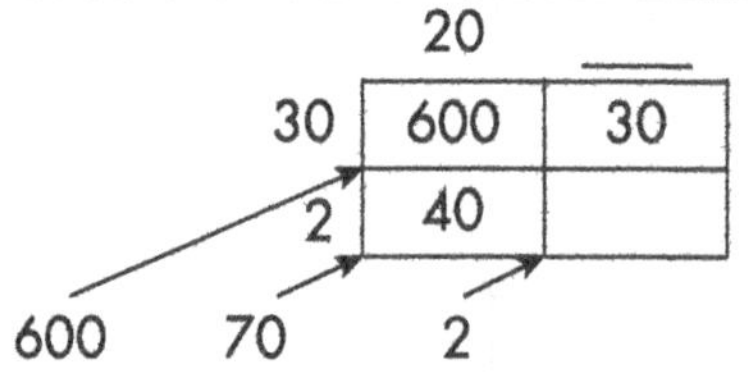20 30 600 30 2 40 600 70 2	
21 32)672 −64 32	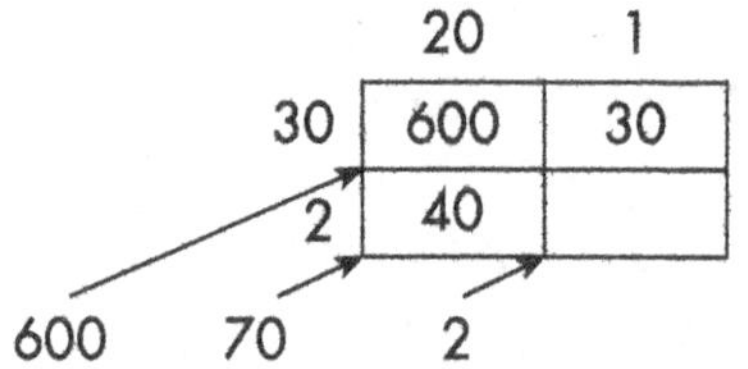20 1 30 600 30 2 40 600 70 2	
21 32)672 −64 32 −32 0	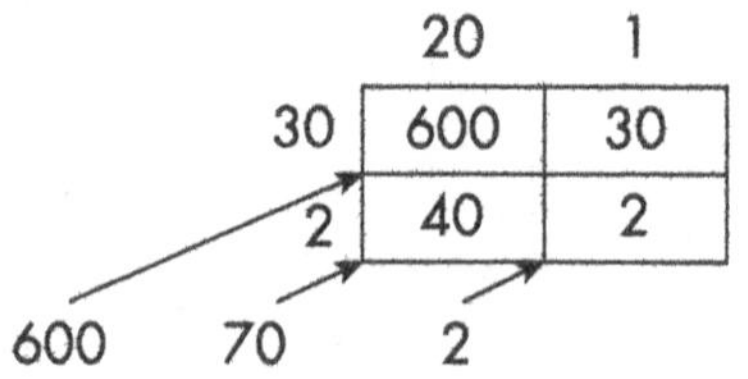20 1 30 600 30 2 40 2 600 70 2	

Figure 5.3 Division with the Area Model

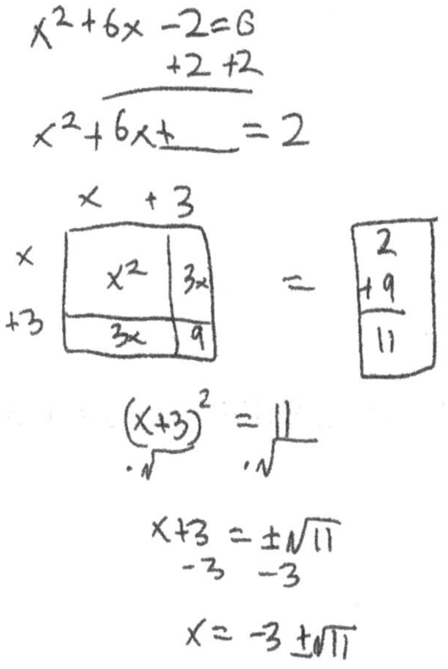

CHAPTER 5: MAKING MATHEMATICAL CONNECTIONS

$$x^2 + 6x - 2 = 6$$
$$ +2 \ +2$$
$$x^2 + 6x \pm \underline{} = 2$$

[Diagram: square divided into 4 regions labeled x, $+3$ on sides; interior cells x^2, $3x$, $3x$, 9; equals column showing 2, $+9$, 11]

$$(x+3)^2 = 11$$
$$\sqrt{} \qquad \sqrt{}$$

$$x + 3 = \pm\sqrt{11}$$
$$-3 \quad -3$$

$$x = -3 \pm \sqrt{11}$$

Figure 5.4 Completing the Square

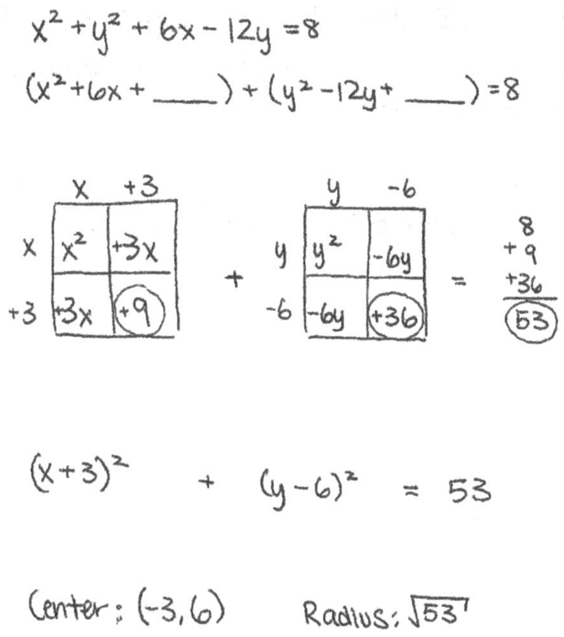

$$x^2 + y^2 + 6x - 12y = 8$$
$$(x^2 + 6x + \underline{}) + (y^2 - 12y + \underline{}) = 8$$

[Diagrams: first square with sides x, $+3$ showing x^2, $+3x$, $+3x$, $+9$; plus second square with sides y, -6 showing y^2, $-6y$, $-6y$, $+36$; equals column 8, $+9$, $+36$, 53]

$$(x+3)^2 + (y-6)^2 = 53$$

Center: $(-3, 6)$ Radius: $\sqrt{53}$

Figure 5.5 Determining the Center and Radius of a Circle

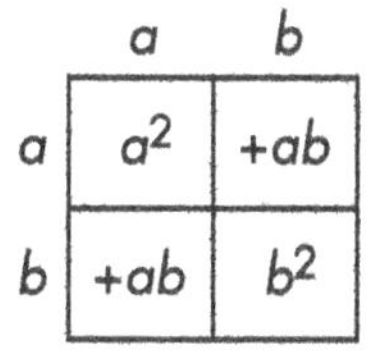

	a	b
a	a^2	$+ab$
b	$+ab$	b^2

$$(a + b)^2 = a^2 + 2ab + b^2 \neq a^2 + b^2$$

Figure 5.6 Why $(a + b)^2 \neq a^2 + b^2$

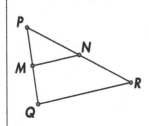

1. In △PQR, $\overline{QMP}$, $\overline{PNR}$, and ∠PMN ≅ ∠PQR. Complete the following statements:
   a. △PMN ~ _____
   b. $\dfrac{PM}{PQ} = \dfrac{}{PR} = \dfrac{}{}$
   c. $\overline{MN}$ ∥ _____
   d. $\overline{QR}$ is a dilation of $\overline{MN}$ with center _____ and scale factor _____ .
2. The statements above illustrate the following theorem (fill in the blanks):
   A line splits two sides of a triangle proportionally if and only if it is _____ to the third side.

Figure 5.7 Ratios and Similarity

1. For each pair of polygons below, the figure in column A is similar to the figure in column B.

   a. Write the ratio *(in simplest form)* of the corresponding sides of each polygon in column A to each figure in column B. Then write the ratio *(in simplest form)* of the corresponding areas of each polygon in A to each figure in column B.

A	B	Ratio of corresponding sides A:B (in simplest form)	Ratio of corresponding areas A:B (in simplest form)
A 9 B, square with sides 9, D C	E 3 F, square with sides 3, H G		
I 4 J, rectangle 4 by 2, L K	M 20 N, rectangle 20 by 10, P O		

   b. Based on your work above, fill in the blanks in the following sentence:
   The ratio of the areas of similar polygons is the _____ of the ratio of the corresponding side lengths.

Figure 5.8 Areas of Similar Polygons

1. For each pair of solids below, the solids in column A are similar to the solids in column B.

  a. Write the ratio *(in simplest form)* of the corresponding sides of each solid in A to each solid in B. Then write the ratio *(in simplest form)* of the corresponding volumes of each figure in A to each figure in B.

A	B	Ratio of corresponding lengths A:B (in simplest form)	Ratio of corresponding volumes A:B (in simplest form)
(box 3 × 2 × 1)	(box 15 × 10 × 5)		
(prism 9 × 6 × 3)	(prism 6 × 4 × 2)		

  b. Based on your work above, fill in the blanks in the following sentence:

The ratio of the volumes of similar solids is the _____ of the ratio of the corresponding side lengths.

Figure 5.9 Volumes of Similar Solids

1. A large circular pizza has a diameter of 16 inches. Calculate the area and circumference of the pizza.

   Area =

   Circumference =

2. Andre, Bonnie, Cheryl, and Deandre are sharing slices of the 16"-diameter pizza. Fill out the table below.

Name	Fraction of pizza	Central angle of slice (degrees)	Amount of pizza crust eaten (inches) (in terms of π)	Amount of pizza eaten (square inches) (in terms of π)
Andre	$\frac{1}{2}$			
Bonnie	$\frac{1}{4}$			
Cheryl	$\frac{1}{3}$			
Deandre	$\frac{1}{10}$			

3. In the problems above, what is the relationship between the amount of *pizza* eaten and the amount of *pizza crust* eaten?

Figure 5.10 Arc Length and Sector

Angle measure	Arc length	Area of sector
In circle O, radius $OA = 1$ and $m\angle AOB = 45$. What is $m\angle AOB$ in radians?	In circle O, radius $OA = 1$ and $m\angle AOB = 45$. What is the length of AB?	In circle O, radius $OA = 1$ and $m\angle AOB = 45$. What is the area of sector AOB?
$\dfrac{45°}{360°} = \dfrac{x}{2\pi} \Rightarrow x = \dfrac{\pi}{4}$	$\dfrac{45°}{360°} = \dfrac{x}{2\pi} \Rightarrow x = \dfrac{\pi}{4}$	$\dfrac{45°}{360°} = \dfrac{x}{\pi} \Rightarrow x = \dfrac{\pi}{8}$

Figure 5.11 Proportional Reasoning in Circles

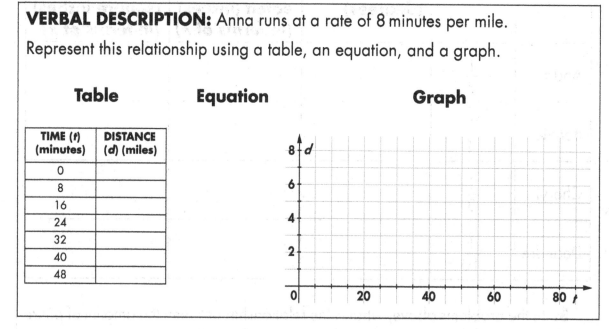

Figure 5.12 Four Views of a Function

1. The accompanying table shows the relationship between the number of square floor tiles (x) and the surface area covered by the tiles (y).

Number of square floor tiles (x)	2	4	6
Area of tiled surface (square inches) (y)	72	144	216

a. Calculate the rate of change in the table from x = 2 to x = 4. Interpret the rate of change in context.

b. Calculate the rate of change in the table from x = 4 to x = 6. Interpret the rate of change in context.

c. Write an equation that relates x and y.

d. On the coordinate plane below, graph the equation that you wrote in part c.

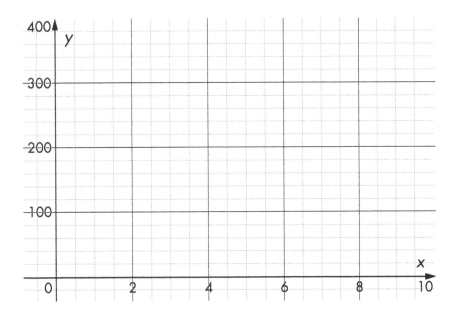

Figure 5.13 Rate of Change

e. Explain how you can find the slope of the line represented in this problem using the table of values, the equation (in part c), and the graph (in part d):

TABLE OF VALUES:

EQUATION:

GRAPH:

Figure 5.13 (Continued)

1. For each function below, complete the table, find its intercepts and zeros, and draw a graph.

   a. $f(x) = (x-4)(x-1) = x^2 - 5x + 4$

x	y
-4	
-3	
-2	
-1	
0	
1	
2	
3	
4	

   x-intercepts: _____

   zeros: _____

   y-intercept: _____

   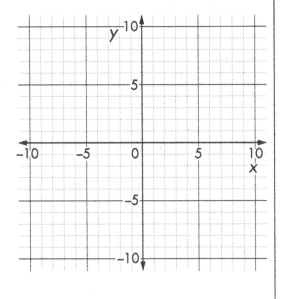

   b. $h(x) = (x-1)(x-2)(x+3) = x^3 - 7x + 6$

x	y
-4	
-3	
-2	
-1	
0	
1	
2	
3	
4	

   x-intercepts: _____

   zeros: _____

   y-intercept: _____

   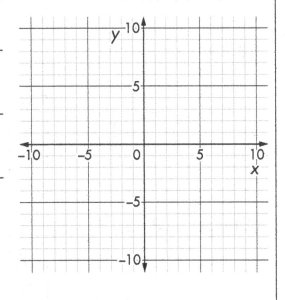

Figure 5.14 Characteristics of Polynomial Functions

2. Based on your work above, explain how you can determine the *x*-intercepts of a function based on the:
   a. table of values:
   b. equation:
   c. graph:

3. Based on your work above, explain how you can determine the *y*-intercept of a function based on the:
   a. table of values:
   b. equation:
   c. graph:

4. What is the relationship between the *x*-intercepts and zeros of a function?

Figure 5.14 (Continued)

Even Functions: For each *even* function below, complete its table and sketch its graph.

$f(x) = x^2 + 1$

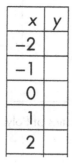

x	y
-2	
-1	
0	
1	
2	

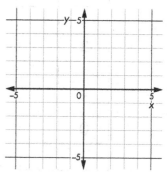

$g(x) = 3x^4 - 5x^2 - 2$

x	y
-2	
-1	
0	
1	
2	

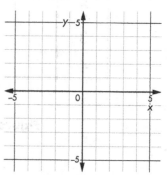

Figure 5.15 Even and Odd Polynomial Functions

Odd Functions: For each *odd* function below, complete its table and sketch its graph.

$j(x) = x^3$

x	y
-2	
-1	
0	
1	
2	

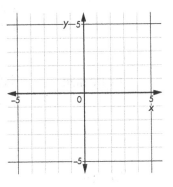

$k(x) = x^3 - x$

x	y
-2	
-1	
0	
1	
2	

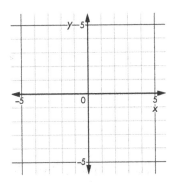

Based on your work above, fill in the blanks below to explain how you can determine whether a polynomial function is even or odd based on its characteristics:

Function type	Evidence from equation	Evidence from table	Evidence from graph
Even	All of the exponents are _____ .		The graph has _____ symmetry over the _____ .
Odd	All of the exponents are _____ .		The graph has _____ symmetry over the _____ .

Figure 5.15 (Continued)

1. Complete the information below for f(x) = sin (x).

   Table

x	$-\pi$	$\frac{-3\pi}{4}$	$\frac{-\pi}{2}$	$\frac{-\pi}{4}$	0	$\frac{\pi}{4}$	$\frac{\pi}{2}$	$\frac{3\pi}{4}$	π
sin (x)									

   Equation   HINT: Use a diagram of the unit circle to help you compare f(x), f(−x), and −f(x).

   Graph
   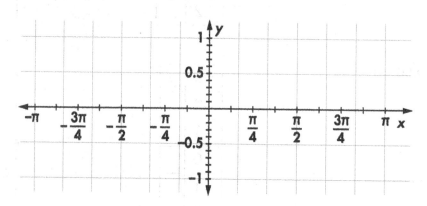

2. Complete the information below for g(x) = cos (x).

   Table

x	$-\pi$	$\frac{-3\pi}{4}$	$\frac{-\pi}{2}$	$\frac{-\pi}{4}$	0	$\frac{\pi}{4}$	$\frac{\pi}{2}$	$\frac{3\pi}{4}$	π
cos (x)									

   Equation   HINT: Use a diagram of the unit circle to help you compare g(x), g(−x), and −g(x).

   Graph
   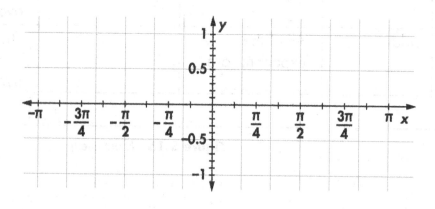

Figure 5.16 Why f(x) = sin (x) Is Odd and g(x) = cos (x) Is Even

Based on your work above, fill in the blanks below to explain how you can determine which function is even and which is odd. Use information from the equation, table, and graph to support your answer.

Function type	Evidence from equation	Evidence from table	Evidence from graph
$f(x) = \sin(x)$ is _____ (even or odd)			The graph has _____ symmetry over the _____ .
$g(x) = \cos(x)$ is _____ (even or odd)			The graph has _____ symmetry over the _____ .

Figure 5.16 (Continued)

1. The accompanying graph shows the relationship between the number of hours that a tool is rented from a home improvement store (x) and the total rental cost (y).

   a. What is the equation of the line shown in the graph? Explain how you got your answer.

   b. Interpret the slope of the line in context.

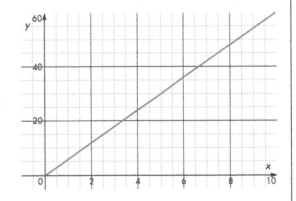

Figure 5.17 Linear Regression

2. The accompanying graph shows the relationship between the number of miles that Isabella runs (x) and the time (in minutes) of her runs (y).

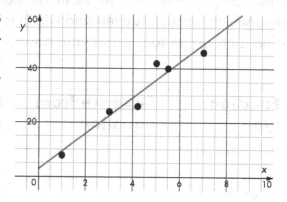

a. The equation of the regression line is $y = 6.64x + 2.54$. Based on the graph, how do you know that the value of the correlation coefficient is *not* 1?

b. Explain why the slope of the line cannot be interpreted in context as follows: "For every increase of 1 mile that Isabella runs, her running time increases by 6.64 minutes."

c. Write a correct interpretation of the slope of the line in context.

Figure 5.17 (Continued)

1. If you flipped a fair coin 50 times, about how many times would you expect to get tails? Explain.

2. Flip a fair coin 50 times. After each flip, record the result. Calculate the total number of times you get tails. Then divide the total number of tails by the total number of flips to find the relative frequency of tails so far. Express this relative frequency as a decimal to the nearest hundredth.

Flip #	Result	Number of tails so far	Rel. freq. (nearest hundredth)	Flip #	Result	Number of tails so far	Rel. freq. (nearest hundredth)
1				26			
2				27			
3				28			
4				29			
5				30			
6				31			
7				32			
8				33			
9				34			
10				35			
11				36			
12				37			
13				38			
14				39			
15				40			
16				41			
17				42			
18				43			
19				44			
20				45			
21				46			
22				47			
23				48			
24				49			
25				50			

Figure 5.18 Long-Run Relative Frequency

3. Graph the relative frequency of the total number of tails.

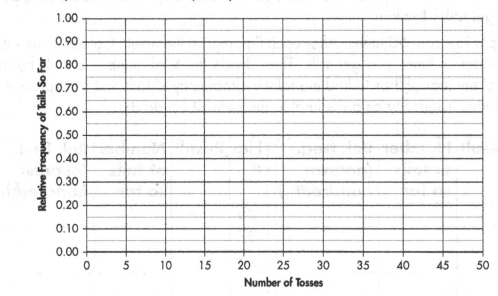

4. Look at the graphs of your classmates. What do you notice about the graphs?

5. Based on your work above, fill in the blanks in the following sentence:
   As the number of flips _____, the relative frequency of tails approaches _____ .

Figure 5.18 (Continued)

CHAPTER 5: MAKING MATHEMATICAL CONNECTIONS

The US Census Bureau measures race and ethnicity by asking individual Americans two questions:

- Are you of Hispanic origin?
- What is your race?

In 2017, the Census Bureau made the following estimates of race and ethnicity for Americans:

- 59 million Americans are of Hispanic origin, while 267 million are not of Hispanic origin.
- Of those that identify as Hispanic, 38 million are white, 1 million are black, 3 million belong to two or more races, and 17 million belong to another racial category.
- Of those that identify as non-Hispanic, 197 million are white, 40 million are black, 8 million belong to two or more races, and 22 million belong to another racial category.

1. Explain why a two-way table is better than a Venn diagram for representing this information.

2. Using the two-way tables below (highlight the appropriate numbers used in your calculations), determine the probability that a randomly selected American is:

Figure 5.19 Two-Way Tables

a. White *and* Hispanic

	Hispanic	Non-Hispanic	TOTAL
White	38	197	235
Black	1	40	41
2+ races	3	8	11
Other	17	22	39
TOTAL	59	267	326

b. White *or* Hispanic

	Hispanic	Non-Hispanic	TOTAL
White	38	197	235
Black	1	40	41
2+ races	3	8	11
Other	17	22	39
TOTAL	59	267	326

c. Black *and* non-Hispanic

	Hispanic	Non-Hispanic	TOTAL
White	38	197	235
Black	1	40	41
2+ races	3	8	11
Other	17	22	39
TOTAL	59	267	326

d. Black *or* non-Hispanic

	Hispanic	Non-Hispanic	TOTAL
White	38	197	235
Black	1	40	41
2+ races	3	8	11
Other	17	22	39
TOTAL	59	267	326

3. Based on your work above, explain how you can use a two-way table to calculate the following:

   a. The probability that events *A* and *B* both occur

   b. The probability that event *A* or *B* occurs

Figure 5.19 (Continued)

1. The US Census Bureau's 2017 estimates of race and ethnicity for Americans can be summarized in the following table. Highlighting the appropriate numbers in the table for each part, find the probability that a randomly selected American will be:

a. White

	Hispanic	Non-Hispanic	TOTAL
White	38	197	235
Black	1	40	41
2+ races	3	8	11
Other	17	22	39
TOTAL	59	267	326

b. Hispanic

	Hispanic	Non-Hispanic	TOTAL
White	38	197	235
Black	1	40	41
2+ races	3	8	11
Other	17	22	39
TOTAL	59	267	326

c. Hispanic, given that the person is White

	Hispanic	Non-Hispanic	TOTAL
White	38	197	235
Black	1	40	41
2+ races	3	8	11
Other	17	22	39
TOTAL	59	267	326

d. White, given that the person is Hispanic

	Hispanic	Non-Hispanic	TOTAL
White	38	197	235
Black	1	40	41
2+ races	3	8	11
Other	17	22	39
TOTAL	59	267	326

Figure 5.20 Conditional Probability

e. white and Hispanic

	Hispanic	Non-Hispanic	TOTAL
White	38	197	235
Black	1	40	41
2+ races	3	8	11
Other	17	22	39
TOTAL	59	267	326

2. Based on your work in #1, explain how you can use a two-way table to find the probability of an event A given that another event B has happened.

Figure 5.20 (Continued)

PART II

How to Plan

CHAPTER 6

How to Plan Units

What Is It?

Unit plans consist of related ideas and objectives that are taught over a period of time, usually several weeks.

A unit plan typically consists of the following:

- Relevant state, local, or Common Core standards
- Major learning goals for the course
- Important vocabulary
- List of lessons in the order they are taught
- Brief summary of the main concepts in the unit (wikiHow, n.d., "How to Write a Unit Plan"; Cunningham, 2009, pp. 118–119)

Unit plans don't have to be long, complex documents. Most of our unit plans are between two and three pages long.

Why We Like It

Planning good lessons can be overwhelming, especially for teachers who are new to a subject. By linking lessons that share common themes and are taught consecutively into a unit plan, teachers can make lessons more coherent and effective. When we create a unit plan, we develop a better understanding of the key concepts in the course and the order in which they should be taught.

In addition, planning a unit helps us implement teaching strategies that take students' cultures and social-emotional well-being into account. Otherwise, we get sucked into the day-to-day pressure of just "covering the standards."

A good unit plan also helps us manage our time more efficiently. From it, we can see how many teaching days we need for each topic. Unit plans that make connections between ideas give students and teachers confidence because they reinforce our belief that math is a logical system. Such planning decreases the stress that can arise from worrying about whether we will have enough time at the end of the year to review.

Most teachers that we know do some long-range planning by using a pacing calendar that lists lessons in the order in which they are taught. We find that expanding on that calendar in ways we suggest in this chapter helps us improve our teaching from year to year.

Supporting Research

Researchers suggest that unit plans help teachers develop a comprehensive understanding of the material being taught (wikiHow, n.d., "How to Write a Unit Plan"; Cunningham, 2009). This is particularly important when teachers use a curriculum written by other authors whose standards, objectives, and teaching styles differ from their own (Taylor, 2016, p. 441). Unit plans also help teachers find gaps in their own content knowledge (Matteson, 2016, p. 40).

Without looking at the big picture, teachers can become too focused on the small details and not see the larger connections in the unit. Creating unit plans helps teachers think about what can be asked to promote higher-order thinking, what prerequisite skills students need, what new content they will learn, and how that new content fits together into a cohesive whole (Matteson, 2016, p. 42).

Common Core Connections

The Common Core standards are organized by groups of related standards called *clusters*, which in turn are grouped together into larger groups called domains. Some of the domains roughly correspond to units, but other clusters and domains span across several units. For example, the function domain covers three different families of relations (linear, quadratic, and exponential), each of which is a major theme that should be put into its own unit (NGA & CCSSO, 2010, pp. 68–71). Furthermore, many standards are broad statements of what should be taught (such as the Common Core Geometry standard G-CO.C.10, "Prove theorems about triangles"), so they often don't tell you what a unit or lesson actually must cover. The bottom line is that the Common Core standards provide some guidance for what goes into a unit, but they often are only a starting point.

Application

Here are the steps that we use to create a unit plan.

GETTING STARTED

We design the unit plan to be read by teachers, not by students. We find that sharing the unit plan with students hinders their ability to connect ideas on their own. Furthermore, allowing students to see the unit plan may be overwhelming, especially since many might be confused by the language used in standards.

When we teach a course, we rarely have to write a unit plan from scratch. Many times, we use the teacher's edition of our textbook or the pacing calendar of our curriculum as a starting point.

Whether we use someone else's unit plan or write our own, we find that we have to think about three things (Posamentier, Smith, & Stepelman, 2010, pp. 14–18):

1. **Identify major themes.** What major themes for the course can you identify from standards, course outlines, textbooks, or websites? Speak to teachers who have taught similar courses to see what others have done. We like to compare several sources to see which topics work best for our students and for us. Each major theme should correspond to a unit.

2. **Get to know your students.** What students will be in the class? What courses have they already completed? What courses will they be expected to take afterwards? Will the class have many English Language Learners (ELLs) or students with learning differences? What does your school expect from these students? What are the cultural backgrounds of the students? What assets (such as a different understanding of number systems) might they bring to the class? Have students taken classes together in the past, or are they strangers to each other? We find that adjusting our unit plans to meet our students' needs is particularly important when we use a curriculum written by someone who is unfamiliar with our students' situations. Keeping our audience in mind is critical while writing unit plans.

3. **Find available resources.** What technology, books, and other resources are available? Will you have easy access to computers, tablets, or calculators? What books or manipulatives can you use? The resources that you have available will help determine pacing.

MAKING CONNECTIONS BETWEEN BIG IDEAS

Once we identify the big themes for a course, we select one theme and begin the unit planning process. If none of the resources that we are currently using has a unit plan, then we create one.

An important part of a unit plan is a *concept map*, a hierarchical plan that shows relationships or links between ideas, with "big ideas" at the top and more specific information below. Concept maps help teachers get a better understanding of the key concepts in the course and the order in which they should be taught. If a unit is not well organized, students won't be able to make connections between topics or understand more complicated ideas (Matteson, 2016, pp. 40, 42).

We start by listing the specific skills that our students have to master in the unit. Figure 6.1: Unit Plan: List of Skills shows the skills for our Geometry unit on angle pairs. To make the concept map, we put each idea on index cards so that we can easily reorganize them. We then group related items together, as shown in Figure 6.2: Unit Plan: Concept Map. Chapter 5: Making Mathematical Connections also discusses ways to connect mathematical ideas.

DEVELOPING A LOGICAL SEQUENCE

After creating a concept map, we organize the ideas into a logical sequence. Units can begin with basic ideas and gradually include more complex problems. Another possibility is to start with a real-world problem and develop the skills necessary to solve it.

Many times, our units have real-world applications. We take these problems from standardized tests, textbooks, or even newspapers or magazines. Word problems can be customized to match student interests or current events. Applications can also take the form of individual or group projects in which students summarize and apply the concepts that they learned. For example, in an Algebra II unit on polynomials, students can use polynomial functions to design a 1-liter can that uses the smallest amount of material possible. In a unit on coordinate geometry, students can create artwork on a coordinate plane and write equations that describe the shapes that they used. We describe more ideas for designing and grading student projects in Chapter 16: Project-Based Learning.

Understanding the prior knowledge for each skill is critical for figuring out the order in which concepts can be taught. In our experience, we've found that determining the best sequence for the ideas in a unit is usually the hardest part of creating a good unit plan. For example, it's clear that addition must be taught before subtraction, but it's not as clear whether solving equations graphically should be presented before solving equations algebraically. We've had countless debates with colleagues over the years about the sequencing of lessons and units! Often, we end up agreeing to disagree, which illustrates an important point: *a unit can often be organized in a variety of equally valid ways*. The method that you pick depends on what makes the most sense to you.

Figure 6.3: Unit Plan: Sequence of Lessons shows one way in which the grouped skills from the previous figure can be organized into a logical sequence.

ORGANIZING TOPICS AND PROBLEMS

After we have a rough idea of how the topics in a unit should be sequenced, we choose appropriate problems and put them into a logical order. We then divide them into lessons based on what we think will fit into one class period. Typically, each lesson will focus on one main concept, with development and practice problems that increase in difficulty.

After organizing the problems and dividing the unit into lessons, we go through the lessons and make a list of all the vocabulary, objectives, and standards that we covered. This provides a handy list that we frequently use as a reference.

SUMMARIZING THE UNIT PLAN

Finally, we write a paragraph that briefly explains the unit's big themes. We find that the unit summary is one of the most important parts of a unit plan. We refer to it frequently while teaching. It helps to remind us of where we are in the unit, what ideas we need to emphasize, and how the unit relates to concepts taught before and afterward. Sometimes, our summary also contains brief suggestions to ourselves that we made in the past for future improvement, such as "try teaching graphing polynomials right after factoring so students see the connection between the two topics."

Figure 6.4: Sample Unit Plan shows an example of a unit plan on angle pairs from our Geometry course. This represents an ideal sequence that we developed after many years of teaching and revising the unit. Not all of our unit plans are this detailed, but at a minimum, we try to include at least a list of lessons and a summary.

BEING FLEXIBLE

Unit plans represent how the lessons would unfold under ideal conditions. However, we find that we always have to adjust them based on what else is going on in our school, our lives, and our students' lives. For example, we try to avoid teaching difficult lessons right before a long holiday, or we may move an interesting lesson to another day if we have visitors coming into our classrooms. If we're absent for an extended period of time, we often have to combine lessons. Being flexible helps maintain sanity!

Flexibility also allows teachers to create space for current events that can be mathematically analyzed, such as unusual election results, the probability of winning a large lottery jackpot, or a controversial new study.

DEVELOPING STUDENTS' SOCIAL AND EMOTIONAL LEARNING

We consider how students' social and emotional learning skills can be developed in each unit. Researchers have found that students who believed that intelligence was fixed became more stressed when their grades declined, while students who believed that their intelligence could be increased were generally not as stressed when their grades declined (Lee, Jamieson, Miu, Josephs, & Yeager, 2018).

We find that many students come to us with math anxiety. They readily say to us, "I've never been good at math," implying that their mathematical knowledge can't be increased. Often, the early lessons of a unit can promote a growth mindset (which we discussed in Chapter 4: Promoting Mathematical Communication).

In our experience, many students are afraid to take Geometry because they have heard that the course requires a great deal of abstract reasoning. We keep this in mind and start the school year by teaching a unit on angle pairs. This unit allows students to solve new problems using familiar concepts that they learned in Algebra I, such as writing and solving linear equations. While thinking about unit plans, teachers can consider ways to help students overcome anxiety and build up their self-confidence, which will help them persist when work becomes more challenging.

INCORPORATING STUDENTS' CULTURES

We also think about ways that we can incorporate students' cultural experiences into our units. For example, when we introduce a unit, we look into the history of the math that we are discussing. We try to show how mathematical concepts developed in different cultures. For example, the Pythagorean theorem, often attributed to the Greek mathematician Pythagoras, was also discovered independently by Mesopotamian, Chinese, and Indian mathematicians.

Adding these types of references is an important example of culturally responsive teaching. Unfortunately, many teachers stop at parachuting these examples into their lessons. In fact, a culturally responsive unit plan should incorporate students' cultural identities and real-world experiences. A unit plan can include a project in which students create problems with contexts familiar to their communities, or share problem-solving techniques used in other countries (Dong, 2016, pp. 538–539). In addition, students can use math to analyze and propose solutions to problems that exist in their communities. Here are some examples:

- Students can compare the density of parks, schools, or supermarkets in their neighborhood to that in other areas.
- Students can analyze accident or congestion patterns to identify locations that could benefit from traffic lights or rerouting.
- Students can write word problems that require converting currency from their country of origin to US dollars or converting from metric to imperial units.

CHAPTER 6: HOW TO PLAN UNITS

We discuss many other strategies for promoting culturally responsive teaching in Chapter 2: Culturally Responsive Teaching.

DIFFERENTIATING FOR ELLS AND STUDENTS WITH LEARNING DIFFERENCES

Often, we adjust the pacing in our unit plans to match our students' needs. We provide additional time for ELLs to learn new vocabulary. If students need extra time to process information, we compact lessons to give them additional time to practice. If we have more advanced students, we combine lessons to build in time for enrichment. We find that these changes often benefit not just the specific groups that we targeted but many other students as well.

When possible, we consult with ELL or special education teachers while working on unit plans. They often have insights that can guide the development of Individualized Education Plans (IEPs) for students in our classes. The unit plan can be used to customize learning objectives that students should master. These objectives can then be incorporated into the student's IEP goals (Matteson, 2016, p. 44). We talk more about working with these colleagues in Chapter 13: Collaborating with Other Teachers.

We address other strategies for unique types of learners in Chapter 15: Differentiating for Students with Unique Needs.

Student Handouts and Examples

Figure 6.1: Unit Plan: List of Skills

Figure 6.2: Unit Plan: Concept Map

Figure 6.3: Unit Plan: Sequence of Lessons

Figure 6.4: Sample Unit Plan

What Could Go Wrong

When using unit plans, we sometimes don't spend enough time understanding the thinking behind a unit's structure. This often happens when we are given material written by someone else. Using other people's unit plans can save us a lot of work in creating or organizing problems, especially when teaching a course that's new to us. However, we find that other people's curricula don't always state all of the connections between lessons.

For example, veteran teachers' lessons often don't explain why they chose the problems contained in their lessons. In many cases, they are so familiar with the material that they see no need to put every detail in writing. (As veteran teachers, we are often guilty of this ourselves!) When using other people's lessons, we know which problems to do but don't always know which ones are most important or which ones we need later. When writing a unit plan that we share with others, we try to remind ourselves to avoid the *curse of knowledge,* the cognitive bias that occurs when teachers unknowingly assume that students have the same background knowledge as they do (Heath & Heath, 2006). Just as we try to consider our students' prior knowledge when we teach a lesson, we try to take into account our colleagues' prior knowledge when we share our unit plans. For example, we might briefly note that certain problems need to be emphasized because they foreshadow concepts that are discussed later in the unit. Sometimes, we write out the exact wording of particularly tricky explanations to model appropriate mathematical language for students.

We've found several strategies that have helped us tremendously when using unit plans written by others. First, we read through as much of the unit as possible before teaching it so that we have an overview of its main themes. When we can, we talk to the original authors of the lessons to understand what they were thinking. If we have time, we rearrange, combine, and add or delete lessons as necessary. Our experience agrees with researcher Megan Westwood Taylor's observation that teachers "will always interpret curricula with their beliefs, experiences, and students in mind" (Taylor, 2016, p. 442).

Sometimes, we stick too rigidly to a long-range plan without making frequent re-evaluations (Posamentier, Smith, & Stepelman, 2010, p. 17). Monitoring student progress helps ensure that our lessons are aligned with their learning. When our students don't understand a lesson, we think of alternative approaches or give students more practice to help them achieve the lesson's objectives. When our unit plan moves too slowly for our students, we have to combine lessons, eliminate lessons, or add more challenging problems.

When we think about the pacing of our unit, we try to avoid lessons that give the impression that nothing new is being taught that day. We find that when students feel that we are spending many days on the same lesson, they pay less attention because they expect the next day's lesson to repeat the concepts. Instead, we try to identify coherent tasks for each lesson so that subsequent aims extend concepts that students learned before.

Finally, we've also made the mistake of not setting aside enough time to revisit previous topics. Even though we include review lessons in our unit, we've often found that some units are harder for students to understand than we originally thought. Again, paying attention to student understanding and being flexible with the calendar both help tremendously.

Technology Connections

To create unit plans, we use the software that our school or district makes available to us. For example, our district has purchased Microsoft Office software. Collaborative software like the free Google Docs (part of the G Suite, http://docs.google.com) can be useful for sharing unit plans among teachers. Microsoft Office also allows collaboration among users, but this requires a subscription.

To type mathematical symbols and equations, we use an equation editor that allows us to insert symbols directly into a document. MathType (http://store.wiris.com/en), and EquatIO (http://www.texthelp.com/en-us/products/equatio) can be integrated into major word processors like Microsoft Word and Google Docs. We recommend trying out each software first to make sure it can be used with your word processor and that it has all of the mathematical symbols that you need. Although they are not free, special pricing may be available for teachers or through your school or district.

To align our unit plan with the current calendar, we use an online calendar program. We look for the ability to export events into a spreadsheet file (useful for creating a compact list of aims) and share calendars (useful when working with other teachers). Current examples include Apple Calendar, Microsoft Outlook, and the free Google Calendar (http://calendar.google.com).

Several unit plans and curricula are available for free online. Illustrative Mathematics (www.illustrativemathematics.org) has an excellent and widely praised Common Core–aligned middle school and high school curriculum. EngageNY (www.engageny.org) has curricula and lessons from pre-kindergarten through pre-calculus. While EngageNY's lessons are often too challenging or confusing for our students, their unit plans provide a good overview of the relevant standards and are a good starting point for developing a unit plan. For high school teachers, eMathInstruction (www.emathinstruction.com) has units for Common Core Algebra I, Geometry, and Algebra II.

Figures

Figure 6.1. Unit Plan: List of Skills
Figure 6.2. Unit Plan: Concept Map
Figure 6.3. Unit Plan: Sequence of Lessons
Figure 6.4. Sample Unit Plan

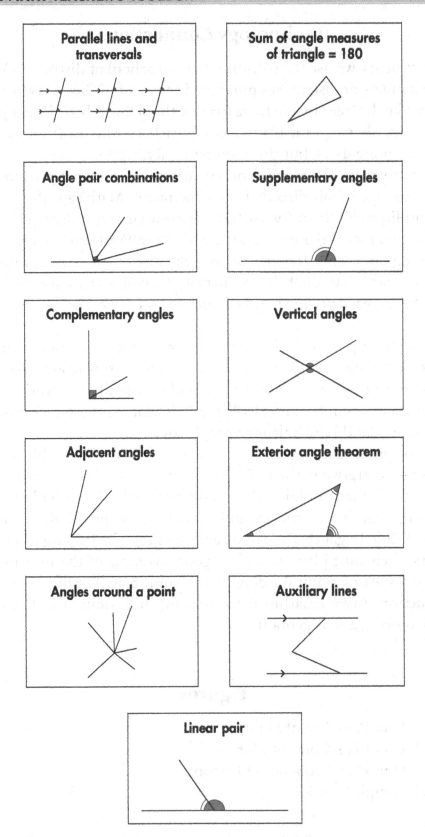

Figure 6.1 Unit Plan: List of Skills

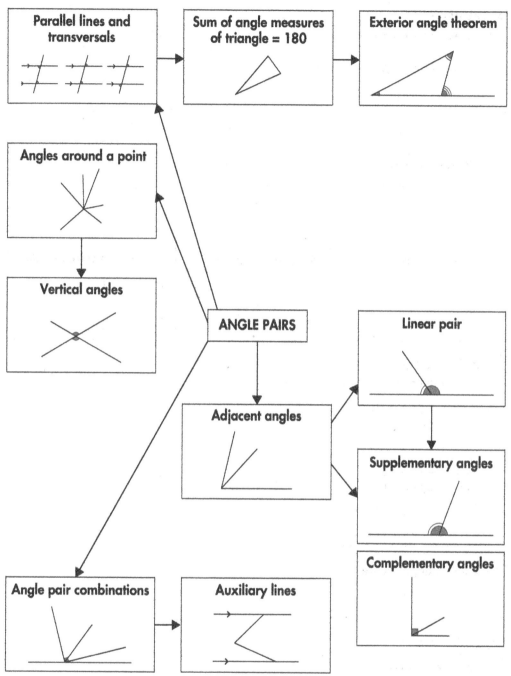

Figure 6.2 Unit Plan: Concept Map

ANGLE PAIRS

PRIOR KNOWLEDGE: Angle types (acute, obtuse, right, straight angles)

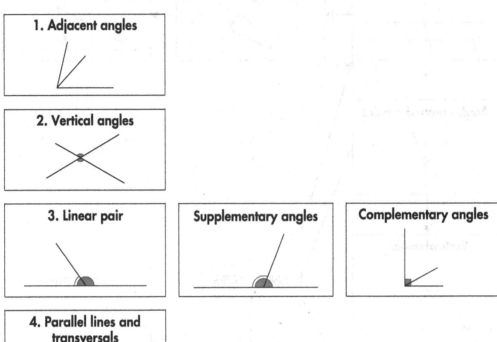

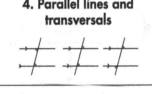

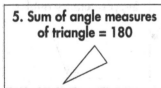

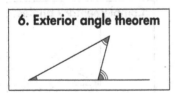

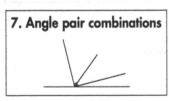

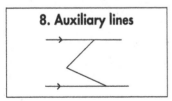

Figure 6.3 Unit Plan: Sequence of Lessons

Geometry Unit 1: Angle Pairs

Summary

This introductory unit serves as a transition between the familiar concept of solving equations in Algebra I and the more abstract writing of formal proofs in Geometry. Using the principles discussed when justifying the solution to equations (A-REI.A.1), students solve algebraic problems related to angle pairs. Throughout the unit, students build their mathematical vocabulary and learn to justify their algebraic solutions using geometric facts, culminating in simple proofs related to angle pairs. This lays a foundation for the next unit, congruence, where students must write formal proofs.

Common Core Standards

- (7.G.B.5) Use facts about supplementary, complementary, vertical, and adjacent angles in a multistep problem to write and solve simple equations for an unknown angle in a figure.
- (8.G.A.5) Use informal arguments to establish facts about the angle sum and exterior angle of triangles, about the angles created when parallel lines are cut by a transversal, and the angle-angle criterion for similarity of triangles. *For example, arrange three copies of the same triangle so that the sum of the three angles appears to form a line, and give an argument in terms of transversals why this is so.*
- (A-REI.A.1) Explain each step in solving a simple equation as following from the equality of numbers asserted at the previous step, starting from the assumption that the original equation has a solution. Construct a viable argument to justify a solution method.
- (G-CO.A.1) Know precise definitions of angle, circle, perpendicular line, parallel line, and line segment, based on the undefined notions of point, line, distance along a line, and distance around a circular arc.
- (G-CO.C.9) Prove theorems about lines and angles. *Theorems include: vertical angles are congruent; when a transversal crosses parallel lines, alternate interior angles are congruent and corresponding angles are congruent; points on a perpendicular bisector of a line segment are exactly those equidistant from the segment's endpoints.*

Figure 6.4 Sample Unit Plan

Goals

- Define adjacent angles, vertical angles, complementary angles, and supplementary angles.
- Name and recognize the angle pairs formed when parallel lines are cut by a transversal.
- Solve problems involving angle pairs.
- State the Exterior angle theorem and apply it in related problems.
- Define isosceles and equilateral triangles.
- Solve problems involving isosceles and equilateral triangles.
- Recognize when an auxiliary line is appropriate for a problem and draw it when necessary.
- Solve problems involving auxiliary lines.
- Define the substitution postulate and transitive property.
- Write simple proofs using the substitution postulate and transitive property.

Vocabulary

- Angle
- Adjacent angles
- Vertical angles
- Complementary angles
- Supplementary angles
- Parallel lines
- Exterior angle theorem
- Isosceles triangle
- Equilateral triangle
- Auxiliary line
- Substitution postulate
- Transitive property

Calendar of Lessons (11 Days)

1. How do we define adjacent and vertical angles?
2. How do we solve problems with adjacent and vertical angles?
3. How do we solve problems involving complementary and supplementary angles?
4. How do we find the measures of angles centered on a point?
5. How do we discover the relationship between pairs of angles formed by parallel lines cut by a transversal?
6. How do we find the measures of angles formed by parallel lines cut by a transversal?
7. How do we apply the Exterior angle theorem?
8. How do we solve problems involving isosceles and equilateral triangles?
9. How do we use auxiliary lines to find the measures of unknown angles?
10. How do we write simple proofs involving angle pairs?
11. How do we use substitution and the transitive property in proofs?

Figure 6.4 (Continued)

CHAPTER 7

How to Plan Lessons

What Is It?

A *lesson plan* is a written description of a teacher's instruction during a class (Meador, 2017). Our lesson plans usually have the following components:

- Explanation of the lesson's scope (aim, goals, related Common Core or state standards, required materials)
- List of required materials (such as manipulatives, graphing calculators, or compasses)
- Introductory activity
- Presentation of new content
- Practice and further development
- Summary

Why We Like It

Lesson plans are the foundation of successful teaching. Planning lessons helps us to address student misconceptions, which keeps students focused on the material. Writing an effective lesson will improve the chances that students will gain knowledge and pay more attention in class—assuming that we follow the plan. Increasing engagement minimizes student misbehavior, boosts our confidence, and reduces stress.

Lesson plans are particularly important in math since mathematical concepts need to be presented in a logical order for students to see how the big ideas in the unit are related to each other.

Supporting Research

Writing lesson plans can help teachers rehearse their questions, analyze tasks, and revise lessons to align with appropriate learning goals. If teachers spend time collaborating on lesson plans, then students are more likely to gain a better conceptual mathematical understanding (Baldinger, Selling, & Virmani, 2016; Boyle & Kaiser, 2017, p. 406).

Researchers recommend a lesson format consisting of an introduction, explanation of new material, practice with progressively more difficult problems, and a summary (Artzt, Armour-Thomas, & Curcio, 2008; Posamentier, Smith, & Stepelman, 2010, pp. 24–25).

Common Core Connections

The Common Core Standards for Mathematical Content provide a starting point for mapping out the content of a lesson plan. In addition, the Common Core Standards for Mathematical Practice outline key mathematical reasoning skills that students should develop, such as modeling with mathematics and finding structure (NGA & CCSSO, 2010).

Application

Here is a description of how we put together each part of a lesson.

DEFINING THE LESSON'S SCOPE

As we explained in Chapter 6: How to Plan Units, we start by creating a unit plan. After we create a concept map that outlines the major themes of the unit and organize them into a logical sequence, we identify and organize problems that correspond to the themes.

We then divide the problems into groups. Each group would ideally focus on one topic, correspond to one lesson, and, in a perfect world, be taught in one day. When topics are closely related (such as complementary and supplementary angles) and roughly equal in difficulty, we often combine them into one lesson since separating them into different days would make the lesson too slow and boring. However, topics that require different procedures, such as addition and subtraction, should generally be taught in separate lessons. Finding the right amount of material to fit into a period can also vary depending on student proficiency, bureaucratic interruptions, block periods, and other factors.

In addition, we include *goals*—the specific skills that we want students to be able to do by the end of the lesson (these goals are for teachers and are *not* written on the board). While we use the standards as a starting point for defining the scope of a lesson, we have found that they often are not specific enough. We usually have to write more specific objectives for each lesson.

We also write an *aim*, sometimes called a *learning objective*, on the board that succinctly states what all students should be able to do by the end of the lesson. Sometimes, we leave blanks in the aim in case we want students to guess what the lesson is about based on the work they did that day. The format of the aim can vary. Some schools require that aims be written as SWBAT ("Students will be able to...") statements. We like to write them in the form of a question (such as "How do we use the distributive property to solve linear equations?") instead of a phrase ("To solve linear equations using the distributive property") in order to encourage students to reflect and assess their own learning during the lesson.

We also include a list of required materials for the lesson (such as graphing calculators or algebra tiles) to help us be prepared for teaching that lesson.

INTRODUCTORY ACTIVITY

Usually called the Do Now or Warm-Up, this introductory activity prepares students for the lesson. We put our Do Now problems on the board before the class begins so that students can work on it immediately.

Our Do Now problems are short—they usually take students no more than 5 minutes to answer. Students work on the problems as we take attendance, return homework, and do other administrative tasks at the beginning of class. After we review the homework, we immediately discuss the Do Now.

The Do Now problems set the tone for mathematical learning from the moment students walk into our rooms. The Do Now questions can quickly review a concept that is needed for the lesson or highlight a gap in student knowledge that will be filled by the lesson. Often, our Do Now questions review a concept that is needed for the lesson or highlight a gap in student knowledge that will be filled by the lesson. Researchers have found that retrieval practice increases the likelihood that students will remember the information later and apply it in new situations (Lemov, 2017; Smith & Weinstein, 2016). Retrieval practice is especially important in math, where many concepts build on previously learned ones. In addition, the Do Now activity immediately engages students in mathematics, creating a bridge between what they already know and what they are about to learn (Wilburne & Peterson, 2007, p. 210). We discuss retrieval practice more in Chapter 4: Promoting Mathematical Communication.

The diagram in Figure 7.1: Do Now Problem shows a Do Now question from our Geometry lesson on drawing auxiliary lines to find unknown angle measures. This activity guides students through two questions that use previously taught concepts (the Exterior angle theorem and angles formed by parallel lines cut by a transversal line). The third question in the Do Now shows a new type of problem that contains parallel lines but neither a triangle (so the Exterior angle theorem can't be applied) nor a transversal (so the theorems related to parallel lines and transversals can't be used). However, the diagram in #3 looks similar enough to the diagrams in #1 and #2 so that students can draw an "auxiliary" or "extra" line into the diagram for #3 to make the problem similar to #2. The resulting class discussion leads directly into a lesson on drawing auxiliary lines to make difficult problems easier to solve.

PRESENTING NEW MATERIAL THROUGH GUIDED QUESTIONS

The presentation of new material is the heart of the lesson. While many teachers we know think of this as the "direct instruction" part of the lesson, we usually don't treat this as a time for lecturing. We like to think of it, instead, as a time to ask students guided questions to elicit the new ideas embedded in the lesson. Usually, we keep these questions to about 10–15 minutes so that students have enough time to practice.

Determining what questions to ask students and what answers we expect is a critical part of our planning for the lesson. Sometimes, we ask these questions orally to the entire class, but many times we also write these questions on paper or on the board. By posing our questions in writing and orally, we give students more time to think about the question. This increased wait time especially helps English Language Learners (who need more time to process the words). By asking students to write, we also encourage students to be more thoughtful in their responses. We discuss these open-ended questions more in Chapter 4: Promoting Mathematical Communication and Chapter 18: Formative Assessment.

Sometimes, we skip our typical Do Now activity and put our guided questions directly at the beginning of the lesson. We may do this if we want to change our routine slightly or if a lesson doesn't require much prior mathematical knowledge. For example, when we introduce the standard deviation, we start with an example that many students are familiar with—grades:

> Ann, Min, and Juan are three students in the same statistics class. Their test scores are listed below. The mean grade for each student is 85. Based on their test scores, are Ann, Min, and Juan the same kind of student? Explain.
>
> Ann's scores: 68, 87, 98, 86, 86
> Min's scores: 79, 98, 68, 94, 86
> Juan's scores: 85, 84, 83, 86, 87

We find that many students can answer this question without much prior knowledge. The resulting discussion—that all three students have the same average score, but Juan's scores are more consistent than the others—leads to a derivation of the formula for the standard deviation, a measure of the spread of a data set (the complete lesson appears in Figure 7.2: Lesson Plan—Standard Deviation).

When we teach a task that can be done in several mathematically correct ways, such as writing the equation of a line (which can be done using slope-intercept form, as done in Figure 7.3: Lesson Plan—Slope-Intercept Form, or point-slope form), we may *temporarily* insist that students use only the method we are teaching in that particular lesson. However, once students have seen all of the methods, we then discuss the advantages and disadvantages of each and recommend that they choose which one they prefer. We try not to prohibit students from using other methods unless we feel that those methods are confusing or error-prone (we discuss this more in the What Could Go Wrong section of Chapter 3: Teaching Math as a Language).

Emphasizing conformity to a particular algorithm without regard to students' understanding rewards only those who can memorize well and discourages original thinking (Witherspoon, 1999, p. 398). This can erode students' confidence by making them think that math consists of arbitrary rules and that success can only come from mindless memorizing.

PRACTICE

Students have to apply what they have learned through practice in order to solidify their understanding. We try to set aside at least 30 minutes of time for practice. Depending on the topic, each problem should generally take students between 2 and 5 minutes to complete and discuss, so we usually plan to complete 5–15 problems in a 45-minute period.

We use both technology and traditional paper-and-pencil methods for student practice, depending on what topic we teach and what resources we have available. The websites and practice methods that we use for homework can also be used for classwork practice. We talk about homework more in Chapter 8: How to Plan Homework and about technology in Chapter 19: Using Technology.

Whenever possible, we differentiate our instruction by dividing practice problems into levels of increasing difficulty. By clearly labeling each level from Level 1 (the easiest problems) to Level 5 (the hardest problems), we help students monitor their progress. We explain this more in Chapter 14: Differentiating Instruction.

DIFFERENTIATING FOR ELLS AND STUDENTS WITH LEARNING DIFFERENCES

As we create a lesson plan, we try to customize our lessons to meet the needs and interests of everyone in our classroom. For advanced students, we reduce the number of simple problems so that they have more time to do more challenging ones. For struggling students, we focus on easier problems so that they can develop skills and build confidence. When we discuss mathematical definitions or theorems, we often express them not just with words but also with pictures and mathematical symbols so that ELLs can improve their vocabulary skills. When possible, we allow students to choose problems that they feel are most appropriate for them.

The methods (as discussed in Chapter 6: How to Plan Units) that we use to make our unit plans culturally responsive and sensitive to students' social and emotional learning can also be applied to writing individual lessons. To build confidence, students can analyze and correct anonymous student work. To take advantage of students' diverse cultural experiences, teachers may ask those who attended school in different countries to share alternate methods of solving a problem. Considering students' cultures and social-emotional learning needs can make lessons more productive by keeping students more engaged and less stressed. In addition, all students can benefit from learning about different cultures and developing their emotional intelligence.

SUMMARY ACTIVITY

When possible, we end our lesson with a summary activity so that students can reflect on what they learned (Wilburne & Peterson, 2007, p. 211). We find that summarizing the lesson helps students remember it. In addition, it helps us see how well students understand the material so that we can adjust future instruction if necessary.

One way to summarize a lesson is to use a Think-Pair-Share strategy, in which we pose a question for students to answer. They can briefly discuss it in pairs before sharing answers with the entire class.

Another option is for students to complete an *exit ticket* (sometimes called an *exit slip*), which they answer on a small piece of paper that we collect to informally assess student understanding. An exit ticket can be a problem that applies a concept from the lesson, a prompt that asks students to explain part of the lesson in their own words, or a question that asks students to compare different methods. Since we write our aims as questions, many times we simply ask students to answer the aim.

Table 7.1: Summary Questions shows some examples of summary questions and the lessons in which they appear.

We often use questions from state or national tests to give students a sense of what to expect on those exams. If the questions are multiple-choice, we ask students to discuss briefly with classmates before answering by holding up the number of fingers that correspond to their answer choice. If we have technology available, we may also use clickers or mobile apps (see the Technology Connections section of this chapter for some examples).

Table 7.1 Summary Questions

Lesson aim	Summary question
How do we solve exponential equations?	To the nearest tenth, solve for x in the equation $30(3)^x = 671$.
What is meant by factoring?	Explain how factoring and dividing are related.
How do we use coordinate geometry to prove that quadrilaterals are parallelograms?	Today, we learned several methods to prove that quadrilaterals are parallelograms. Which one do you prefer? Explain.

Another possibility for summarizing a lesson is to ask students to write their solutions on the board or chart paper. Then the class could do a "gallery walk" around the room, making comments or asking questions on sticky notes.

Many of the techniques that we describe in Chapter 17: Cooperative Learning and Chapter 18: Formative Assessment can also be used as summary activities.

Student Handouts and Examples

Figure 7.2: Lesson Plan—Standard Deviation

Figure 7.3: Lesson Plan—Slope-Intercept Form

What Could Go Wrong

Sometimes, we teach a lesson without thinking enough about how it fits into the larger unit. This can happen when we use an unfamiliar curriculum or get so overwhelmed with day-to-day work that we don't spend time thinking about the "big picture." When we teach "day-to-day," we don't know which concepts are most important, so we often wind up not making connections to future lessons.

Another mistake we've made is to not write out important definitions, theorems, and solutions in our lesson plans. We've found that as we've gotten more experienced, we write less-detailed lesson plans since we know many of the steps. However, classrooms can often be unpredictable. An unexpected event, such as a student outburst or a bird flying in through an open window, can rattle even the most seasoned teacher. Writing down the answers and the precise wording of key concepts helps us stay calm no matter what happens in class.

Many times, we are so concerned with finishing our lesson that we don't pay enough attention to how our students react. This can often make students frustrated and demoralized. We find that checking with students frequently during the lesson and adjusting it as necessary improves our instruction, makes it more culturally responsive, and helps send the message that we care about their learning. We discuss ways to monitor student learning during the lesson in Chapter 18: Formative Assessment.

Focusing too much on making lessons "fun" can also lead to problems. Ideally, lessons should be both educational and entertaining, but it's easy to get so caught up in making beautiful animations or planning elaborate activities that we neglect to spend enough time thinking about what students need to learn. If students can't meet our learning objectives, then the lesson *probably* has failed, no matter how pretty the presentation looks. While teaching any lesson, we sometimes take the time to teach other important concepts not included in our original goals. In these cases, we don't consider our lesson a "failure."

As we said in Chapter 2: Culturally Responsive Teaching, the classwork problems in the lesson should not reflect any cultural, gender, or ethnic bias. For example, a question about buying something on credit can be challenging for a student from a culture that relies on a cash economy (The Education Alliance, n.d.). In addition, questions should not require any specialized knowledge (Leith & King, 2016, p. 673). Instead, any contextual information that is necessary to answer the question should be provided. For example, the word problem below requires knowledge of baseball to answer correctly:

> On a diamond, the distance from the pitcher's mound to home plate is 60.5 feet. What is the distance from the pitcher's mound to first base, to the nearest foot?

The question is supposed to assess knowledge of the law of cosines. To answer, though, students need to be familiar with the layout of a baseball diamond. They must realize that the pitcher's mound is not located at the center of the diamond. This problem can be challenging not just to students of other cultures but also to anyone who is unfamiliar with baseball. ELLs may have difficulty interpreting the word *diamond*. Not only could they associate the word with jewelry instead of sports, but they may not realize that a diamond is actually a square. Many students may spend more time processing the language than the math.

To make the question clearer, the question should clearly describe the shape of a baseball field and clarify what is meant by confusing words like *diamond*. It should include the straight-line distance from home plate to first base and clearly show that the angle formed by home plate, the pitcher's mound, and first base is not a right angle. A labeled diagram not only helps students at all levels process the information but makes the problem less wordy by allowing references to labeled points.

Figure 7.4 is a revised version of the word problem.

Although it is longer than the original version, it provides the necessary background information and a visual aid needed to solve the problem. Many students, particularly ELLs, may still have difficulty understanding phrases like "pitcher's mound," "home plate," and "first base." However, since the distances are stated in the diagram and the distance that needs to be found is labeled *PF*, students can answer the question even if they are unfamiliar with the baseball terminology. We discuss other strategies for solving word problems in Chapter 4: Promoting Mathematical Communication.

Not thinking about the lesson after teaching it can lead to problems in the future. Immediately after teaching a lesson, we try to take a few minutes to reflect on how it went and how we can make it better. Many times, we discover improvements to the lesson while we're teaching it. When these moments occur, we try to make a note on our lesson plan so that we can revise it later. For example, a student may see another way to solve a problem or explain an idea in words that is clearer than what's in our lesson. Looking at exit tickets also helps us assess how much students understood. Writing these ideas immediately after teaching the lesson, while it's still fresh in our minds, increases the likelihood that the lesson will be better the next time we teach it.

Of course, we don't always have the time or energy to revise a lesson "on the spot." We find that writing down even just a few short notes, like "rephrase #4—wording is confusing" or "add more problems with negative numbers—too repetitive" can be helpful for the next time we teach this lesson. Otherwise, we'll wind up making the same mistakes the next period or the following year!

Technology Connections

Many websites contain resources that are useful for writing lessons. We list some of them in Chapter 6: How to Plan Units. The National Council of Teachers of Mathematics website (http://nctm.org) contains many lesson plans and student activities for grades K–12. Membership is required to access the material. In addition, well-established websites like Math Bits Notebook (www.mathbitsnotebook.com) and Khan Academy (www.khanacademy.com) often provide explanations or problems that can be incorporated into a lesson.

Two websites that we have found especially useful for writing math lessons deserve special mention. Desmos Classroom Activities (http://teacher.desmos.com) contains many free online ready-made student activities that can be incorporated into lessons. Geogebra (http://geogebra.org) has free online math tools (including many animations and guided discovery activities) for graphing and geometry.

Websites like Kahoot! (http://kahoot.com) or Plickers (http://plickers.com) allow students to record and analyze student responses instantly.

When making lessons, we often use a snipping tool (such as the free Preview software that comes with Apple computers or the free Snipping Tool on Windows 10 computers) to take a partial screenshot from a computer and paste it into a document.

From our experience, we know that standards and curricula change all the time. To facilitate revising, reordering, and finding lessons, we try to put all of the lessons for a unit into one computer file. Having one file for the unit also makes reordering lessons and searching for content much easier. Sometimes, these files can get quite large, so navigating through them can be tedious. To find things more easily, we create a table of contents or put pages into groups, which allows us to have an overview of the topics in the unit.

When we write lessons, we use both word processing and presentation software.

Word processors like Microsoft Word or the free Google Docs (http://docs.google.com) are great for creating handouts. Teachers who are new to a course may find word processors helpful for writing more detailed lesson plans that contain such things as the problems we plan to cover, solutions to classwork problems, anticipated student errors and ways we plan to address them, and questions we plan to ask students. (We discuss using word processors and equation editors in the Technology Connections section of Chapter 6: How to Plan Units.) When we work with other teachers, we find that word processors (such as Google Docs) that allow teachers to work simultaneously on the same document can help facilitate collaboration.

However, formatting word documents to appear nicely on a large projector or interactive whiteboard can often be difficult. Instead, we use presentation software, such as Microsoft PowerPoint or the free Google Slides (http://slides.google.com), to display our lessons on projectors or interactive whiteboards. Each slide contains a small part of the lesson, such as one definition or theorem, one or two problems, key points on a particular topic, or an illustration. Google Slides also can provide automatic closed captioning if a teacher wears a microphone, which can be particularly helpful for English Language Learners. Interactive whiteboards typically come with their own lesson creation and delivery software, which often has features unique to the board. We suggest checking with your school to see what software is available and what works best for you.

Many teachers we know don't write formal lesson plans and instead rely on their presentation files to guide them in class. Often, they print out their presentations and write notes on them, such as solutions to problems or reminders of what to emphasize. This method has the advantage of taking less time than typing up a lesson plan and then creating a separate file for presentations. However, teachers who want digital versions of their lessons or who are required by their schools to have written lesson plans may find a word processor easier.

When deciding what software to use for presenting lessons, we look for the following features:

- **Can it save handwritten work from the whiteboard?** We often encourage students to write solutions to problems on the whiteboard. The software we use should have the ability to save handwritten work for each problem so that we can go back and refer to it if necessary.

- **Can it convert files to a format that students can access?** We often convert our class notes to a commonly available format like PDF so that students can download and view them.

- **Can we type mathematical symbols?** Software like EquatIO (www.equatio.com) and MathType (www.mathtype.com) work with most presentation and word processing software applications.

- **What interactive tools are included?** We often use mathematical tools like protractors, rulers, and compasses in our lessons, so having an electronic version on the board is helpful. Electronic dice or timers are also useful for managing classwork.

Figures

DO NOW: Find the measure of each angle. Justify your answer.

1.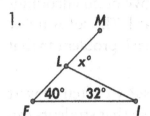
2.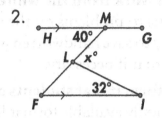
3.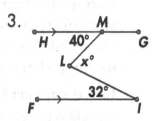

Figure 7.1 Do Now Problem

Algebra I

Aim: How do we calculate and interpret the standard deviation?

Goals:
- Define and distinguish between a population and a sample.
- Derive the formula for the variance and standard deviation.
- Define and calculate the range, variance, and standard deviation.
- Interpret the standard deviation.

Standards: (S-ID.A.2): Use statistics appropriate to the shape of the data distribution to compare center (median, mean) and spread (interquartile range, standard deviation) of two or more different data sets.

Materials: Classwork sheets

Do Now

1. Ann, Min, and Juan are three students in the same statistics class. Their test scores are listed below. The mean grade for each student is 85. Based on their test scores, are Ann, Min, and Juan the same kind of student? Explain.

   Ann's scores: 68, 87, 98, 86, 86
   Min's scores: 79, 98, 68, 94, 86
   Juan's scores: 85, 84, 83, 86, 87

   Juan's scores are generally in the 80s while Min's are less consistent.

   Measures of dispersion are measures of the spread or variation in the data.

Figure 7.2 Lesson Plan: Standard Deviation

Lesson

2. The range is one measure of dispersion. The ranges for each student's scores are as follows: Ann = 30, Min = 30, Juan = 4. Based on the student scores, is the range a good measure of dispersion for this data?

No, it can be deceiving. Ann's scores are in the 80s more often than Min's, yet they have the same range. However, both are clearly different from Juan's scores, which have a range of 4.

Discuss the following definitions:
A **population** is the collection of all outcomes, responses, measurements, or counts that are of interest.
A **sample** is a subset of a population.

One way to measure dispersion is to consider how far each score is from the mean.

The **deviation** of a value in a population is the difference between the value and the mean, μ, of the population. Deviation of $x = x - \mu$.

3. Find the deviations for each score and calculate the mean of the deviations for each student. Why is the mean deviation a poor measure of dispersion for the scores?

Make a table for each student and fill in the first three columns. (Leave the fourth column blank for now.)
The mean deviation always sums up to 0.

Ann's Scores

x_i	μ	$x_i - \mu$	$(x_i - \mu)^2$
68	85	−17	289
87	85	2	4
98	85	13	169
86	85	1	1
86	85	1	1

Variance = $\sigma^2 = \dfrac{289 + 4 + 169 + 1 + 1}{5}$
= 92.8
Standard deviation = $\sigma = \sqrt{92.8} \approx 9.6$

Figure 7.2 (Continued)

Min's Scores

x_i	μ	$x_i - \mu$	$(x_i - \mu)^2$
79	85	−6	36
98	85	13	169
68	85	−17	289
94	85	9	81
86	85	1	1

Variance $= \sigma^2 =$
$\dfrac{36 + 169 + 289 + 81 + 1}{5} = 115.2,$
Standard deviation $= \sigma = \sqrt{115.2} \approx 10.7$

Juan's Scores

x_i	μ	$x_i - \mu$	$(x_i - \mu)^2$
85	85	0	0
84	85	−1	1
83	85	−2	4
86	85	1	1
87	85	2	4

Variance $= \sigma^2 = \dfrac{0 + 1 + 4 + 1 + 4}{5} = 2$
Standard deviation $= \sigma = \sqrt{2} \approx 1.4$

4. To eliminate the issue that you discovered in the previous problem, we can square each of the deviations. Calculate the variance (the average of the squares of the deviations from the mean) for Ann's, Min's, and Juan's scores.

   Show on board:
   The **variance**, σ^2, of a set of data is the average of the squares of the deviations from the mean.

   $$\sigma^2 = \frac{1}{n} \sum_{i=1}^{n} (x_i - \mu)^2$$

5. Since the units in the variance are "square units," which are different from that of the original data (e.g., "square points"), we take the square root of the variance to get our original units. Calculate the standard deviation for each student's scores.

Figure 7.2 (Continued)

Show on board:
The **standard deviation of a population**, σ (lowercase sigma), is the square root of the variance:

$$\sigma = \sqrt{\sigma^2} = \sqrt{\frac{\sum_{i=1}^{n}(x_i - \mu)^2}{n}} = \sqrt{\frac{1}{n}\sum_{i=1}^{n}(x_i - \mu)^2}$$

Complete the fourth column for each table.

6. Interpret the standard deviation for each student using the conditions of the problem.

Show on board:
The standard deviation measures the average amount by which individual items of data deviate from the mean of the data.
Ann's scores typically vary from the mean by 9.6 points.
Min's scores typically vary from the mean by 10.7 points.
Juan's scores typically vary from the mean by 1.4 points.

Summary

7. The times, in seconds, needed by five athletes to run 400 meters were as follows: 45.5, 57.1, 50.3, 55.0, 48.3. Find the mean and the population standard deviation to the nearest tenth. Interpret the results.

Mean = 51.2; standard deviation = 4.3. The runners' times typically varied from the mean by 4.3 s. (Most of the runners ran between 46.9 and 55.5 s.)

Figure 7.2 (Continued)

Aim: How do we write an equation of a line in slope-intercept form?

Goals
- Relate the points on a graph of an equation to the equation's solutions.
- Write an equation of a line given its slope and y-intercept.
- Write an equation of a line given its slope and the coordinate of one point on the line.
- Write an equation of a line given the coordinates of two points on the line.

STD (8.EE.B.6) Use similar triangles to explain why the slope m is the same between any two distinct points on a nonvertical line in the coordinate plane; derive the equation $y = mx$ for a line through the origin and the equation $y = mx + b$ for a line intercepting the vertical axis at b.

Do Now

Identify the slope and y-intercept of the line shown in each graph.

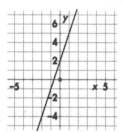 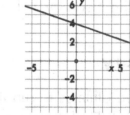 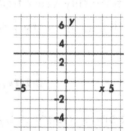 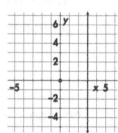

1. Slope = 3
   y-intercept = 2

2. Slope = $-\frac{1}{3}$
   y-intercept = 4

3. Slope = 0
   y-intercept = 3

4. Slope: undefined
   y-intercept: none

Lesson

Review from previous lessons:
- The graph of an equation is the graph of all ordered pairs that are solutions to that equation.
- If a line passes through a point, then the coordinates of that point are a solution to the equation of that line.

Elicit from #1 that the points (−1, −1), (0, 2), (1, 5), etc., are all solutions to the equation $y = 3x + 2$.

Figure 7.3 Lesson Plan: Slope-Intercept Form

The slope-intercept form allows us to see the slope (rate of change) and y-intercept (starting value, which is when $x = 0$) quickly. Write on the board:

SLOPE-INTERCEPT FORM of a linear equation: $y = mx + b$ (m = slope, b = y-intercept)

Elicit the equations of the lines in the Do Now.

1. $y = 3x + 2$ **2.** $y = -\frac{1}{3}x + 4$ **3.** $y = 3$ **4.** $x = 3$

Classwork

5. Identify the slope and y-intercept of each equation.
   a. $y = -4x + 5$ $m = -4, b = 5$
   b. $y = 6 - 7x$ $m = -7, b = 6$
   c. $2y + 8 = x$ $y = \frac{1}{2}x - 4 \Rightarrow m = \frac{1}{2}, b = -4$

6. Determine whether the given equation passes through the given point:
   a. $y = 2x - 4$, $(1, -2)$ Yes since $2(1) - 4 = -2$.
   b. $y = \frac{3}{4}x + 2$, $(4, 4)$ No, since $\frac{3}{4}(4) + 2 \neq 4$.
   c. $16x + 57y = -66$, $(60, -18)$ Yes, since $16(60) + 57(-18) = -66$.
   NOTE: Questions in #6 and 7 can be solved using point-slope form. Don't mention it yet—it will be covered in a later lesson—but encourage students who may know point-slope form to use slope-intercept form today. Assure them that once they have learned both methods, they may choose.

7. Write an equation of a line that has:
   a. a slope of $\frac{1}{3}$ and passes through the point $(0, 4)$.
      Find the y-intercept: $4 = \frac{1}{3}(0) + b \Rightarrow b = 4$, so the equation is $y = \frac{1}{3}x + 4$.
   b. a slope of -2 and passes through the point $(-26, 41)$.
      From slope–intercept form $y = mx + b$, substitute coordinates of a point and the slope to solve for the y-intercept:
      $$41 = -2(-26) + b \Rightarrow 41 = 52 + b \Rightarrow b = -11$$
      $$y = -2x - 11$$
   c. a slope of -18 and passes through the point $(-37, 59)$.
      $m = \dfrac{y_2 - y_1}{x_2 - x_1} \Rightarrow -18 = \dfrac{y - 59}{x + 37} \Rightarrow b = -607$, so the equation is $y = -18x - 607$.

Figure 7.3 (Continued)

8. Write an equation of the line that passes through the given points:
   a. (−1, −1) and (2, 5)
   $m = \dfrac{5-(-1)}{2-(-1)} = \dfrac{6}{3} = 2$, $b = 1$, so the equation is $y = 2x + 1$.
   b. (12, 18) and (15, 12)
   $m = \dfrac{18-12}{12-15} = \dfrac{6}{-3} = -2$, $b = 42$, so the equation is $y = -2x + 42$.
   c. (−8, 7) and (8, 15)
   $m = \dfrac{15-7}{8-(-8)} = \dfrac{8}{16} = \dfrac{1}{2}$, $b = 11$, so the equation is $y = \tfrac{1}{2}x + 11$.

Summary

List all of the information that you need to write an equation of a line.

- Graph of equation
- Slope and *y*-intercept
- Slope and coordinates of one point on the line (find the *y*-intercept)
- Coordinates of two points on the line (find the slope, then the *y*-intercept)

Figure 7.3 (Continued)

As shown in the accompanying diagram, a baseball field contains a square, known as the *diamond*. The pitcher's mound (point *P*) is 60.5 ft from home plate (point *H*). The distance *HF* between home plate and first base (point *F*) is 90 ft. The measure of ∠*PHF* is 45°. To the nearest foot, what is *PF*, the distance from the pitcher's mound to first base?

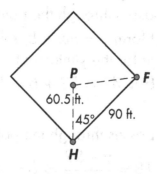

Figure 7.4 Revised Baseball Field Word Problem

CHAPTER 8

How to Plan Homework

What Is It?

Homework is an assignment, one that we give every day, that students do outside of class. Typically, we assign homework to reinforce the lesson through practice. Sometimes, we use homework to help students discover ideas independently.

We recognize, and often experience, that many students often don't do homework regularly. In the What Could Go Wrong section, we share our strategies for dealing with this issue, including creating opportunities during class for students to complete it.

Why We Like It

We find that daily homework is an essential part of our class. In addition, we use homework to monitor student work informally. Students can assess their own work without publicly acknowledging weakness (Jackson, 2014, p. 532). When students do homework, they can review the concepts discussed in class and apply what they learned to different problems. Students who don't do homework not only miss an opportunity to practice independently but may also forget the lesson.

When we give students work to do outside of class, we don't always give the traditional set of problems from a textbook or worksheet. We sometimes assign other types of work, such as projects. We discuss these strategies more in Chapter 16: Project-Based Learning.

Supporting Research

Experts offer conflicting evidence on the value of homework.

Some argue that homework can hinder student learning. Many studies indicate that assigning homework to elementary or middle school students has no positive benefit (Kohn, 2008). Homework can increase negative attitudes toward school while taking time away from play, independent activities, and spending time with family (Bennett & Kalish, 2006, p. 22; Shumaker, 2016). Many students feel that their homework doesn't reinforce their learning and takes too much time and effort to complete (Burriss & Snead, 2017, pp. 203–204).

Others argue that homework has several important benefits. Homework allows students to reflect on the lessons and work at their own pace. Teachers can use homework assignments to review previously learned material without taking up valuable instructional time (Posamentier, Smith, & Stepelman, 2010, pp. 26–27). Furthermore, homework can help students recognize what topics they don't understand and when they need help (Jackson, 2014, p. 528). Some research indicates that homework can help older students more than younger students (Cooper, 2006, p. 50) and that students who complete high-quality homework scored better on assessments (Dettmers, Trautwein, Lüdtke, Kunter, & Baumert, 2010, p. 478).

In our experience, homework, when done right, builds student confidence and fosters conceptual understanding. Effective homework is directly tied to what students need to know, respects students' time by being efficient (usually about 15–20 minutes per night), and has an appropriate level of difficulty for all students (Vatterott, 2010). In addition, reducing the number of homework problems and including a mix of straightforward and challenging problems can help make homework more meaningful and manageable for students (Walk & Lassak, 2017, p. 549).

Common Core Connections

Doing homework can help students develop the expertise they need to meet the goals expressed in the Common Core's Standards for Mathematical Practice. In order to "make sense of problems and persevere in solving them" (NGA & CCSSO, 2010, p. 6), students can benefit from doing additional work outside of class.

Application

Our students have busy lives outside of our classroom. As a result, we try to do everything we can to make homework a manageable and meaningful routine, which encourages students to complete it. In this section, we discuss our techniques for assigning, discussing, and grading homework.

SOURCES

We use many sources for homework problems. If the course that we teach uses a textbook, we usually assign questions from it, sometimes supplementing it with questions from other places.

Most of the time, we find that no one textbook fits all of our students' needs. For example, the textbook we use may not have useful problems on a particular topic, so we have to supplement it with material from another book or a website.

When possible, we also like to use online practice sites as another source of homework questions (we list some of our favorites in the Technology Connections section of this chapter). Most of them allow students to practice topics repeatedly and automatically grade their work, saving us a great deal of time. Unfortunately, the problems on these sites tend to be limited to grid-in or multiple-choice questions since proofs or other problems that require explanations cannot be graded accurately by a computer. Sites like these are more appropriate for practicing skills or preparing students to take computer-based tests than for deep mathematical thinking (we discuss how technology can reinforce practice in Chapter 19: Using Technology).

We like to use a combination of both paper-based and technology-based homework in our classes. Paper-based homework helps students learn how to show their work and gives them examples that they can use for studying. Technology-based homework allows students to practice more and receive instant feedback (Cox & Singer, 2011, p. 518). Not relying exclusively on online homework also enables us to accommodate students who don't have reliable internet access at home.

HOMEWORK FORMAT

We assign homework every day except for the day before and day of a test.

Since many of our students can get online at least briefly every day, we prefer to post our homework online rather than write the questions on the board. Many schools have content management systems that allow teachers to post assignments and other related material online for students to access. Although we generally post homework online, we always remind students in class of upcoming assignments. For students who don't have internet access, we print out copies of our homework and give it to them or write the assignment on the board.

Since we like to plan ahead, we often have several homework assignments ready at any given time. This practice gives us some flexibility in case unexpected events change our schedule. However, we typically post only one assignment at a time to avoid student confusion.

When we assign questions from a textbook for homework, we ask students to copy the questions from the textbook with all diagrams and directions, even though they may have access to the book outside of school. We find that homework works best as a study guide when questions are right next to the answers. In many cases, the work required to solve a math problem only makes sense when seen next to the original question. If we have time, we try to create files for each homework assignment and put them online for students to download. This avoids problems that arise when students copy the wrong questions or travel between different homes during the week. We also ask students to annotate problems as part of the problem-solving process (as we explain in Chapter 4: Promoting Mathematical Communication), which they can only do if they can write on the assignment itself.

With every assignment, we include references that students can use. If we use a textbook, then we include relevant pages for students to read. In addition, we also try to provide links to two or three websites or online videos that reinforce the lesson so that students may learn from different sources.

As we stated earlier, we generally prefer that students write their homework solutions by hand. Many math problems require writing special symbols or drawing accurate diagrams that can sometimes take more time to construct on a computer or phone than students realize. Students who take paper-based tests should get used to writing math by hand. Writing homework on paper is usually cheaper and easier for students than writing homework on a computer.

However, having students type their homework also has several advantages. Some students, such as those with physical disabilities, may not be able to write easily by hand. Students can solve many math problems with the help of online graphing tools, so typing homework enables students to paste work into a document instead of transfer work onto paper by hand. In addition, students who have to take computer-based state tests should have some homework assignments that mimic testing conditions (we suggest some useful resources for computer-based state test practice in the Technology Connections section of this chapter).

HOMEWORK AS PRACTICE

Homework assignments that reinforce skills learned in class generally have between four and six problems, depending on the lesson. If the assignment contains a more complicated problem, such as a word problem or a proof, we usually make the other problems shorter and easier. Two of the problems typically relate to that day's lesson, while the others deal with previous topics. As we discussed in Chapter 7: How to Plan Lessons, retrieval practice helps students use previously learned concepts in unfamiliar situations, thus strengthening their learning (Lemov, 2017; Smith & Weinstein, 2016).

We try to select problems in such a way that students will spend 15–20 minutes on each homework assignment. The amount of time may vary according to your school's culture. When possible, we talk to our students and check with teachers in other departments to see how much homework is given in other classes. Making homework predictable in length and difficulty helps students set aside time in their schedules to do it.

Figure 8.1: Homework as Practice shows a typical homework assignment with practice problems.

HOMEWORK AS DISCOVERY

Sometimes, we use homework to help students find a pattern that would take too much time in class. These assignments usually consist of a series of exercises that will allow students to discover a new idea in the next lesson (Posamentier et al., 2010, p. 26). We find that discovery homework is most effective when it includes tasks that are straightforward enough to be done at home with minimal assistance, such as calculations that students have previously learned or questions that can be answered with some simple research online or in widely available sources like newspapers. Ideally, students should do enough work at home so that we can focus our classwork discussion on eliciting the new idea. Many times, guided questions used to introduce a lesson (which we discuss more in Chapter 7: How to Plan Lessons) can also be assigned as homework.

Figure 8.2: Homework as Discovery—Ratios shows a homework assignment that is used to introduce ratios. Students deal with a simple real-world problem that relates the production of large cars to small cars at a factory.

Figure 8.3: Homework as Discovery—Mean Proportional Theorem shows a homework task in which students explore the relationship among similar triangles formed when an altitude is drawn to the hypotenuse of a right triangle.

Figure 8.4: Homework as Discovery—Parabolas shows an example of an Algebra II assignment in which students solve quadratic equations, graph the corresponding functions, and state the characteristics of their roots. In previous lessons, students already learned how to complete these tasks. This homework leads into a lesson on how the graph of the function is related to the nature of its roots.

HOMEWORK AS TRANSFER

We also use homework to help students transfer learning from one situation to another. In math, we often ask students to apply their knowledge of a particular topic to real-world situations through word problems. For example, after learning how to solve linear equations, students typically learn how to use them to solve word problems. (We talk more about strategies for solving word problems in Chapter 4: Promoting Mathematical Communication.)

Sometimes, we assign homework in which students react to a prompt or research a question. Figure 8.5: Homework as Transfer—Similarity contains an example that we assign at the beginning of our unit on similarity. In this assignment, we give students a picture and ask them to list all of the mathematical terms that they can see. Students should be able to see terms like *point, line, similarity, parallel,* and *dilation,* all of which are relevant to this unit. Another example appears in Figure 8.6: Homework as Transfer—Bank Accounts, which we assign at the beginning of our unit on exponential functions. In this assignment, students look through newspapers or the internet to find information about three different savings accounts, rank them from best to worst, and use appropriate math to justify their answer. Since we assign this lesson before we teach the concept of the annual percentage yield, it provides a powerful motivation to devise a way to easily compare interest rates.

We like to give these types of assignments periodically because they provide a welcome change of pace from our typical homework. In addition, these "low floor, high ceiling" tasks allow all students to do meaningful work. Finally, these kinds of "transfer" assignments can enhance students' intrinsic motivation (discussed further in Chapter 1: Motivating Students) by helping them see how their math skills can be applied outside of school.

DISCUSSING HOMEWORK

When we discuss homework in class, we try to let students do most of the talking. We have found that this helps promote an atmosphere of community among students and encourages them to take more responsibility for their work. Here are some methods that we use to talk about homework (we discuss other strategies for fostering student discussion in Chapter 4: Promoting Mathematical Communication).

Selecting Students to Share Work

When we require all students to put up homework problems or select one row every day to put up homework, students often resist. We prefer to give extra credit to students who volunteer to put up problems on the blackboard. Another option to help create a safe space for sharing their work is to tell the class that some students may occasionally make mistakes on purpose to see if the class is paying attention. We may privately speak to students beforehand to make mistakes that we want to highlight during the homework discussion. If we have students who don't want to share a solution and lead a discussion about it, we encourage all students to talk with their neighbors about the homework as their classmates put up problems. We want to help all students have meaningful mathematical conversations without feeling embarrassed or socially awkward, which can reinforce negative feelings about math.

Whole-group Review

In our default method for reviewing homework, individual students write the solution to a problem on the classroom board at the beginning of the period while other students compare and correct their work. If we want to keep the work for future reference, we sometimes ask students to put their work on chart paper and tape the papers to the board. Each student who puts up a problem then leads a discussion on it, fielding questions from classmates and correcting his or her work on the blackboard. Typically, this method takes no more than 10 minutes.

We find that this whole-group review method works best when most students do homework and the problems are short and straightforward. To make sure that all problems are covered, we divide up the space on the board so that only one person puts up each problem.

Online Review

Sometimes, we review by sharing student pictures of their homework assignments. We use this method when writing homework solutions on the board would be too difficult and time-consuming. For example, we may teach in a room with inadequate board space, or we may be doing questions with solutions that are too long or complicated to write on the board quickly (such as geometric proofs).

Many content management systems, such as Google Classroom, allow students to upload their work so that their classmates can see it. Ideally, students would look at the uploaded problems before coming to class so that we can reduce the amount of time spent discussing homework in class. However, we find that students often don't look at uploaded work before class, so we also insert the pictures into our lesson presentation files and project the students' work on the classroom's whiteboard or projector screen.

Another possibility is to ask for student volunteers to email their homework to us beforehand. This method can be particularly helpful for students who want feedback on a particular question or method but want to submit their work anonymously. When we put their work into our lesson presentation files, we crop out their names so that students may speak freely without embarrassing anyone.

COLLECTING HOMEWORK

To collect homework that is written on paper, we ask students to pass their work to the first person in their row (or a designated person in each group if seats are arranged by group), who folds the papers in half and writes a preassigned row or group number. We keep the papers in these groups when grading and return them folded to the designated monitor for each group. Students then take their work and pass the remaining assignments to others in the group until all papers are returned.

Some teachers may prefer to collect and comment on homework electronically. In these situations, students can take pictures of their homework and submit it online using a site like Google Classroom.

GRADING HOMEWORK

Grading homework allows us to hold students accountable, give feedback, and informally monitor student progress. We usually grade homework for both completion and accuracy. Since we usually review homework in class, students should have a complete and correct solution on their papers, which can serve as model examples for studying. We emphasize to our students that they are graded not on whether they originally got the problems right or wrong but to what extent they correct their work. We ask students to grade their own work clearly using a different-colored pen or pencil so that we know what they were able to complete independently.

The top of Figure 8.1: Homework as Practice contains a scoring guide that we use for grading homework. We put the scoring guide at the top of our handouts to make our expectations clear and facilitate communication about the grade. We occasionally add other comments directly on the paper. While grading homework, we note common mistakes and use that to improve our instruction. If many students are making the same mistake, we try to point them out right after returning the assignment.

If we don't have the time to grade homework ourselves, we try to recruit a former student who has already taken the course or a teacher assistant (student assigned by the school to help a teacher) to help us grade. We find that many students enjoy being our monitors because they can get service credit or they just like to help us. Another possibility is to ask students to switch homework with a partner and grade each other's work. In this method, teachers must clearly explain to students how the homework is scored. As teachers show answers on the classroom projector, students could grade each question and point out common mistakes. Teachers could check student grading by collecting a sample of student homework to ensure that it is accurate and respectful.

Sometimes, we get so busy that we can't keep up with grading every homework assignment every day and we don't have student monitors who can help us. In these situations, we grade only a sample of student work. For example, we may roll a die in class and grade the assignments of the rows whose numbers come up. We may flip a coin in class to determine if we collect that day's homework. We find that when we don't grade every student's work every day, students generally are less careful with their homework and less likely to complete it properly. While not ideal, it sometimes is our only option.

DIFFERENTIATING FOR ELLS AND STUDENTS WITH LEARNING DIFFERENCES

Ideally, we would like to differentiate homework so that students receive assignments that match their individual needs and interests. However, this can be hard to do on a regular basis. Here are some tips to customize homework but keep the workload manageable for teachers.

One simple way to differentiate homework is to employ the same methods that we use to differentiate our lessons (see Chapter 14: Differentiating Instruction and Chapter 15: Differentiating for Students with Unique Needs for more information). For example, we sometimes divide our homework into three levels of difficulty and ask students to choose two levels to complete. Each level of homework problems would roughly correspond to a level of classwork problems. When reviewing homework, we ask students to put up all of the problems. This method does take more time to review, however.

Many English Language Learners can find homework particularly challenging because they often lack the supports that we provide in class. When possible, we incorporate links to websites in different languages, such as Khan Academy (we list some of these sites in the Technology Connections section of this chapter). Students who don't have reliable internet access often find printed glossaries to be a valuable resource. We typically provide these links on the day that an assignment is given because we find that if we post many homework assignments simultaneously, students get confused about which assignment to do. However, ELLs may benefit from the *preview-view-review* strategy, in which they get instruction in their home language before and after receiving instruction in English. Teachers can use a modified preview-view-review strategy for ELLs by posting multilingual language summaries or videos online one or two days before the lesson. Students could then watch the videos in their free time at home or in class before the lesson (Ferlazzo & Sypnieski, 2018b).

Another technique that can be helpful for all students is to encourage them to form study groups. ELLs may benefit from discussing homework in English, another language, or both, depending on their preference. As we get to know students, we foster the formation of these study groups by having students who speak a common language sit next to each other. We encourage them to exchange contact information and ask each other questions about the homework. If we have time, we sometimes allow study groups to meet during class (for example, while we discuss homework or work on classwork practice).

Homework may also be customized for students with particular needs. For example, some students need additional help or can't complete assignments at home. We offer some suggestions to address inclusion in the What Could Go Wrong section in this chapter.

Student Handouts and Examples

Figure 8.1: Homework as Practice
Figure 8.2: Homework as Discovery—Ratios
Figure 8.3: Homework as Discovery—Mean Proportional Theorem
Figure 8.4: Homework as Discovery—Parabolas
Figure 8.5: Homework as Transfer—Similarity
Figure 8.6: Homework as Transfer—Bank Accounts

What Could Go Wrong

STUDENTS WHO DON'T DO HOMEWORK

One of the most serious problems that we encounter is that some students don't do homework. We try not to take this personally or conclude that students aren't trying hard enough. Instead, we try to identify factors that prevent them from completing assignments. Students may not have a quiet space or time at home, they may be traveling between different homes during the week, or they may have part-time jobs or family obligations.

One of the most effective ways that we have learned to deal with this is to reach out to parents, guardians, or school personnel (such as a guidance counselor, social worker, or psychologist) to determine what is going on with their child and what we can do to help. Many times, we find that parents are unaware of the situation and are happy to encourage their children to complete assignments. We also try to encourage students who don't do homework at home to complete it in school during a free period or at a local library or tutoring center if available. Many schools have study hall periods or tutoring available before or after school. Some libraries or school districts offer homework help after school. Check with your district or neighborhood to see what resources may be available.

We sometimes allow students who don't do homework at home to work on it during our homework review in class. The advantage of this method is that they can get additional help from us or their peers if they need it. The disadvantage of this strategy is that students may not have time in class to finish the homework. In addition, if students see that they don't need to do homework outside of class because we give them time in class to do it, they may stop doing it altogether, forcing us to take more time during our lessons to review.

As a last resort, we may set aside time at the beginning of class for students to practice independently. This "quiet time" then replaces what would be the homework review in class. To help students focus, we sometimes play quiet music in the background as students work. Since we need time at the end of this practice to briefly discuss what students did, students get less practice time than they would in a regular homework assignment. However, this strategy may be the only option for classes in which most students don't do homework at home. Teachers could also "flip" their classrooms by assigning instructional videos to watch at home and doing practice in class (we discuss flipped classrooms in Chapter 19: Using Technology).

We find that building good relationships with students and being flexible helps ensure that they complete homework and use it to learn important mathematical concepts. This can be as simple as giving copies of homework to a student without a printer at home, praising someone who does a good job on a problem, or having a private conversation with a student who is struggling. In our experience, small steps like these can pay big dividends in the long run.

MISMANAGING CLASS TIME

We have also made the mistake of not spending enough time reviewing homework in class. We've done this when we feel pressed for time or when we use online tools that automatically grade student work. We find that when we don't spend time talking about the homework in class, students can feel frustrated because they can't figure out if their work is right or why they got a problem wrong. In our experience, when our homework review confirms students' thinking as well as corrects their mistakes, then they feel more validated as mathematicians.

Other times, we don't manage our time properly at the beginning of class, resulting in homework review that takes much longer than 10 minutes and leaves less time for learning new material. As a result, we wind up not being able to finish our lesson, which means that students may be assigned homework problems that they didn't learn how to solve in class. The following day, we then have to spend more time explaining homework problems than we had originally planned. This can lead to a vicious cycle of extended homework review and unfinished lessons.

To prevent this cycle from occurring, we try to monitor our time at the beginning of class very carefully. Sometimes we even use a stopwatch to help us ensure that we move student conversations along. We also adjust homework immediately after teaching a lesson, but before students start it, if we feel that they won't be able to answer certain questions. Of course, we don't want to cut off valuable mathematical conversations. However, if we find that our homework review is consistently running too long, then we shorten or simplify our homework so that our students can engage in deeper mathematical thinking during the lesson.

HOMEWORK REVIEW CHALLENGES

Sometimes, we lead the homework review process ourselves by taking student questions and explaining the correct answer. This happens when we are covering a topic that is particularly challenging to the class or when we're just not getting many students willing to come up to the board, even when we offer extra credit.

We try to avoid just reading off the answers on a regular basis. Our experience confirms existing research that when teachers simply read off homework answers, they miss opportunities to assess students formatively and help those who make errors (Otten, Cirillo, & Herbel-Eisenmann, 2015, p. 101). Instead, we focus on other methods, such as making questions more appropriate for students' levels of understanding, giving them extra credit for putting up problems, or having them review homework in small groups, which can help make homework review more meaningful and effective.

CHOOSING THE WRONG PROBLEMS

We have found that giving students unfamiliar problems in homework usually backfires. When our homework includes questions that we haven't covered in class, students tend to get frustrated and lose motivation to do future assignments. Instead of trying to teach through homework, we generally try to teach during the lesson, except when we design homework as discovery or transfer, as we describe in the Application section of this chapter.

Another mistake we've made is to assign homework questions without first working through them ourselves. Sometimes a problem may require additional knowledge that we haven't discussed in class yet or may be too difficult or too easy for our students. This can confuse students and ultimately discourage them from doing homework. Not creating an answer key can be particularly tempting when we use a textbook or website that already provides the answers. We've found that the best way to avoid this issue is to complete the problems the way we would expect our students to do them. Doing the problems ourselves also helps us to identify potential student pitfalls that can be addressed during the homework review.

Technology Connections

The technology available for students to do homework changes constantly. Since many schools or districts purchase student accounts for online homework resources, we recommend checking with your administrator to see what is possible.

We have found that the online homework tools that work best have certain features in common. Here's what we look for in a technology solution for homework:

- **Does it grade student work automatically?** We should not have to grade student work on a homework website. The site should also provide an item analysis of student responses so that we can determine the topics that students struggled with.

- **Can we download grades easily?** Many schools now use online gradebooks, so having the ability to download grades from an online homework tool saves us a lot of time.

- **Can we monitor student work?** Teachers should be able to see what questions students have answered and which ones they got correct.

- **Does it give students immediate feedback?** Our students have told us that they like finding out whether they got a question right or wrong immediately after answering it. Such immediate feedback helps students correct their mistakes more quickly, especially since we are not there to help them. The best sites explain how to obtain the correct answer and why other choices (if the question is multiple-choice) are incorrect.

- **Does it provide many examples for student practice?** Websites that provide many computer-generated examples for each topic help students develop procedural fluency. In contrast, sites that simply republish items from state tests aren't as helpful, especially if students recognize the questions from our lessons, test prep books, or the internet.

For online practice, one free site that stands out because it offers all of the characteristics above is DeltaMath (www.deltamath.com). On DeltaMath, teachers create assignments that students complete online. DeltaMath can generate hundreds of similar questions for middle and high school math, allowing students to do many problems on a particular topic. ZooWhiz (www.zoowhiz.com) has thousands of online activities for elementary and middle school math.

Many sites provide ready-made worksheets or allow teachers to create their own. Kuta Software (www.kutasoftware.com) has a paid version that allows teachers to customize worksheets for topics ranging from pre-algebra to calculus. Kuta Software also offers dozens of pre-made worksheets for free. Dad's Worksheets (www.dadsworksheets.com) provides thousands of free worksheets, most of which deal with K–6 math. Homeschoolmath.net (www.homeschoolmath.net) allows teachers to generate free worksheets for topics ranging from addition to linear equations.

To find homework questions and student tutorials, teachers can use many of the resources that we discuss in Chapter 7: How to Plan Lessons. Khan Academy (www.khanacademy.com) is particularly useful since it contains free written explanations and videos. Khan Academy also offers much of its material in several other languages, including Chinese, Korean, Spanish, Swahili, and Hindi. The list of available languages is currently accessible by selecting the appropriate language at the bottom of the Khan Academy webpage.

Many states have mathematical glossaries in different languages that can be printed and given to English Language Learners who have limited internet access at home. Examples include the New York State Education Department (http://www.p12.nysed.gov/assessment/nysaa/2013-14/glossaries/glossarymathrev.pdf), the Texas Education Agency (http://www.texasgateway.org/resource/interactive-math-glossary), and the California Department of Education (www.cde.ca.gov/ci/ma/cf/documents/mathfwglossary.pdf).

Many states and testing organizations that administer computer-based assessments provide sample questions online that students can use. Some of these organizations allow teachers to create customized tests online for students. Check with your school or district to see what resources are available. In addition, sites like Google Classroom (http://classroom.google.com), Kahoot! (www.kahoot.com), and Quizlet (www.quizlet.com) enable teachers to create online assignments that can be given for homework.

Many tests allow students to use the Desmos (www.desmos.com) graphing calculator. We recommend that teachers encourage students to download the Desmos Test Mode app, which restricts certain features on devices that students may not use during assessments. Geogebra (www.geogebra.org), a free interactive algebra, statistics, geometry, and calculus app, also has a test mode.

Figures

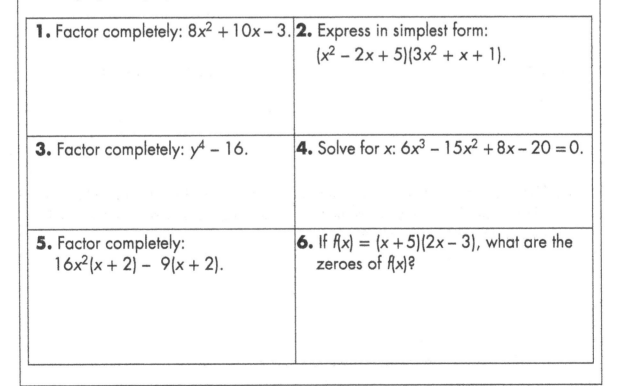

Figure 8.1 Homework as Practice

HOMEWORK # _____   **Name** _____

A factory produces 5 small cars and 2 large cars per hour.

1. Draw a picture illustrating what the factory produces after 1 hr. How many cars does the factory produce in all?

2. Draw a picture illustrating what the factory produces after 2 hr. How many cars does the factory produce in all?

3. Draw a picture illustrating what the factory produces after 3 hr. How many cars does the factory produce in all?

4. What patterns do you see in answers to questions #1–3 above?

5. Use your answer to #4 to determine how many cars the factory will produce after 12 hr. You may draw a picture, make a table, or use any other method that you think is reasonable.

Figure 8.2   Homework as Discovery—Ratios

CHAPTER 8: HOW TO PLAN HOMEWORK   185

GEOMETRY: HW #2.24   Name: _____ Date: _____

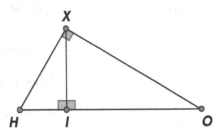

As shown in the diagram above, △HOX, $\overline{XI} \perp \overline{HO}$, and $\overline{HX} \perp \overline{OX}$.

1. If $m\angle XHI = 40$, find the measures of all of the acute angles in the diagram and label the measures in the diagram.

2. How many triangles are in this diagram? Name them.

3. Name all of the pairs of similar triangles in this diagram. Redraw the similar triangles so that they are ordered from smallest to largest and are oriented in the same direction. Write the appropriate label and angle measure for each vertex.

Figure 8.3 Homework as Discovery—Mean Proportional Theorem

186 THE MATH TEACHER'S TOOLBOX

Algebra II: HW #33		Name
Solution	Graph	Circle everything that applies to the roots of each example
1. $2x^2 + 3x - 2 = 0$	$y = 2x^2 + 3x - 2$	**The roots are** 1. Real 2. Rational 3. Irrational 4. Unequal 5. Equal 6. Complex
2. $3x^2 + x - 1 = 0$	$y = 3x^2 + x - 1$	**The roots are** 1. Real 2. Rational 3. Irrational 4. Unequal 5. Equal 6. Complex
3. $9x^2 + 6x + 1 = 0$	$y = 9x^2 + 6x + 1$	**The roots are** 1. Real 2. Rational 3. Irrational 4. Unequal 5. Equal 6. Complex

Figure 8.4 Homework as Discovery—Parabolas

Algebra II: HW #33		Name _____
Solution	**Graph**	**Circle everything that applies to the roots of each example**
4. $2x^2 + 3x + 4 = 0$	$y = 2x^2 + 3x + 4$	The roots are 1. Real 2. Rational 3. Irrational 4. Unequal 5. Equal 6. Complex

Figure 8.4 (Continued)

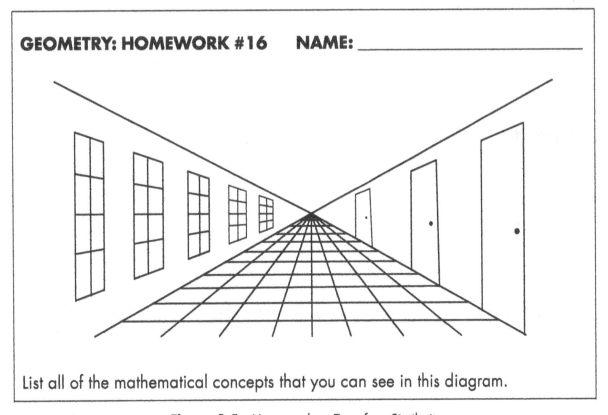

List all of the mathematical concepts that you can see in this diagram.

Figure 8.5 Homework as Transfer—Similarity

Algebra II: HW #2.20

Name: _____

DIRECTIONS: Find the best savings account!

Look through newspapers, magazines, or the internet to find current information about any three different savings accounts. Your job is to rank the three accounts from best to worst. Attach or summarize the information about each account in the spaces below.

1. (Best)

2.

3. (Worst)

In the space below, explain why you ordered the accounts in this way.

Figure 8.6 Homework as Transfer—Bank Accounts

CHAPTER 9

How to Plan Tests and Quizzes

What Is It?

An effective teacher constantly assesses students' learning. These assessments can be diagnostic (typically done before instruction to identify students' current knowledge), formative (done in class to adjust learning while it is happening), or summative (given after instruction has occurred, often to compare student understanding with content standards) (Garrison & Ehringhaus, n.d.; Posamentier, Smith, & Stepelman, 2010, p. 182).

Unfortunately, in today's world, assessments are often associated with the "high-stakes" tests used to make decisions about the effectiveness of educators, schools, and districts. Many teachers feel pressure to "teach to the test," often because their evaluations depend on how well their students perform. *Tests* (assessments of student learning after a significant part of a unit has been taught) and *quizzes* (shorter assessments that typically cover several days' worth of lessons) are a necessary part of a math teacher's toolbox.

The tests and quizzes that we talk about in this chapter represent only some of the ways that we assess students. We discuss other methods in Chapter 16: Project-Based Learning and Chapter 18: Formative Assessment. We also discuss the relative values of assessments in our discussion of grading practices and philosophy in Chapter 10: How to Develop an Effective Grading Policy.

Why We Like It

Well-designed tests and quizzes help to hold students accountable for instruction and measures their skills, not simply their effort and practice (Garrison & Ehringhaus, n.d.). When used properly as one of many assessment methods, good tests and quizzes can improve our instruction and help students prepare for the high-stakes tests that they need to take.

Supporting Research

Research indicates that teachers should evaluate students using many different methods of assessment, including tests, quizzes, classwork, homework, and projects (Clarke, 1997, p. 21). Since knowledge is often constructed individually and tied to a particular context, no one measurement tool can measure that knowledge objectively (Romagnano, 2001, p. 35).

In addition, assessments should not simply measure students' content knowledge but also their reasoning, problem-solving, and communication skills (Hunsader, Zorin, & Thompson, 2015, p. 71; Swinford, 2016, p. 518). Assessment should not simply be used to rank students (Posamentier et al., 2010, p. 159). Teachers should also use assessments to improve future instruction and monitor student progress (Brendefur, Strother, Rich, & Appleton, 2016, p. 177).

Common Core Connections

The Common Core contains both standards for mathematical content (which list the procedures for a particular grade or subject that students should master) and standards for mathematical practice (which describe general skills, such as constructing viable arguments or making sense of structure, that students at all levels should master) (NGA & CCSSO, 2010). Ideally, assessments should address both content and practice standards.

Application

In this section, we outline the steps involved in creating an in-class test or quiz and analyzing its results.

We present many different strategies here. Obviously, we can't implement all of them all of the time. We decide whether to use each of these techniques based on our students' individual gifts and challenges as well as our own priorities and limitations.

TYPES OF QUESTIONS

Ideally, tests should allow students to communicate mathematically (through symbols, pictures, or words), connect to other mathematical ideas or a real-world context, and translate between different representations (such as verbal, symbolic, and graphical) (Hunsader et al., 2015, p. 75). Questions should test various skills, such as using basic procedures, choosing an appropriate method to solve problems, showing conceptual understanding, and justifying a solution method (Brendefur et al., 2016, p. 176).

Here are some of the question types that we use most often on our tests.

Multiple-choice Questions

In *multiple-choice questions*, students choose one best answer out of four or five possible choices. Multiple-choice items are easy to grade and can allow teachers to cover more questions on assessments, which can improve the reliability of tests (Butler, 2017). They are more objective since graders are not negatively influenced by poor grammar or inconsistent human graders (Xu, Kauer, & Tupy, 2016, p. 149).

However, multiple-choice questions have several disadvantages. They are the most difficult and time-consuming questions for teachers to write. If a student gets a question incorrect, teachers often have no way of knowing whether they guessed, used partially correct reasoning, or simply bubbled in the incorrect choice. Students may also think that multiple-choice questions require less higher-order thinking and may not think as deeply through them as a result (Xu et al., 2016, p. 149).

Since multiple-choice questions occur on many tests, knowing what makes them effective is important. Good multiple-choice questions should have the following characteristics:

- Have plausible alternatives (Brame, 2013). If a question asks students to calculate the volume of a cylinder, a negative number is not a reasonable choice. Ideally, a good distractor correctly answers another question (Butler, 2017) or incorporates a common mistake.

- Avoid negative statements like "all are true *except*...," which can easily confuse students (Xu et al., 2016, p. 148). Unfortunately, many standardized tests have these types of questions. When we have to prepare students for these exams, we include them in our assessments.

- Avoid clues in the grammar that hint at the correct answer (Mueller, n.d.). A question that asks students to "find the solutions" to an equation implies that the equation has more than one answer. Rephrasing the directions to "solve for *x*" avoids this problem.

- Have a roughly similar length and structure for each choice (Xu et al., 2016, p. 148). If one choice is more detailed than the rest, students may select it for that reason.

Teachers can modify multiple-choice questions to get more insights into what students are thinking, although these changes will make tests harder to grade. Students could earn partial credit for explaining their reasoning for incorrect answer choices or eliminating a certain number of answer choices (Cleveland, 2018). They could also attach a confidence level to an answer choice (Xu et al., 2016, p. 150). Three-point multiple-choice questions could be graded according to Table 9.1: Confidence Level Scoring Guidelines:

These scoring guidelines reward students who get the correct answer with a high degree of confidence but also value correctness over confidence. This discourages students from simply expressing high confidence for questions they guessed on.

Grid-in Questions

Often given on standardized tests like the SAT, *grid-in questions* have no answer choices, and students must provide a numerical answer. This format discourages students from guessing since there are hundreds or thousands of possible answers, so teachers are less likely to misinterpret a correctly answered question.

Unfortunately, we find that grid-in questions, like multiple-choice questions, often have limitations. They don't give students a way to show partial understanding. In addition, they only allow for numerical answers, so students can't show how they think.

True-false and Always-sometimes-never Questions

True-false questions are a simpler version of multiple-choice questions since they typically have two possible choices. Many students find them challenging since they require very close reading—altering one word can make a true statement false or vice versa.

One variation that we have used in the past is *always-sometimes-never questions*, in which students must determine whether a statement is always, sometimes, or never true. Students can answer these questions to help them compare definitions. For example, an *identity* is a mathematical statement that is true for *all* defined values of its variables, but an *equation* is a mathematical statement that is true only for *some* values of the values (i.e., the *solution set* of the equation) (MARS, 2015, pp. T12–T14).

Table 9.1 Confidence Level Scoring Guidelines

Correct/Incorrect	Student confidence level	Point value
Correct	High	3
Correct	Low	2
Incorrect	High	1
Incorrect	Low	0

We sometimes use always-sometimes-never questions when comparing families of quadrilaterals. The statement "All squares are rectangles" is *always* true but the statement "All rectangles are squares" is *sometimes* true since some rectangles have sides that are not congruent.

Matching and Fill-in-the-blank Questions

Sometimes, *matching questions* can be helpful for units that have a great deal of vocabulary, such as geometry. To minimize student guessing, we often make sure that the number of choices in one group is not the same as the number of choices in another group. Many English Language Learners (ELLs) may find matching questions less intimidating than multiple-choice questions.

Fill-in-the-blank questions, in which students must provide the correct word or phrase to complete a sentence, can reduce student guessing and be more challenging than matching questions since students have no terms to choose from. Teachers can also provide a word bank with answers. This modification can help not just ELLs but everyone in an inclusive classroom. One way to make a "word bank" more challenging is to include more words than blanks.

Constructed-response Questions

Constructed-response questions (sometimes called *free-response questions*) are open-ended questions with no answer choices, so students must write in a numerical or verbal response. This format allows students to show their thinking (Swinford, 2016, p. 518). Students can also earn partial credit, which we find can be particularly helpful for struggling learners.

However, in our experience, constructed-response questions can be difficult to write and much more difficult to grade than multiple-choice questions. Even good constructed-response questions have their limitations. For example, they don't measure students' ability to perform valuable real-world tasks like making effective oral arguments and performing research (Darling-Hammond & Adamson, 2010, p. 4). We discuss other methods of real-world assessment, such as performance-based and project-based assessments, in Chapter 16: Project-Based Learning.

TEST FORMAT

Like many teachers, we've found that tests can induce frustration and anxiety, which may discourage students from doing well. To make tests less stressful for our students, we try to make our test format as predictable as possible so that they can focus on the content of the questions instead of their organization. Here are some strategies we use to structure our tests.

Topics

To determine what topics should appear on the test, we look at our unit and lesson plans, standards, and sample questions from any standardized tests that our students will take. Most of the test focuses on topics that we recently taught (typically 70–80 percent—the exact percentage depends on the difficulty of the topics). We also include a few questions covered in previous tests. If the topic of the test is particularly challenging, adding easier questions from previous units improves the chances that students will do well. Adding questions from previous topics builds confidence, helps students prepare for end-of-year tests, and promotes the retrieval practice that we incorporate into our lessons and homework (see Chapter 7: How to Plan Lessons and Chapter 8: How to Plan Homework for more details).

Number of Questions

Many times, we teach a course that culminates in an end-of-year test, such as a state standardized assessment or a school-wide test that determines whether students advance to the next grade level or course. In these situations, we set the number of questions and the format of our tests to be proportional to those on the end-of-year tests. Figure 9.1: Algebra I Test shows a test given in a 45-minute period. (The layout has been compressed slightly from its original legal-size format.) This test has four parts that roughly approximate the format of our Algebra I state exam: nine multiple-choice questions, and six constructed-response questions with various point values.

When we teach a course that doesn't culminate in an end-of-year standardized test, we often have more flexibility in how we design our assessments. In these situations, we frequently omit multiple-choice questions and include questions that enable students to show more of their thinking. Figure 9.2: Precalculus Test shows a test without multiple-choice questions. The test contains eight 4-point constructed-response questions and three longer constructed-response questions.

No matter what structure we choose for our tests, we group questions with the same type and point value into a separate part. Doing so helps students navigate through the test more easily to determine which questions they should focus on.

Order of Questions

We order test questions from easiest to hardest for several reasons:

- Students can answer questions in the order that they prefer. Those who like to start with easy problems to build their confidence would start at the beginning, while those who prefer starting with the hardest problems would start at the end.

- Students can tap into what psychologists call the *Zeigarnik Effect*—the idea that once people start something, they are more inclined to finish it (Dean, 2011).
- Teachers can get a better overall sense of a test's difficulty, reducing the chances that it will be too easy or too hard.

To help students make an informed decision, we describe our test structure to the class before the first test. All of our tests have the same format so that students know what to expect.

When we give paper tests, we use the layout of the page to give students a visual clue about their level of difficulty. Questions that have a lower point value or require less work get less space. We put a box around each question to help students organize their work.

Skills in Questions

Our assessments contain questions that require four different types of skills: *factual* (recall basic information), *procedural* (use a procedure or algorithm), *conceptual* (generalize or explain their reasoning), and *application* (solve a real-world problem in context) (Achieve, 2018b, Independent Analysis, p. 53; Achieve, 2018a, A Framework, p. 2). We try to ensure that all four skill types appear on our tests. (Doing so for quizzes can be difficult since they have fewer questions.) Including all question types also helps align our assessment with our instruction (Murphy, 1999, p. 248).

For example, the test in Figure 9.1: Algebra I Test has one factual (#3) and seven procedural questions (#1–2, 4–5, 10, 11, 13) worth about 50 percent of the test, five conceptual questions worth about 30 percent (#6, 8, 9, 12, 14), and two application questions (#7, 15) worth about 20 percent.

The test in Figure 9.2: Precalculus Test has no factual questions, three procedural questions (#1–3) worth about 15 percent of the exam, six conceptual questions (#4–10) worth about 60 percent of the exam, and two application questions (#11, 12) worth about 25 percent of the exam. Although some of the conceptual questions here appear procedural since they lead to a numerical calculation, they also require a great deal of conceptual understanding.

The ratio of factual, procedural, conceptual, and application questions can vary depending on the difficulty level of the course, the unit being taught, and student performance. We find that sticking to a rigid ratio for each test makes tests too hard to create—what may work for students 1 year or even for one test might not work for another.

In our experience, more advanced courses tend to have fewer factual and procedural questions and more conceptual and application questions. We don't use this as a justification to limit students in lower-level courses. No matter what course we teach, we try to give all students a mix of factual, procedural, conceptual, and application questions. We recommend that teachers be aware of the skill types required for test questions and adjust questions as necessary to make more balanced assessments.

QUIZ FORMAT

Our quizzes are shorter and more focused than tests. We use quizzes as a checkpoint for both students and for us to see how students are doing in the unit—worth more than formative assessments (which we don't count as part of students' grades) but less than larger summative assessments like tests or projects.

Usually, our quizzes only take up a part of the period so that we can have a lesson afterward. We try to design our quizzes so that they take up about as much time as we typically would spend reviewing homework—about 10–15 minutes. In reality, of course, quizzes usually take more time than homework review, so we often have to adjust our lesson slightly to accommodate. As a result, we try to schedule quizzes on days that we have easier or shorter lessons. We usually give quizzes at the beginning of the period since we don't review homework when we give quizzes in a 45-minute class.

As we do with our tests, we try to make our quiz formats predictable so that students know what to expect. Since quizzes are shorter than tests, we tend to be more flexible in the number and point values of questions. We give between two and four questions on each quiz, typically ranging from two to eight points each, depending on the type of question and its level of difficulty.

Figure 9.3 shows a quiz from an Algebra I class. This quiz is given during a unit on linear functions, and has questions that deal with writing the equation of a line and determining the characteristics of functions. The paper provides space for students to solve either algebraically or graphically since we discussed both methods in class—a simple example of how assessment should be aligned with instruction. The four questions on this quiz should take students no more than about 15 minutes to answer.

REVIEWING FOR ASSESSMENTS

Students who take tests often need considerable modeling and practice before they can write about their reasoning and answer mathematical questions properly (Swinford, 2016, p. 517). We try not to limit test review to the day before the test. Instead, we incorporate review questions in classwork and homework throughout the unit. Sometimes we label these questions as "Test Prep" to increase students' focus and make their test expectations more accurate. We find that when students are more familiar with the format and content of assessment items, they are less likely to feel anxious about exams and are more likely to be engaged.

We don't spend much time reviewing for low-stakes quizzes. However, we do spend class time preparing students for tests.

Here are some methods that we use to review for them.

Mini-whiteboards

One of our favorite ways of reviewing for tests is to have students solve problems on mini-whiteboards. Typically, we do this activity on the day before the test as a final review. To keep everyone focused, we don't give out the problems on a sheet. Instead, we show the problems one at a time on our classroom projector. Of course, we don't use whiteboards only for test review. Using them regularly in class helps promote *whole-class engagement* or *total participation*, in which as many students as possible are actively participating in and cognitively engaged with the lesson (Himmele & Himmele, 2017, p. 4).

When we use whiteboards, we divide the class into pairs. Each pair of students gets a whiteboard, dry-erase marker, and eraser cloth. We show each question on the board one at a time, so students work simultaneously on the same question on their whiteboards. Students who don't have access to whiteboards can use markers to write their answers on scratch paper.

After a certain amount of time (usually 1–4 minutes, depending on the question), we ask students to hold up their answers for the entire class to see. This allows us to quickly assess student work. We then show the correct solution. We either reveal it on our classroom projector or we show student boards. If we share student work, we always ask for permission first. We also discuss common mistakes and alternate solution methods. Occasionally, some pairs of students need more time to discuss a particular question, so we allow them to continue their conversation as necessary while the rest of the class moves on to another problem.

Technology

Teachers can now use many technology tools to create review activities for students. We describe some of them in more detail in the Technology Connections section of this chapter.

Jeopardy

Many times, we play a quiz game modeled after the *Jeopardy!* television game show, where contestants must guess the question that corresponds to a given answer. Like the TV game show, we use a game board consisting of six categories with six clues in each category. We display the questions on our classroom projector, but Jeopardy doesn't require technology. It can be played using an over-the-door shoe rack with each question written on an index card and placed in a pocket. It could also be drawn on a classroom whiteboard. Each clue has increasing values (we use the same values as the current game show, but these can vary depending on your preference) that correspond to their level of difficulty, so clues worth more are harder. Students must provide a question that is answered by the clue. If the clue is, "This is a comparison of two numbers using division," students must answer, "What is a ratio?" Like the game show, we insist that students respond in the form of a question, so simply responding "Ratio!" would be considered incorrect. We find that insisting on this form playfully reminds students that they should speak and write precisely in our class!

Teachers have a great deal of flexibility in running a Jeopardy activity. We divide students into groups of three or four students (groups that are larger tend to be harder to manage and often wind up making students less focused). Each group needs its own unique identification, such as a number or name. Students can even name their groups themselves. We prefer to have students respond using whiteboards, but if whiteboards aren't available, we use paper.

To start the game, we randomly select a group who chooses a category and number amount from anywhere on the board. We reveal and read the clue to the class. Like the game show, teachers *could* use electronic buzzers to determine which group gets to answer first. We find that giving points to the group that chooses to answer first makes the game more competitive and potentially more fun for some. However, it can also put pressure on students to finish quickly and doesn't keep everyone engaged. Since we typically value accuracy over speed, we use a different strategy. We prefer that groups work simultaneously for each clue and reveal their responses simultaneously on whiteboards or pieces of paper that they hold up. We then give points to each group that provides the correct response. We keep track of the points on chart paper or the board.

A game like Jeopardy usually keeps students engaged in meaningful work and allows them to practice many problems on a variety of topics. Unfortunately, it often requires a lot of time and energy to prepare. Teachers can easily find premade Jeopardy activities online or find a customizable template. We list some resources for making a Jeopardy activity in the Technology Connections section of this chapter.

Student-generated Test Questions

Another way that we review for tests is to allow students to create their own test questions. We typically do this only when students feel sufficiently comfortable with the format of test questions. Making test items and creating an answer key helps students deepen their content knowledge and strengthens the link in their minds between assessment and understanding (Rapke, 2017, p. 615).

We encourage students to write their own problems from scratch or modify someone else's problem. We give extra credit when students show creativity and originality, such as when they incorporate their own personal interests into a problem (we discuss this more in Chapter 1: Motivating Students). Students may use existing problems as long as they cite sources, include a variety of sources, and include a certain percentage of problems that can clearly be identified as original (for example, word problems that incorporate their names and interests). The percentage of original work can vary, depending on the unit and our students.

To introduce student creation of test items, we plan a separate lesson about a week before the test. We typically wait until students have taken a few tests so that they have some experience knowing how difficult questions should be. We review the format of the test and the importance of creating a test that contains a mix of easy and challenging questions. We also discuss how we make scoring guidelines for a constructed-response item so that students can keep these expectations in mind when creating their questions.

After going through the list of topics with students, we allow them to work in groups to go through their notes, various textbooks (which we bring to class on that day), and internet resources so that they can find questions on their own.

Each group is responsible for creating one question for each section of a test and working out a complete solution. Doing this activity several days before the test allows us to look over our students' work and select some items that will appear on the actual exam. Once students are familiar with the process, they usually no longer have to spend much time in class creating questions (when students are unable to collaborate outside of class, we allot small amounts of time periodically during the unit for them to work in groups). For subsequent tests, we simply provide them with the format and the list of topics and allow them to work outside of class to create items. We then review for tests by allowing students to work in groups to answer each other's questions. Our experience matches existing research that students are much more engaged in analyzing their mistakes when they make their own questions (Clarke, 1997, p. 50).

CREATING SCORING GUIDELINES FOR ASSESSMENTS

To grade a test or quiz in which students can earn partial credit on calculations and explanations, teachers need to use scoring guidelines. Scoring guidelines contain *benchmarks*—specific tasks that should be accomplished in order to solve a problem.

Scoring guidelines differ from *rubrics*, which describe levels of performance for a task. We typically use scoring guidelines for mathematical problems that lead to a specific answer and rubrics for open-ended tasks such as projects (which we discuss more in Chapter 16: Project-Based Learning).

Figure 9.4: Creating Scoring Guidelines shows how we generate scoring guidelines for a four-point test question on quadratic equations. This figure represents an idealized method. In everyday practice, we write our solution, highlight benchmarks, and assign point values without creating a three-column table.

Here are the steps that we use to make scoring guidelines:

1. **Identify the point value for the problem.** When possible, we use assessments from other sources, such as colleagues, textbooks, websites, or state tests, to help us determine an appropriate point value.

2. **Solve the problem.** When solving problems, we write all of the work that we expect students to show.

3. **Create benchmarks.** We highlight or mark the most important steps in the solution. If we feel that a step is particularly important (such as checking for extraneous solutions or labeling the diagram), we make it a benchmark. We usually combine steps that students should know from previous courses. Table 9.2: Comparing Scoring Guidelines shows how different content is emphasized in benchmarks for Grade 6, Algebra I, and Algebra II. In the Grade 6 problem, each step in solving $x + 2 = 0$ is a benchmark since students are learning how to solve one-step linear equations. In Algebra I and II, solving linear equations is considered prior knowledge, so we don't consider it a benchmark. Similarly, each step in solving $x + 6 = x^2 + 8x + 16$ is a benchmark in Algebra I (where students learn how to solve quadratic equations) but is worth only two points when it is part of the more complicated process of solving radical equations.

4. **Assign point values to the benchmarks.** Once we identify the benchmarks, we assign point values to each one. If necessary, we modify the question's total point value if the question has more or fewer benchmarks than we originally anticipated.

5. **Add additional comments to the scoring guidelines if necessary.** We typically only do this step if we share our scoring guidelines with others or if we want to make short explanatory or additional notes to ourselves, such as "look for a question mark above the equal sign" or "if they don't cross out the additive inverse, don't take off points."

Creating scoring guidelines can take a great deal of time and effort. We find that with experience, we can streamline this process a great deal (unless we're writing guidelines that someone else will use to grade) by just writing the solution to problems, circling benchmarks in our solution, and writing appropriate point values next to the corresponding steps.

Table 9.2 Comparing Scoring Guidelines

Grade 6	Algebra I	Algebra II
Solve for x in the equation $x + 2 = 0$. Check your answer.	**Algebraically solve for x in the equation $x + 6 = x^2 + 8x + 16$.**	**Algebraically solve for x in the equation $\sqrt{x + 6} - 4 = x$.**
[1 pt.] Subtracts 2 from both sides. $x + 2 - 2 = 0 - 2$ [1 pt.] Simplifies both sides. $x + (2 - 2) = 0 - 2$ $x + 0 = -2$ [1 pt.] Uses additive identity property to find final answer: $x = -2$ [1 pt.] Checks answer in original equation. $-2 + 2 = 0$	[1 pt.] Writes a quadratic equation in standard form: $x^2 + 7x + 10 = 0$ [2 pts.] Factors trinomial and sets each factor equal to 0 OR completes the square: $(x + 5)(x + 2) = 0$ $x + 5 = 0$ or $x + 2 = 0$ [1 pt.] States final answer: $x = \{-5, -2\}$	[1 pt.] Isolates the radical and squares both sides. $\sqrt{x + 6} = x + 4$ $x + 6 = (x + 4)^2$ [1 pt.] Writes a quadratic equation in standard form: $x + 6 = x^2 + 8x + 16$ $x^2 + 7x + 10 = 0$ [1 pt.] Solves the quadratic equation by factoring or completing the square: $(x + 5)(x + 2) = 0$ $x + 5 = 0$ or $x + 2 = 0$ $x = \{-5, -2\}$ [1 pt.] Check for extraneous solutions and writes final answer: $x = -2$ $\sqrt{-2 + 6} = -2 + 4$ $2 = 2$ $x = -5$ $\sqrt{-5 + 6} = -5 + 4$ $1 \neq -1$ REJECT $x = -5$ ANSWER: $x = -2$

GRADING ASSESSMENTS

We freely admit that we don't like to grade tests. Unfortunately, we can't avoid it! In this section, we discuss some techniques that help make grading assessments more manageable.

Formatting

First, we try to fit each test onto one sheet of legal-size paper. In our experience, single-period tests that are several pages long intimidate students and heighten their math anxiety. Multipage tests are also harder for us to grade since we constantly have to turn pages while grading. We prefer that students not use separate sheets to write since they can easily get lost or passed on to students taking our test later in the day. However, we make exceptions when necessary. Some students may have very large handwriting, or some tests may have several questions whose responses require a lot of space.

Understanding the Scoring Guidelines

Before grading a constructed-response question, we start by reading through the scoring guidelines carefully, locating the benchmarks for the problem. We then tentatively grade a few papers to give us an idea of how the scoring guidelines work and see what possible ambiguities arise. At this stage, we write scores in pencil so that we can change them if necessary. We also make notes if we encounter situations that weren't immediately clear from the scoring guidelines. Once we feel that we understand the scoring guidelines, we proceed to grade all of the papers.

We find that scoring guidelines usually can't account for all possible student errors. When students make a mistake in their work, we have to determine the error is *conceptual* (an error in thinking) or *computational* (an error in calculation or rounding). We then deduct more points for conceptual than for computational errors. On a four-point question, we typically deduct two points for conceptual errors and one point for computational errors. Often, the same error can be either conceptual or computational depending on the course. Our general rule is that conceptual errors stem from misunderstanding an idea that students learn in the course they are currently taking, while computational errors stem from ideas that should be prior knowledge. Adding 2 instead of subtracting 2 from both sides of the equation $x + 2 = 0$ may be considered a conceptual error in Grade 6 but a computational error in Algebra II.

After finding a student mistake, we then read through the work to see if it is consistent or if other mistakes were made. If a student makes a mistake at the beginning of the work but follows through to get an answer that differs from the correct one, we deduct only for the error and give points for the rest of the work. Students can receive partial credit for an incorrect answer as long as they show correct work after making a mistake. We discuss ways for students to recognize the difference between conceptual and computational errors in Chapter 4: Promoting Mathematical Communication.

We try to grade only a few constructed response questions at a time so that we don't get confused between questions. Over time, we've gotten better at keeping track of the scoring guidelines of several questions in our heads simultaneously, so we can grade five or six constructed-response questions at once. Less-experienced teachers may only want to grade one or two constructed-response questions at a time.

Peer Grading

One possible way to save time when grading is to have students grade the work of other students. We recommend this only in situations when students can be trusted to grade fairly. Peer grading is easiest when used for questions that require little or no interpretation, such as true-false, fill-in-the-blank, multiple-choice questions, or constructed-response questions with simple scoring guidelines (like a two-point question).

Commenting on Student Work

We do not correct or write detailed notes on every student mistake. Often, we just circle an error. We sometimes write a brief comment like "What does x represent?" or "Wrong formula" to help students think about what they did wrong.

When possible, we also try to make positive comments on student work, such as "Nice solution!" or "Well done!" We especially like to give praise when students come up with solution methods that are more elegant than the ones we discussed in class. Many times, we ask these students to share their methods with the rest of the class when we hand back tests.

ANALYZING TEST RESULTS

After grading a test, we try to get an overall picture of how students did on each question. We look through test papers to see which items many students seemed to answer incorrectly or incompletely. Sometimes, we take a representative sample of student test papers to see what mistakes occurred most often.

If we have time, we sometimes do a more detailed item analysis. We record each response to see what percentage of students got each question correct. This process is much easier when we use scanners to grade machine-readable student test forms and generate summary statistics. An item analysis enables us to see which questions students struggled with to make sure that we can address them later. When possible, we use technology tools such as a spreadsheet or website to help us understand our test results. In addition to calculating statistics like the mean, we use these tools to generate a stemplot or histogram. This gives us an overall picture of our students' performance and identifies trends in the grades (see the Technology Connections section of this chapter for more details).

Here are some of the questions we ask when analyzing our test scores:

- **Did one class do significantly better or worse than another?** If a class's performance is noticeably different, we may have to adjust our instruction accordingly. If that class has more absentees, we may need to reach out to parents and guardians.

- **Did students generally do significantly better or worse on one test compared to another?** If test scores dropped noticeably from one test to another, we may have made the test too hard, or our instruction might not have been as clear as before.

- **Did students do particularly well or poorly on a particular topic?** If we see that students struggled on a particular topic, we may need to revisit it in future instruction and homework.

- **Are there any unusual features, such as clusters or outliers, in the test score distribution?** If some students did much worse than the rest of the class, we reflect on how we can improve our instruction for them. In some cases, they may simply need more practice. However, we often find that providing "more of the same" instruction doesn't change the result, so we may need to try another strategy. We may pair them with students who can provide regular help in class, find better instructional materials for them, or even invite a colleague to observe our teaching and suggest ways that we can improve.

RETURNING TESTS

As we mentioned earlier, we typically don't write detailed comments on student papers. We prefer to have students figure out their mistakes on their own. We find that giving students time to think about their mistakes helps them better understand what they did wrong and avoid the same errors in the future. Here are some techniques that we use to review tests after we return to students.

Group Discussion

We find that allowing students to discuss their papers in small groups can often be effective. This activity typically takes about 15 minutes. We do this either at the beginning of class (in place of homework review) or at the end of class if we feel that students would not be focused after getting their tests back. Working in groups enables most students to find and correct their mistakes. We also bring groups together at the end to discuss questions that almost everyone got incorrect.

Correcting Mistakes

After allowing students to look over their tests, we sometimes invite them to explain in writing what they did wrong and what they should have done. Thinking about a mistake as "the right answer to a different question" (Meyer, 2018) and giving students an opportunity to show their thinking can help them learn from their mistakes.

We ask students to write up this work using the template shown in Figure 9.5: Blank Test Corrections Sheet. We provide a blank copy of the sheet on our class website or print them out for students. We find that giving students a template keeps their work more organized.

The sheet has three columns. In the first column, students write the question number. In the second column, they explain their error as specifically as possible. For example, they can say, "I forgot to isolate the radical before squaring both sides," or, "I added when I should have subtracted." In the third column, they write the correct work for the problem.

To make test corrections more manageable for us, we set a grade above which students may not write test corrections. Often, we set this grade to be relatively high, like 80, since we feel that these students typically can figure out their mistakes by themselves and will take appropriate steps to avoid similar mistakes in the future.

We grade these written corrections as extra credit to motivate students to learn from their mistakes. One relatively simple way to calculate the extra credit is to allow students to get back half of the points that they lost. Students who originally earned only one point on a four-point question and successfully correct their work would then earn a total of $1 + 0.5(3) = 2.5$ points for that question.

To discourage students from simply copying someone else's work, we explain the importance of learning from their mistakes. When we have time, we encourage students to meet with us to correct their tests so that we can correct errors in their thinking and improve our future instruction.

We find that allowing students to review and correct their mistakes promotes metacognition (which we discuss in Chapter 4: Promoting Mathematical Communication). When students think and write about their mistakes, they often reflect more thoughtfully about how they do math. Students may see that they selected only the first answer that seemed reasonable instead of the best choice, or they failed to check that their final answer made sense in the context of the problem.

Figure 9.6: Completed Test Corrections Sheet shows a sample of student work for test corrections.

Reflecting on Mistakes

Teachers can also ask students to think about their mistakes using a test reflection form, shown in Figure 9.7: Test Reflection Form. It is based on the Test Aftermath Form devised by Scott Brown, a test reflection that facilitates metacognition and improves student learning (Brown, 2005, p. 69). Students can complete this assignment for homework immediately after getting the test back in class.

Our test reflection form has four questions. The first question asks students to identify two mathematical strengths that they showed on the test. The second asks students to identify two mathematical weaknesses that they showed on the test. Students should use correct mathematical language and avoid comments that deal with test-taking strategies, such as, "I ran out of time" (Brown, 2005, p. 70). The third question asks students to show that they know a question that wasn't asked on the test. The final question asks students to show that they now know how to solve a problem that they got incorrect on the test.

The test reflection form has the advantage of being shorter and easier to grade than a complete test correction. Teachers who don't have time to grade test corrections may find test reflections easier to manage. In addition, all students can reflect on their tests, not just students who did poorly on the test. However, test reflections allow students to write about only a very small number of mistakes on their test. A combination of test retakes, corrections, and reflections can give students opportunities to learn from their mistakes while giving teachers flexibility throughout the year.

Retaking Tests

Sometimes, we allow students to retake tests. Giving students a chance to redo a test on which they did poorly can prepare students for the real world, where adults can often retake high-stakes assessments like the bar exam or a driver's license test for full credit (Wormeli, 2011, p. 24). Furthermore, many students learn at different rates and may improve their skills with more practice. In contrast, setting an arbitrary deadline for students to learn can discourage them (Wormeli, 2011, p. 23).

In our experience, though, allowing students to retake tests can often be more trouble than it's worth. First, we have to find time for students to retake the test. Some students may only need a few days to learn the material, while others may need weeks. So that we don't lose instructional time, we usually ask students to meet with us outside of their school day or during a free period. We also have to make up multiple versions of retakes since students can simply memorize the answers to a previous

CHAPTER 9: HOW TO PLAN TESTS AND QUIZZES

test. To make retakes easier for us to create, our versions are shorter than the original tests, typically consisting of constructed-response questions. When possible, we use online sources for test questions (which we outline in the Technology Connections section of this chapter) to help us.

Test retakes can be effective in certain situations. If students did badly on a test because our instruction wasn't clear, we may reteach the topic and give students another chance to demonstrate their skills. Sometimes, personal issues in students' lives may prevent them from studying for a test. However, other methods of allowing students to learn from their mistakes, which we outline in this section, are often more effective and more sustainable than test retakes.

DIFFERENTIATING FOR ELLS AND STUDENTS WITH LEARNING DIFFERENCES

When creating a test, we keep the diverse needs of our students in mind. We often take test questions from other sources, such as state assessments or exams written by other teachers. We frequently modify them when necessary to meet our students' needs.

Sometimes, the questions we find from other sources don't give students enough opportunities to make connections or provide their reasoning. A question like, "If the legs of a right triangle are 5 and 7, find the exact length of the hypotenuse," is solved procedurally by substituting into the Pythagorean theorem. To give students more of a challenge, this question can be modified so that students must explain their thinking: "Two sides of a right triangle are 5 and 7. Find all possible lengths of the third side. Justify your answer." Students who struggle to remember a procedure can experience more success with open-ended questions, where they can receive partial credit.

To ensure that ELLs have a better chance at doing well on tests, we try to use clear and concise language in questions. When we follow Strunk & White's (1979, p. 23) writing rule of "Omit needless words," students can focus on the math without getting confused by the wording. For example, the question "Given the polynomial function $f(x) = 3x^{10} - 7x^2 + 4x$, determine the remainder when $f(x)$ is divided by the linear polynomial $3x + 5$" can be shortened to "What is the remainder when $f(x) = 3x^{10} - 7x^2 + 4x$ is divided by $3x + 5$?" without losing any mathematical precision. In our experience, verbosity can be particularly challenging in word problems.

Many times, students are given special accommodations on assessments. In our state, some ELLs are allowed to use glossaries that translate mathematical terms into their first language. Some students are allowed extra time or the use of calculators on assessments. Students with vision problems may have difficulty reading small print, so we print large-print versions of tests on several pages. We modify our in-class assessments accordingly.

When we have classes with wide variations in ability, we often differentiate our instruction by dividing our classwork problems into clearly labeled levels of increasing difficulty and allowing students to choose what questions to answer (Wong & Bukalov, 2013, p. 56). We differentiate our tests in the same way (we discuss this more in Chapter 14: Differentiating Instruction).

ALTERNATE FORMS OF ASSESSMENT

We can't simply rely on traditional tests and quizzes as our only method of evaluating students. When possible, we balance in-class tests and quizzes with other types of summative assessments like projects. In a geometry unit on graphing, we may ask students to create artwork on a coordinate plane and write the equations of all lines and curves that they used. We may also give in-class group quizzes or take-home tests as an occasional alternative to in-class assessments (we discuss projects and alternate forms of assessment in more detail in Chapter 16: Project-Based Learning and Chapter 17: Cooperative Learning). Giving students different ways to show mastery makes each individual assessment less "high-stakes" and reduces student anxiety about testing (Murdock, 1999, p. 589).

Whenever we give assessments, we also try to be sensitive to the attention levels of our students, particularly at the end of the year when we often feel pressured to finish our curricula. Sometimes, we replace tests with shorter quizzes or make tests cover only the most recent topics. We may do this when we don't have time to write or grade tests, or if we want students to focus on a smaller amount of material (especially if a unit is particularly difficult for students).

Student Handouts and Examples

Figure 9.1: Algebra I Test
Figure 9.2: Precalculus Test
Figure 9.3: Quiz
Figure 9.4: Creating Scoring Guidelines
Figure 9.5: Blank Test Corrections Sheet
Figure 9.6: Completed Test Corrections Sheet
Figure 9.7: Test Reflection Form

What Could Go Wrong

When we give tests and quizzes, many things can go wrong.

POOR SCHEDULING AND PREPARATION

Giving too many or too few assessments can cause problems. Testing too often eats into instructional time, makes tests too repetitive, and overwhelms us with grading. Testing infrequently forces students to study too much material.

We also try to be flexible enough to change our assessment schedule depending on what else is going on in our lives or our students' lives. We usually avoid giving an assessment the day before a long break because many students either leave early or are thinking so much about their vacation that they are unable to focus on studying. In addition, we don't like to spend our vacations grading! Returning assessments after a long vacation makes them almost impossible to discuss since students often forget what the test covered or what they did on it.

When we don't do enough review in class, we sometimes assume that students will just study at home. That option may be a challenge for many students. They may have to travel a long distance to and from school or they may not have a quiet place to study at home. We incorporate some review in class through homework (which we discuss in Chapter 8: How to Plan Homework) and test review, which we usually do the day before a test. In addition, we make connections in our lessons to previous topics to help students reinforce those ideas.

Occasionally, we forget to remind students about tests and quizzes. Simply posting a notice on a class website or on the board isn't enough. We hear questions like, "Wait, we have a test tomorrow?" too often in our teaching careers! We remind students verbally every day for several days before an assessment. In addition, putting a reminder on the top of every screen of our presentation file (where we also put the aim or learning objective), which we display on our classroom projector during the lesson, helps students. Encouraging students to put reminders on their phones or posting events on online calendars can also help students to remember. We give students at least a week's notice before a test so that they can budget their time accordingly. This also usually allows them to study over the weekend. We give students at least a few days' notice before a quiz.

Spacing assessments evenly throughout the year makes them more predictable, which reduces student anxiety. In our experience, giving a test about once every 2 or 3 weeks strikes an appropriate balance between testing too frequently and testing too rarely. We also give at least one quiz in between each test. However, this schedule can vary depending on our students' ability to handle tests, our ability to grade quickly, school culture, work that students have in other classes, and events in students' lives. Some classes may function better with fewer tests and more quizzes because they get anxious about full-period tests. Classes that have other forms of assessments, such as projects and presentations, may have fewer tests. Some schools mandate a certain minimum or maximum number of assessments.

One big mistake is to reduce test review to a single sheet of questions that we give a day or two before the test. Relying only on review sheets can lead some students to believe the test will be the review sheet—but with different numbers. Even if we include topics on tests that aren't on the review sheets, students tend to study only what is on them. Review sheets also don't prepare students for the end-of-year standardized tests that most of our students take, when nobody knows what questions will appear. Instead, we incorporate test review throughout the unit with similar questions given in our classwork and homework. Our test review in class is simply equivalent to a longer homework assignment with more questions and topics.

ASSESSMENTS AS CLASSROOM MANAGEMENT

One big mistake we've made over the years is to use tests or quizzes to threaten classes that misbehave. Punitive measures like these devalue assessments, which should measure student learning, not student behavior. Giving an assessment to misbehaving students may quiet them at that moment. Unfortunately, they rarely do well on these tests, which lowers both their grade and their motivation for learning. Such disciplinary measures often make future student behavior worse, leading to a downward spiral of more assessments and grading with no benefit. In the long term, testing or quizzing students to control them does more harm than good.

Instead of punishing students with assessments, we suggest more positive measures of classroom management, such as asking students to consider the impact of their actions on others, having private conversations with misbehaving students, or giving students more autonomy in the classroom (Ferlazzo, 2013). Many times, students act out because our lessons are too easy, difficult, or boring, so we reflect on how we can improve them. We discuss better ways to motivate students in Chapter 1: Motivating Students.

POORLY CHOSEN QUESTIONS

Sometimes, we make an exam that doesn't reflect what we taught in class. This often happens when we use an assessment written by someone else and fail to adjust it based on what we did in class. In our experience, the best assessments contain a mixture of familiar items that we've done in class and somewhat unfamiliar items that can be solved using the methods discussed in class.

Making all assessment items exactly like classwork or homework questions may seem like a good thing, but we find that it actually hurts students in the long run. Not training students to expect the unexpected can make them more nervous when they take tests that we didn't write. It fails to prepare students adequately to solve difficult questions like word problems or proofs, where anticipating every single possible context is virtually impossible. It also makes students think that math is simply a collection of rote procedures rather than a way of understanding how the world behaves. Instead, we incorporate many different types of problems with varying levels of difficulty into our lessons (see Chapter 7: How to Plan Lessons for more details).

Sometimes, we make the opposite mistake by choosing assessment items that are *too* different from what we discussed in class to be attainable. This makes students more frustrated and less engaged because they feel that what we do in class has little to do with what they will see on a test.

Another mistake we've made is putting multiple questions on the same topic on a test or quiz. This often happens when we feel an obligation to make an exam worth a certain number of points or we don't have enough topics for an assessment. When we ask the same question multiple times, students who don't know the topic will be punished repeatedly, while students who know only that topic will be disproportionately rewarded. To avoid this problem, we determine the topic that corresponds to each question to ensure that there is no repetition. If necessary, we adjust the timing of our tests to ensure that enough topics can be covered.

Assessment items, like classwork questions, should also not reflect any cultural, gender, or ethnic bias. As we said in the What Could Go Wrong section of Chapter 7: How to Plan Lessons, any specialized nonmathematical knowledge (such as the number of yards required to make a first down in football) should be provided. Making these questions as unbiased as possible is particularly important for tests since students may not feel comfortable asking for clarification during an assessment.

MISTAKES ON ASSESSMENTS

One of the most serious errors we've made is not making an answer key before giving an assessment. Working through the problems ourselves helps us find typos, ambiguous test items, and other issues. If our tests constantly contain mistakes, students don't know whether they have difficulty answering a question because the question is flawed or they don't know how to solve it, which leaves them frustrated and anxious.

Sometimes, despite our best efforts, we don't find a mistake until after the assessment is over. How we respond depends on the type of mistake we made. If a multiple-choice question has several possible correct answers, then we accept all of them. If a question has no possible answer, then we don't count it. In rare cases, we may also allow students to retake a question as a short quiz. We also publicly acknowledge our mistakes to students and apologize for them. We believe that modeling this kind of behavior is an important life lesson for students to see.

STUDENT CHEATING

Unfortunately, we sometimes encounter students who try to cheat on our assessments. Students cheat for a variety of reasons—they are afraid of failing, they feel pressure from parents or guardians, they may need good grades for an extracurricular activity, or they feel that the assessments aren't fair. Teachers may not be clearly stating instructional objectives or may be putting material on tests that was not covered in class. Conversely, when students value the material being taught and when they see that they can control their outcomes on assessments, they are less likely to cheat (Murdock, 1999, p. 588).

To limit student cheating, we first make sure that our assessments are aligned with our instruction. If we allow students to use calculators in class, then we allow them to use calculators on assessments (Murdock, 1999, p. 588). We also try to make a clear connection between what is covered in class and what is covered on tests. We believe that assessments should be challenging enough so that students are prepared for the unexpected but accessible enough so that students who study can achieve reasonable success.

Another source of potential cheating are apps like PhotoMath, Mathway, and MathPapa that allow them to take pictures of math problems on their phones and see step-by-step solutions. We point out that these camera calculator apps, like calculators and formulas, are tools designed to make procedural work easier (Webel & Otten, 2015, p. 370). We discourage the use of camera calculator apps in class so that students can understand the concepts behind the skills and work independently. Students may then use these apps once they have mastered the skill, just as they would be allowed to use a formula or a calculator (Webel & Otten, 2015, p. 371). However, since we don't allow the use of phones on tests, students must be able to solve problems using permissible tools like calculators. In addition, we point out that relying too much on camera calculator apps may give students a false sense of confidence and prevent them from developing the deep conceptual understanding required to solve word problems, which these apps can't solve.

We point out that since math skills are cumulative, cheating on one test often requires cheating on future tests, which will become increasingly harder as the topics get more challenging (Murdock, 1999, p. 590).

Despite our best efforts, we often have to resort to other measures to discourage cheating. We circulate around the room constantly during assessments—not just to limit cheating, but also to see if we've made a mistake on a question, if students have written their name on the paper, or if students are struggling with a particular question (this often gives us valuable feedback as we analyze assessment results).

We also collect students' cell phones, smart watches, and any other electronic devices. To make the collection process quicker and easier, we hang an over-the-door shoe rack in our room and ask students to put their devices in one of the pockets.

CHAPTER 9: HOW TO PLAN TESTS AND QUIZZES

However, this method may not be possible or practical in all schools. Another option is to ask everyone to take out their phones, turn them off, and place them in backpacks.

In addition to these pragmatic responses to potential cheating, we also periodically discuss ethical questions with our students throughout the year: What harm can cheating cause to ourselves and others? What self-image and what public image do our ethical decisions promote? What values do our present families encourage, and what do we want to teach to our future families? Referring back to these conversations when we have tests helps reduce incidents of cheating. In addition, research indicates that having students write about values that are important to them can improve academic achievement and motivation (Borman, Grigg, Rozek, Hanselman, & Dewey, 2018, p. 9).

DIFFERENT VERSIONS OF TESTS

Many times, we make different versions of an assessment. However, we often have failed to check to see if some versions are harder than others. We find that the best way to make different versions is for each item to have a similar context but whose solutions involve the same steps with equal levels of difficulty. In Table 9.3: Two Test Questions with Unequal Difficulty, students are asked to solve a quadratic equation. Even though both questions appear to be similar since they deal with the same topic, the second is substantially harder than the first, as shown by the solution.

In the first question, the roots are real and the radical has no perfect square factors, so no simplification is necessary. In contrast, the roots of the second equation

Table 9.3 Two Test Questions with Unequal Difficulty

Solve for x in the equation $x^2 - 5x + 3 = 0$. Express the solution in simplest $a + bi$ form. Solution: $$x = \frac{-(-5) \pm \sqrt{(-5)^2 - 4(1)(3)}}{2(1)}$$ $$= \frac{5 \pm \sqrt{25 - 12}}{2}$$ $$= \frac{5 \pm \sqrt{13}}{2}$$	Solve for x in the equation $x^2 - 8x + 21 = 0$. Express the solution in simplest $a + bi$ form. Solution: $$x = \frac{-(-8) \pm \sqrt{(-8)^2 - 4(1)(21)}}{2(1)}$$ $$= \frac{8 \pm \sqrt{64 - 84}}{2}$$ $$= \frac{8 \pm \sqrt{-20}}{2}$$ $$= \frac{8 \pm 2i\sqrt{5}}{2}$$ $$= 4 \pm i\sqrt{5}$$

Table 9.4 Two Test Questions with Similar Difficulty

Solve for x in the equation $x^2 - 5x + 3 = 0$. Express the solution in simplest $a + bi$ form. Solution: $$x = \frac{-(-5) \pm \sqrt{(-5)^2 - 4(1)(3)}}{2(1)}$$ $$= \frac{5 \pm \sqrt{25 - 12}}{2}$$ $$= \frac{5 \pm \sqrt{13}}{2}$$	Solve for x in the equation $x^2 - 7x + 8 = 0$. Express the solution in simplest $a + bi$ form. Solution: $$x = \frac{-(-7) \pm \sqrt{(-7)^2 - 4(1)(8)}}{2(1)}$$ $$= \frac{7 \pm \sqrt{49 - 32}}{2}$$ $$= \frac{7 \pm \sqrt{17}}{2}$$

require simplifying $\sqrt{20}$ and rewriting the expression in $a + bi$ form. Solving the second problem requires two more steps than solving the first, so giving the two questions an equal point value would be unfair. To avoid this problem, both solutions should be roughly similar but have different numbers, as shown in Table 9.4: Two Test Questions with Similar Difficulty.

Making different versions requires more time and creates more opportunities for us to make mistakes. One shortcut that we've tried is to make different versions by taking the same questions and scrambling their order. However, this disrupts the flow of the exam since we prefer to order questions by increasing difficulty, as we explained earlier in this chapter. Using the resources that we describe in the Technology Connections section of this chapter makes the creation of different versions easier. In addition, we find that after we make up different versions of tests for the first few assessments of the year, students discover that there is little or no benefit to cheating. As a result, we don't feel like we have to make alternate versions for every test throughout the year.

Many times we have to give a single-period test over 2 days to students who are allowed extra time on assessments. Students may be tempted to look up information or ask friends after they leave class. To prevent this from happening, we give these students only a part of the test each day. To avoid any misunderstanding, we clearly explain this procedure to students and parents before the test.

GRADING AND RETURNING ASSESSMENTS

Many times, we get so overwhelmed with schoolwork or life that we fail to grade or neglect to return a test promptly. In these situations, by the time we return a test or quiz, many students have forgotten what they did, so correcting their thinking becomes much harder. We try to return assessments within a week after we give them. When possible, we use machine-readable answer sheets or student monitors to help grade multiple-choice questions. For constructed-response items, we use scoring guidelines that are modeled on district, state, or national assessments that we know.

We have often made the mistake of giving back assessments and going over student mistakes in a whole-group discussion. When students get their papers back, they usually compare their work with their classmates' and try to find their mistakes, so they often don't listen to what we're trying to say anyway! Instead of fighting this, we've learned to encourage it by allowing students to work in groups to discover and correct their errors.

TEST RETAKES AND TEST CORRECTIONS

Sometimes, students depend on test retakes or test corrections so much that they wind up not taking the original tests seriously enough since they think they can just take it again and get a better grade. This can lead to an endless cycle of writing, grading, and regrading tests. One solution is to limit the number of retakes or corrections that students are allowed to do. If we find that most of our students do poorly on the initial tests, then we reflect on our instruction to see what we could do differently. We may not be aligning our assessments with our instruction, or we may not be giving students enough time to understand the material.

Technology Connections

Many websites with test and quiz questions come and go, so we try to rely on resources that have been around for several years. The resources that we use to write lessons and unit plans usually have examples that we use for assessments.

TEST QUESTIONS, ANSWERS, AND SCORING GUIDELINES

Many districts, states, and testing companies publish test items for free online. For example, New York State publishes questions, scoring guidelines, and model student responses from its Regents exams dating back to 2000 at http://nysedregents.org. New York State also has selected questions and student responses for its grades 3–8 exams at its EngageNY site (www.engageny.org). JMAP (www.jmap.org) has published New York State math Regents exams going as far back as 1866. The College Board has sample questions for the PSAT and SAT (http://collegereadiness.collegeboard.org/sample-questions). For AP classes, the College Board releases free-response questions and their scoring guidelines every year on the web pages for each course (http://apcentral.collegeboard.org/courses). PARCC, which publishes the Common Core-aligned exams used by many states, has sample questions (http://parcc.pearson.com/practice-tests). Smarter Balanced, another testing company, has sample questions (http://www.smarterbalanced.org/assessments/samples).

The Problem-Attic website (http://www.problem-attic.com) has thousands of questions from several sources, including state (including New York), Smarter Balanced, PARCC, TIMMS, and PISA tests.

Another excellent source of questions for assessments is the free DeltaMath website (www.deltamath.com), which we also use for student practice and homework. DeltaMath generates hundreds of similarly structured questions on the same topic, which makes creating different versions of tests much easier.

ZipGrade (http://www.zipgrade.com) allows teachers to scan and grade multiple-choice papers using their mobile phone. The website has a monthly fee, but a limited number of free scans per month are available.

TEST REVIEW

Online resources like Kahoot! (www.kahoot.com) or Quizlet (www.quizlet.com) can be useful tools for test review. Many of these sites not only have ready-made quizzes but allow you to add your own questions and collate and analyze student responses. Google Classroom (http://classroom.google.com) also has the ability to create an online quiz with multiple-choice or dropdown questions using Google Forms. We also use DeltaMath, which we mentioned earlier, to review for tests.

Another low-tech option that requires no student technology is Plickers (www.plickers.com). Students respond to teacher-created multiple-choice questions by holding up special cards that can be printed from the Plickers website, so students don't need special technology. Teachers can use an app on their mobile phone to scan the student-response cards and instantly record student responses.

Doing an online search for "Jeopardy activity" for a particular topic usually yields many results. Since these activities are often done using presentation software like Microsoft PowerPoint or Google Slides, you can easily customize them to suit your needs.

TEST ANALYSIS

To help us generate histograms and summary statistics for analyzing our test scores, we use a spreadsheet program like Microsoft Excel or the free Google Sheets. Many sites provide spreadsheet templates or online tools for making histograms. One example is Social Science Statistics (https://www.socscistatistics.com/descriptive/histograms), which will create a histogram after you enter data. An online search using a phrase like "spreadsheet histogram template" or "automatically generate histogram from data" should yield many tools that you can use.

Figures

ALGEBRA I: TEST #2   DATE: _____

NAME: _____   **TOTAL:** _____ = **OUT OF 50 POINTS** = _____%

Part I. Answer all questions in this part. Each correct answer will receive 3 points. No partial credit will be given. For each question, write in the space provided the numeral preceding the word or expression that best completes the statement or answers the question. [27 points]

1. What is the solution to the equation $2m + 10 = 50$?
   - (1) $m = 20$
   - (2) $m = 25$
   - (3) $m = 30$
   - (4) $m = 40$

   1_____

2. If $f(x) = x^2 + 7$, what is $f(3)$?
   - (1) 13
   - (2) 16
   - (3) 39
   - (4) 63

   2_____

3. Which equation illustrates the distributive property?
   - (1) $a(b + c) = ab + ac$
   - (2) $a + (b + c) = (a + b) + c$
   - (3) $b + c = c + b$
   - (4) $a + (b - b) = a$

   3_____

4. Which number is in the solution set to $5p + 6 < 3p + 10$?
   - (1) $p = 1$
   - (2) $p = 2$
   - (3) $p = 3$
   - (4) $p = 4$

   4_____

5. Which of the following correctly expresses the set of numbers less than or equal to 15?
   - (1) $(15, \infty)$
   - (2) $[15, \infty]$
   - (3) $(-\infty, 15)$
   - (4) $(-\infty, 15]$

   5_____

6. To prove that $(x + 4) + 5x = 4 + (x + 5x)$, what properties must be used?
   - (1) both the commutative property and associative property
   - (2) commutative property, only
   - (3) associative property, only
   - (4) both the distributive property and commutative property

   6_____

Figure 9.1   Algebra I Test

7. A ball hit during a tennis match travels 50.1 feet in 0.36 seconds. What is the ball's approximate average speed, in feet per second?
  (1) 18.0
  (2) 50.4
  (3) 69.6
  (4) 139.2

  7____

8. If $x = a + b$ and $a + b = 17$, which equation must be true?
  (1) $x = 17$
  (2) $a = 9$
  (3) $b = 8$
  (4) $b = a + x$

  8____

9. Which equation has the same solution set as $25k + 9 = 49$?
  (1) $50k + 18 = 98$
  (2) $25k - 9 = 49$
  (3) $5k + 3 = 7$
  (4) $34k = 49$

  9____

Part II. Answer all questions. Each correct answer will receive 3 points. Clearly indicate the necessary steps, including appropriate formula substitutions, diagrams, graphs, charts, etc. For all questions in this part, a correct numerical answer with no work shown will receive no credit. [9 points]

10. Graph the inequality $-2 \leq x < 6$ on the number line below.

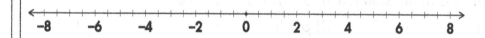

11. Algebraically solve the equation $12 - (6m - 7) = 3(m + 1) - 5m$.

12. Is the relation {(1, 5), (2, 3), (4, −9), (−9, 4)} a function? Explain your answer.

Figure 9.1 (Continued)

Part III. Answer all questions. Each correct answer will receive 4 points. Clearly indicate the necessary steps, including appropriate formula substitutions, diagrams, graphs, charts, etc. For all questions in this part, a correct numerical answer with no work shown will receive no credit. [8 points]

13. Solve for x in the inequality $-2x + 16 \geq 4$ and $3x + 15 \geq 21$.

14. Prove that $(x + y) + z = (z + y) + x$.

Part IV. The correct answer will receive 6 points. Clearly indicate the necessary steps, including appropriate formula substitutions, diagrams, graphs, charts, etc. For all questions in this part, a correct numerical answer with no work shown will receive no credit. [6 points]

15. An electronics store sells only tablets and mobile phones. The store makes a $60 profit on the sale of each tablet and a $25 profit on the sale of each phone. The store wants to make a total profit of at least $290 from its combined sales. To reach this goal, the store must sell t tablets and p phones.

   a Write an inequality that represents the given information.

   b If the store sold two tablets, at least how many phones must the store sell to reach its goal?

Figure 9.1 (Continued)

PRECALCULUS EXAM #3

NAME: _____   SCORE: _____ out of 46 = _____ %

Part I. Show all work in the space provided. Each correct answer will receive 2 points. A correct answer with no work shown will receive only one point. [8 points]

1. Convert $(3, -3)$ into polar form.	2. Convert $\left(8, -\frac{3\pi}{2}\right)$ into rectangular form.
3. Express $\cos(-135°)$ as a function of a positive acute angle.	4. Write the equation in polar form that is represented by the graph below. 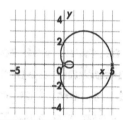

Part II. Show all work in the space provided. Each correct answer will receive 4 points. A correct answer with no work shown will receive only one point. [20 points]

5. If $f(x) = \cos(x) + \sin\left(\frac{x}{3}\right)$, find the exact value of $f(\pi)$.

6. If $\sec(B) = -\sqrt{5}$ and B is an angle in Quadrant III, find the exact value of $\csc(B)$.

Figure 9.2 Precalculus Test

7. Solve the following system of equations for r and x:
   $r = \cos^2(x)$
   $r = 2\cos(x)$

8. Write the equation $4x^2 + 9y^2 = 36$ in parametric form.

9. For the equation $r = 2\sin(2\theta)$:

   a. Sketch its graph below.

   b. Write the equation in rectangular form.

   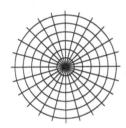

Figure 9.2 (Continued)

Part III. Show all work in the space provided. Each correct answer will receive 6 points. A correct answer with no work shown will receive only one point. [18 points]

10. In the interval $0° \leq A \leq 360$, solve for all values of A in the equation $\cos(2A) = -3\sin(A) - 1$.

11. Two equal forces act on a body at an angle of 80°. If the resultant force is 100 N, find, *to the nearest hundredth of a newton*, the value of one of the two equal forces.

12. A spider and a fly crawl on a coordinate plane so that their positions at time t (in seconds) are:
    Spider: $(x, y) = (3, -2) + t(2, 1)$
    Fly: $(x, y) = (-1, 6) + t(4, -3)$

Figure 9.2 (Continued)

a. On the accompanying coordinate plane, sketch the positions of the fly and spider at different times.

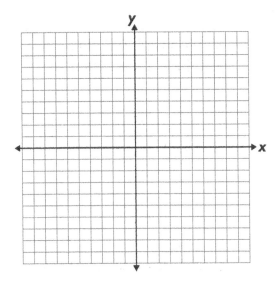

b. Determine *when* and *where* they meet. Justify your answer.

Figure 9.2 (Continued)

224 THE MATH TEACHER'S TOOLBOX

Algebra I Quiz Name: _____ Date: _____

Grade: _____ out of 16 points = _____ %

Answer all questions. Each correct answer is worth 4 points. Show all appropriate work.

1. Write an equation that could represent the line whose graph is shown here.

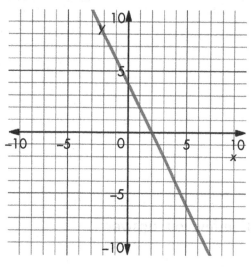

2. The graph of the function $y = f(x)$ is shown below.

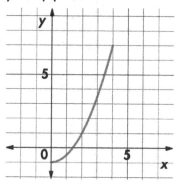

a. State the domain of the function.
b. State the range of the function.

3. If $f(x) = 5x - 2$, then evaluate $f(-1)$. *(The use of the accompanying grid is optional.)*

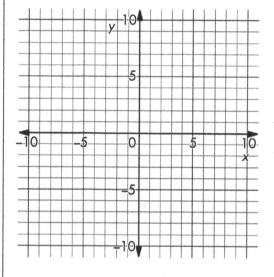

4. Write an equation of the line that has a slope of $\frac{3}{4}$ and passes through the point (2, 1). *(The use of the accompanying grid is optional.)*

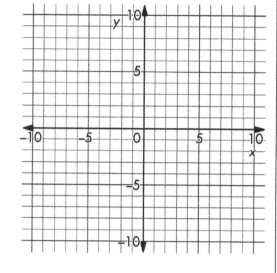

Figure 9.3 Quiz

PROBLEM: Algebraically solve for x in the equation $x + 6 = x^2 + 8x + 16$.

Work	Benchmark	Point value
$x + 6 = x^2 + 8x + 16$ $-x - 6 \quad\quad -x - 6$ $\overline{0 = x^2 + 7x + 10}$	← Write quadratic equation in standard form	1
$0 = (x + 5)(x + 2)$ $x + 5 = 0$ or $x + 2 = 0$	← Factors trinomial and sets each factor equal to 0 (or completes the square) (NOTE: A graphical solution receives no credit for this benchmark since the question asks for an algebraic solution.)	2
$x = -5$ or $x = -2$ $x = \{-5, -2\}$	← States final answer	1

Figure 9.4 Creating Scoring Guidelines

Test # _____ **Corrections**   **Name:** _____

Directions: Use the following sheet to correct your test. If necessary, use multiple copies of this sheet.

- **Number:** write the question number. *If your work for a question was correct, you may skip it here.*
- **Explain your error:** Explain in detail what you did wrong. Be specific: "I thought the angles were supplementary, but they're actually complementary," "I subtracted 2x when I should have added," etc.
- **Correct your work:** Write a complete solution to the question. Show all work. Do *not* copy the question. *Draw a horizontal line to separate each question.*
- **Staple this sheet to your original test and submit both papers.** (Put this sheet on top.) Use as many copies of this sheet as necessary. Do not write on the back of this sheet. *Test corrections that are not stapled will not be accepted.*

Figure 9.5 Blank Test Corrections Sheet

Number	Explain what you did wrong...	Correct your work (Show all work)...

Figure 9.5 (Continued)

Test # __4.1__ Corrections   Name: _____

Directions: Use the following sheet to correct your test. If necessary, use multiple copies of this sheet.

- **Number:** write the question number. *If your work for a question was correct, you may skip it here.*
- **Explain your error:** Explain in detail what you did wrong. Be specific: "I thought the angles were supplementary, but they're actually complementary," "I subtracted 2x when I should have added," etc.
- **Correct your work:** Write a complete solution to the question. Show all work. Do *not* copy the question. *Draw a horizontal line to separate each question.*
- **Staple this sheet to your original test and submit both papers.** (Put this sheet on top.) Use as many copies of this sheet as necessary. Do not write on the back of this sheet. *Test corrections that are not stapled will not be accepted.*

Number	Explain what you did wrong…	Correct your work (Show all work)…
13	I took the wrong information from the two-way table that I've made.	left: 10+ years (4.5%), −10 years 4.5%, total 9%; right: 19.5%, 71.5%, 91%; Total (24%), 70%, 100%. 0.045/0.24 = 0.1875 $\boxed{0.19}$
14.	I did not explain properly and did not show work.	Phys 100, Chem 60, Tot 160; Male (100) 60 (160), Fem 71 61 132, Tot 171 121 (292). 160/292 = 0.548. 100/160 = 0.625. Dependent because they are not equal since P(phys|male) is .548 and probability of physics is 0.625.
15 b	I was looking for the z-score.	0.773 (520,000) = 402,140 ≈ 402,000 normCDF
15 c	I didn't answer the question	inverseNorm (0.9, 72300, 20000, left) = 98,571.81 ≈ 98,600

Figure 9.6 Completed Test Corrections Sheet

Name: _____   Test #: _____

Test Reflection Form

DIRECTIONS: Think about how you did on the test you just got back. Use this form to summarize your thoughts. Make sure you think carefully about your performance. This form is for you. This form is due on the date and time listed on the class webpage.

1. List two *mathematical* strengths that you showed on this test. (For example, "I know how to add fractions with unlike denominators.") *(2 pts.)*

2. List two *mathematical* weaknesses that you showed on this test. (For example, "I don't know how to solve equations with variables on both sides of the equal sign." Avoid test-taking mistakes like "I ran out of time.") *(2 pts.)*

3. You may have studied a topic that wasn't on this test. In the space below, **show that you know** it. Write a problem that you thought would be on the test but wasn't. Then show a complete solution with all appropriate work. Don't just copy a problem from classwork or homework. Explain your work carefully so that you show that you really understand this topic. *(3 pts.)*

4. Find a problem from this test that you thought you knew how to do but you messed up on it. Here's your chance to redo it. Write the question number below and show that you can do it by yourself. Show all appropriate work. *(3 pts.)*
   Question #: _____
   Work:

Figure 9.7 Test Reflection Form

CHAPTER 10

How to Develop an Effective Grading Policy

What Is It?

After teachers present lessons, assign homework, and give tests, quizzes, or other assessments, they must summarize all of this information by assessing students, typically in the form of report card grades. A *grading policy* communicates to administrators, teachers, parents, and students the overall expectations for student performance and the calculations used to derive report card grades.

Developing an effective grading policy can be challenging since various groups have different goals for grades (Guskey & Bailey, 2010, pp. 1–2):

- Parents want to know how their children are doing.
- Administrators want consistency in grading among teachers and across subjects.
- Teachers want grades that can be easily managed so that grading doesn't take time away from providing high-quality instruction.
- Students want to know how they are doing and what they have to do to get the grade they want.

In this chapter, we discuss strategies for giving summative report card grades. We discuss how to grade the different components that go into report card grades (such as homework, tests, and projects) in Chapter 8: How to Plan Homework, Chapter 9: How to Plan Tests and Quizzes, and Chapter 16: Project-Based Learning.

Why We Like It

We don't like dealing with grades, but creating a clear, fair, and equitable grading policy can reduce stress and help us and our students focus on learning.

Supporting Research

The following research informs how we use grades with our students, which we discuss in the Application section of this chapter.

Extensive research dating back over a century shows that grading can vary considerably because teachers weigh distinct criteria differently (Marzano, 2010, p. 41). Landmark studies by Daniel Starch and William Elliott (1912, 1913) showed wide variation in English and geometry papers scored on a numerical scale. Even the use of rubrics may not improve the reliability of grading (Anderson, 2018, p. 12). A recent replication of Starch and Elliott's study found similar variability in grading English papers, even after graders received a week of training using a rubric (Brimi, 2011, p. 6).

In addition, researchers have found that excessively low grades can harm students by encouraging them to give up instead of trying harder (Guskey, 2011). Low grades at the beginning of the year can seriously damage student self-confidence and give students few options to recover (Carifio & Carey, 2010, p. 219).

Common Core Connections

An effective grading policy should reflect student mastery of learning objectives embodied by documents like the Common Core Standards.

Application

In a traditional report card grading system, students receive a single grade based on their performance on different types of assessments, such as homework, classwork, tests, quizzes, and projects. However, these grades can often be confusing and unclear (Kunnath, 2017, p. 53). A report card grade of 85 says nothing specific about a student's effort, content knowledge, or participation.

Here are some strategies that attempt to address some of the shortcomings of traditional grading systems.

STANDARDS-BASED GRADING

Since grades can vary considerably, some researchers have proposed the more rubric-based approach of *standards-based grading*. In a standards-based grading system, students receive several grades, one for their progress on each learning objective in the course (Scriffiny, 2008, p. 70).

To implement a standards-based grading system, teachers must first create or find a list of required skills for the course. This list, typically based on state or national standards, could be provided by a teacher's school or district and possibly modified by teachers. Teachers then must design lessons that meet those goals. Students take periodic assessments that contain several test items aligned to a specific skill (Scarlett, 2018, p. 60). To grade student work, teachers can use scoring guidelines like the one in Table 10.1: Standards-Based Grading. In this table, which is based on work by Texas teacher Dale Ehlert (2015), scores range from 1 (shows insufficient evidence of procedural and conceptual understanding) to 5 (shows an understanding beyond what is expected). Teachers then give a standards-based score for each learning objective that is an average of the scores for each skill.

Standards-based grading can foster a growth mindset (which we discuss in Chapter 4: Promoting Mathematical Communication) since students' grades are based not on how many points they earn on tests but on how well they demonstrate mastery (Ehlert, 2015). Students don't receive zeros for late work since they can retake assessments until they are proficient in a particular skill. Since students receive more specific feedback than a single numerical grade, they are more likely to think about improving their skills than improving their grades (Pinkin, 2016). Standards-based grading can be particularly helpful for struggling students since teachers can provide specific information on gaps in student learning, which in turn helps teachers determine appropriate support for them (Guskey & Jung, 2012, p. 27).

Though we find standards-based grading attractive, we believe it may be impractical for many teachers. First, standards-based grading requires significant support

Table 10.1 Standards-Based Grading

Score	Explanation
5 (Mastery)	Student has shown an understanding that is beyond what is expected.
4 (Proficient)	Student has shown a complete and thorough procedural and conceptual understanding.
3 (Developing)	Student has shown some procedural and conceptual understanding.
2 (Basic)	Student has shown a basic procedural understanding but not yet shown a conceptual understanding.
1 (Incomplete)	Student has not yet shown a procedural and conceptual understanding.

from schools or districts. Since it differs greatly from traditional grading, it requires substantial professional development and explanation in order for teachers, students, and parents to understand and support it. Schools must rework their report cards to determine what reporting standards will be included for each course, what performance levels will be reported for each standard, and how those levels will be labeled (Guskey & Bailey, 2010, pp. 58, 118). In addition, reporting levels of student understanding can be difficult if teachers have to use a school or district gradebook that requires traditional numerical grades for assignments. Furthermore, since students need multiple opportunities to show mastery, teachers must spend more time writing and administering assessments. Grading standards-based assessments can take more time since teachers must think about how proficient students are in each skill instead of simply adding up points (Scarlett, 2018, p. 67). As a result, many teachers may not have the time or support necessary to implement standards-based grading successfully.

We like the idea embedded in standards-based grading that students should focus on how they can improve their content knowledge instead of how they can raise their grade. We believe that all teachers can benefit from incorporating elements of standards-based grading into their existing grading system. We find that dividing classwork and assessment questions into levels of increasing complexity (which we discuss in Chapter 14: Differentiating Instruction) can communicate more clearly what students know and what they can do to improve their understanding.

MINIMUM GRADING POLICY

To minimize the effect of low grades, some researchers have challenged the notion of giving zeros for missing assignments (Reeves, 2004, p. 325; Wormeli, 2006, p. 21). Some schools have experimented with a minimum grade policy, in which any grade below a certain threshold is replaced with a preset number, usually between 40 and 60. Supporters of minimum grades argue that these modifications keep grades within a reasonable zone of student control, enabling more students to keep working and eventually pass their courses. Opponents claim that minimum grades give students unearned assistance and can lead to overall grade inflation, which erodes the quality of the information provided by a grade (Friess, 2008).

In our experience, low grades can have a catastrophic effect not just on students' averages but also on their motivation. A student who receives test grades of 0, 80, and 80 has a test average of 53%. This failing grade doesn't represent the 80s that the student usually received. We find that not giving students another chance to show mastery after receiving a low grade can lead them to give up quickly, especially at the beginning of the year. Often, students have events in their lives that can adversely affect their schoolwork. Not giving them other opportunities to succeed can make them feel unappreciated and discourage their desire to learn.

Unfortunately, we also find that a minimum grading policy can have unintended negative consequences for the struggling students that it is supposed to benefit. In our experience, minimum grades often *discourage* our lowest-performing students from improving. Many students who get very low grades on tests tell us that they see no reason to study more or get additional help if they will get a minimum grade of 50. Minimum grades can also hide growth for the lowest-performing students. A student who improves from a grade of 20 to a grade of 40 shows substantial improvement that would not be reflected if the minimum grade were 40. We find that encouraging students at the early stages of improvement is easier when they can see that their grade increases.

In addition, many grading policies essentially create minimum grades for students who put in maximum effort but have limited content mastery. A policy that sets an excessively high minimum grade for math classes can have disastrous consequences for the struggling students that it is designed to benefit. Math is a cumulative discipline—students must understand basic concepts before they can apply them in more advanced situations (Fleming, 2019). In order to solve rational equations in Algebra II, students must know how to solve linear equations, write equivalent fractions, and perform operations with signed numbers—all of which are learned in previous courses. We find that students who show mastery of 20% or less of content often lack the prerequisite skills necessary to succeed and require a great deal of remedial instruction. A minimum grading policy that allows students to move to a more advanced course without demonstrating sufficient content proficiency sets them up for future failure and encourages them to resent math even more.

Table 10.2: Sample Report Card Grades for a Student with 20% Content Mastery shows how minimum grades can send confusing messages about students' performance. This table shows the grades that a student with minimal content knowledge (20% content mastery) and maximum effort (100%) would receive under three grading policies with different minimum grades. For the sake of simplicity, grades are calculated using two overall categories—content mastery (60%) and effort (40%).

A minimum grade of 20 enables students who show only 20% content mastery and maximum effort to receive an overall grade of 52, which is below the 65 that is usually considered passing. However, a minimum grade of 40 would enable these students to receive an overall grade of 64, which is close to passing. A minimum grade of 50 enables these students to receive a grade of 70, which sends a false message to students, parents, and teachers that they have more than a minimal amount of understanding.

In short, we recognize that rigidly giving zeros for missing assignments harms students. To avoid the problems that arise from giving zeros, we suggest giving students reasonable opportunities to make up missing work, dropping the lowest

Table 10.2 Sample Report Card Grades for a Student with 20% Content Mastery

Category	Minimum grade = 20	Minimum grade = 40	Minimum grade = 50
Demonstrated content knowledge	20%	20%	20%
Content mastery (60%)	20% of 60% = 12%	(Under this policy, a 40 is recorded in place of a 20.) 40% of 60% = 24%	(Under this policy, a 50 is recorded in place of a 20.) 50% of 60% = 30%
Effort (40%)	100% of 40% = 40%	100% of 40% = 40%	100% of 40% = 40%
Overall grade:	52	64	70

assignment score, or setting a minimum grade of 20. We believe that setting a minimum grade that is greater than 20 can create even more problems for struggling students.

POINT ACCUMULATION SYSTEM FOR GRADING

Many grading policies assign percentage weights to each type of assessment, reflecting an estimate of the relative value of each category. However, these percentages can often give a disproportionate weight to an assignment. If quizzes are 20% of a grade and two quizzes are given, then each quiz is worth 10% of the overall grade. If tests are worth 30% of a grade and three tests are given, then each test is worth 10%. In this example, a short quiz has the same value as a full-period test!

One way to reduce problems that can occur with disproportionate weights in grades is to use a point accumulation system. In this method, teachers give each assignment (homework, test, quiz, project) a point value. Classwork can also be given a point value. Student grades are calculated by adding up the total number of points earned and dividing by the total number of possible points.

A point accumulation system has several advantages. It can simplify the grading process since teachers no longer need to determine the appropriate percentages for each type of assessment. Students can calculate their grade by themselves more easily since they no longer need to calculate a weighted average, which in our experience is challenging for many. In addition, this system mitigates the effect of low grades or missing assignments since teachers can give extra credit assignments that add to the total number of points earned but not to the total number of possible points.

CHAPTER 10: HOW TO DEVELOP AN EFFECTIVE GRADING POLICY

A point accumulation system may be difficult to implement if teachers must follow a school or district grading policy that assigns specific weights for each type of assessment. We use a hybrid system in which students accumulate points within each category. We assign extra credit as we would under a point accumulation system—points added to the total number of points earned—but try to give extra credit for several categories.

To make a point accumulation system work successfully, we make sure that the point values of each assignment are proportional to its relative value. A 15-minute quiz worth 10 points should not have the same value as a week-long project graded from a 10-point rubric. In this case, the project grade (out of 10 points) can be multiplied by a number to give it a more appropriate total point value (such as 50 points). While we usually convert the grade for each project to a percentage to give students a clearer idea of how they did, we weigh each assignment in our grade book according to its weight.

We distribute a sheet similar to Figure 10.1: Grade Calculation Sheet at the beginning of the year. We use this sheet to help students understand how their grades are calculated. We find that a sheet like this is useful for giving them a visual illustration of the weighted average and an overall sense of what they need to do to get the grade that they want.

Students could use a grade calculation sheet to keep track of their grades on paper. However, they would have to recalculate their overall grades after entering assignment grades, which can get tedious and messy on paper. A computerized version of this sheet (which teachers could put online so students can download it) can help students calculate their grade more easily and estimate their grade.

We find that grade calculation sheets are especially useful when we use computer-based systems to record and calculate grades. In our experience, when we use computerized systems, students often tend to view them as "black boxes" that magically produce grades. Using a grade calculation sheet can help demystify the grade calculation process, help them monitor their learning, and ultimately empower them to take more control over their learning.

Figure 10.2: Completed Grade Calculation Sheet shows a completed grade calculation using a policy in which assessments are 60% of the overall grade, homework is 20%, and classwork is 20%.

DIFFERENTIATING FOR ELLS AND STUDENTS WITH LEARNING DIFFERENCES

Grading English Language Learners (ELLs) and students with learning differences can present unique challenges. Often, these students don't participate in class activities in the same way that other students do. However, teachers are expected to hold all students to the same high standards (Jung & Guskey, 2010). We find that the techniques we describe below can be effective not just for ELLs and students with learning differences but for *all* students.

Teachers can calculate grades using a combination of *product* (mastery of course objectives), *process* (non-academic factors such as effort and attendance), and *progress* (amount of improvement over a period of time) (Guskey & Jung, 2012, pp. 24–25). To provide a more detailed and meaningful picture of students, teachers can report each of these grades separately. However, since many schools allow students to receive only one grade for each class, teachers may have to weigh product, process, and progress as categories for a student grade.

Teachers can identify specific content learning objectives for ELLs or students with learning differences (Fenner, Kester, & Snyder, n.d.). ELL students may also have language learning objectives, which often can be derived from state or national standards (Himmel, n.d.). For example, the Common Core State Standard "Use similar triangles to explain why the slope m is the same between any two distinct points on a non-vertical line in the coordinate plane" (8.EE.B.6) (NGA & CCSSO, 2010, p. 54) can be converted to the language objective "Use similar triangles to explain *orally* why the slope m is the same between any two distinct points on a non-vertical line in the coordinate plane." Students with learning differences often have learning objectives in their Individualized Education Plans (IEPs) or 504 plans.

Teachers then have to determine what accommodations or modifications are necessary for their students to meet those standards. Often, this information can be found from students' IEPs, 504 plans, ELL teachers, or guidance departments. Some students may need assessment questions read to them or may need access to a glossary. Students with exceptional oral skills but weaker writing skills may not be able to explain grade-level standards in writing, so they may be expected to master standards from a lower grade level (Jung & Guskey, 2010).

Finally, the grades of ELLs and students with learning differences should reflect the customized standards for students (Jung & Guskey, 2010). If possible, teachers can make a notation that the grade reflects modified learning objectives.

MORE THAN JUST A GRADE

We find that no one policy can address all of the problems that can arise with grading. In our experience, our grades are more effective and accurate when we use a combination of strategies, including (but not limited to):

- Dropping the lowest grade to reduce the effect of outliers
- Teaching students how their grades are calculated so that they can monitor their grades
- Aligning assessment items to specific skills to determine students' strengths and weaknesses

- Giving students opportunities to make up missing points through test retakes, test corrections, or extra credit assignments (we discuss test retakes and corrections in Chapter 9: How to Plan Tests and Quizzes)
- Balancing the high standards that students are held to with reasonable accommodations for unexpected events in their lives
- Setting a reasonable minimum grade that is low enough to enable students to show growth but high enough to avoid eroding student morale
- Simplifying the grading process by reducing the number of grade categories

We strongly believe that meeting students where they are and building them up toward mastery is more important than improving our passing rate. We feel that our job as educators is to give our students the academic and social-emotional support necessary for the high-quality instruction that they deserve. A fair and equitable grading policy supports this goal.

Most importantly, we frequently tell students that their grades don't reflect everything that we think about them, their value as individuals, or what they have learned. To reassure them that they are more than just a number or letter, we try to acknowledge their humanity throughout the year by getting to know them personally, praising them when they make progress, and encouraging them when they encounter setbacks. We find that these encounters make a more memorable and lasting impact on students' lives than the grades we give them.

What Could Go Wrong

Sometimes, we rely on our computer-based grading system or grading policy and ignore qualitative factors not captured by grades. A student who received grades of 60, 60, 90, 100, and 100 shows clear improvement over the marking period but may have had personal or medical issues at the beginning that prevented the student from doing well. However, neither the student's mean grade of 82 nor the median of 90 incorporates these factors. In these situations, we try to adjust students' grades by a few points if possible to give students a grade that better reflects what we feel they deserve. Using a statistical algorithm to calculate grades is fair and accurate for most students but can be inappropriate for others (Guskey & Jung, 2016). We find that balancing our professional judgment with concrete evidence to support our decision works best.

Another mistake we've made is using grades punitively, such as lowering a student's grade due to behavioral issues. Grades should measure student learning, not regulate student behavior (Kunnath, 2017, p. 54).

Sometimes, we get so busy that we fail to put grades into our gradebook. We find that when students don't know how they're doing, they get frustrated and lose motivation. We try to keep updates manageable by either inputting grades into our gradebook immediately or at regular intervals (such as at the end of the week) so that we can develop good habits.

Similarly, we sometimes don't keep parents informed about how their children are doing or what modifications or accommodations we have made for them. We find that getting support from parents leads to fewer problems in the long run. We talk more about how to keep parents informed in Chapter 12: Building Relationships with Parents.

As mentioned in the More Than Just a Grade section, creating a classroom environment where students become obsessed with higher grades can be problematic. Instead, we try to foster an atmosphere in which students feel motivated to learn because they want to acquire new knowledge. We discuss some of these efforts in Chapter 1: Motivating Students.

Student Handouts and Examples

Figure 10.1: Grade Calculation Sheet
Figure 10.2: Completed Grade Calculation Sheet

Technology Connections

Since many schools and districts use online grading tools, we recommend checking with them first to see what is available. Few stand-alone grading software applications now exist since they have been replaced by online grading sites. Spreadsheet tools like the free Google Sheets (http://sheets.google.com) can be used to calculate student grades. Templates and tips for creating grade calculator spreadsheets in Microsoft Excel can be found on the Microsoft Office Templates site (http://templates.office.com).

The Association for Supervision and Curriculum Development website (http://www.ascd.org) has many free articles related to grading policy and grading practices, including standards-based grading.

Many math teachers who use standards-based grading have blog posts that describe their grading practices. We find that Dale Ehlert (http://whenmathhappens.com/standards-based-grading), Kate Owens (http://blogs.ams.org/matheducation/2015/11/20/a-beginners-guide-to-standards-based-grading), and Yelena Weinstein's (http://questformasteryblog.wordpress.com) blog posts have been particularly useful. The Credits for Teachers website (http://creditsforteachers.com/K12-Standard-Based-Grading-Resources) has many activities and other resources available. In addition, websites like DeltaMath (http://www.deltamath.com) that organize questions by topic and grade them automatically can help teachers make standards-based assessments.

Figures

Name:	Course:	Term:	Year:
CATEGORY	CATEGORY GRADE %	CATEGORY WEIGHT	CATEGORY TOTAL
Assessments (__%)		× 0.___	=
Homework (__%)		× 0.___	=
Classwork (__%)		× 0.___	=
Add the category totals in the right column to calculate **YOUR OVERALL GRADE**:			

ASSESSMENTS (__ % of grade)

Date	Assessment	Pts. Earned	Possible Pts.

EXTRA CREDIT

Totals for Points Earned and Possible Points:		÷	Pts.
Convert to % (divide numbers above and multiply by 100) to get YOUR TEST GRADE %			

HOMEWORK (__ % of grade)

HW #	Points	HW #	Points	HW #	Points
1		29		57	
2		30		58	
3		31		59	
4		32		60	
5		33		61	
6		34		62	
7		35		63	
8		36		64	
9		37		65	
10		38		66	
11		39		67	
12		40		68	
13		41		69	
14		42		70	
15		43		71	
16		44		72	
17		45		73	
18		46		74	
19		47		75	
20		48		76	
21		49		77	
22		50		78	
23		51		79	
24		52		80	
25		53			
26		54			
27		55			
28		56			
Totals for Points Earned and Possible Points:			÷	Pts.	
Convert to % (divide numbers above and multiply by 100) to get YOUR HW GRADE %					

CLASSWORK (__ % of grade)

Classwork grade:	÷	Pts.
Convert decimal above to % (multiply by 100) to find YOUR HW GRADE %		

Figure 10.1 Grade Calculation Sheet

Name: Edwin Gonzalez **Course:** Algebra II **Term:** 1 **Year:** 2020–2021

CATEGORY	CATEGORY GRADE %		CATEGORY WEIGHT		CATEGORY TOTAL
Assessments (60%)	73.62	×	0.60	=	44.17
Homework (20%)	96.38	×	0.20	=	19.28
Classwork (20%)	100	×	0.20	=	20
Add the category totals in the right column to calculate **YOUR OVERALL GRADE: 83.45**					

ASSESSMENTS (60% of grade)			
Date	Assessment	Pts. Earned	Possible Pts.
9/21	Quiz #1	12	16
9/25	Test #1	40	50
10/2	Quiz #2	8	16
10/16	Test #2	36	50
10/22	Project #1	15	20
10/26	Quiz #3	9	16
10/30	Test #3	35	50
11/20	Test #4	15	50
12/7	Quiz #5	13	16
12/14	Test #5	27	50
1/4	Project #2	14	20
1/8	Quiz #6	12	16
1/15	Final	75	100

EXTRA CREDIT			
10/18	Extra Credit #1	8	
1/7	Extra Credit #2	12	
1/24	HW Review	15	
Totals for Points Earned and Possible Points:		346 ÷ 470 Pts.	
Convert to % (divide numbers above and multiply by 100) to get YOUR TEST GRADE %		73.62%	

HOMEWORK (20% of grade)					
HW #	Points	HW #	Points	HW #	Points
1	1	29	1	57	1
2	1	30	1	58	0.75
3	1	31	1	59	0.95
4	1	32	1	60	1
5	1	33	1	61	1
6	1	34	1	62	1
7	1	35	1	63	1
8	1	36	1	64	1
9	1	37	1	65	0
10	1	38	1	66	1
11	1	39	0.75	67	1
12	1	40	1	68	1
13	0.75	41	1	69	1
14	1	42	1	70	1
15	1	43	1	71	1
16	1	44	1	72	1
17	1	45	1	73	1
18	1	46	1	74	1
19	1	47	1	75	1
20	1	48	1	76	1
21	0.95	49	0	77	1
22	0.95	50	1	78	1
23	1	51	1	79	1
24	1	52	1	80	1
25	1	53	1		
26	1	54	1		
27	1	55	1		
28	1	56	1		
Totals for Points Earned and Possible Points:		77.1 ÷ 80 Pts.			
Convert to % (divide numbers above and multiply by 100) to get YOUR HW GRADE %		96.38%			

CLASSWORK (20% of grade)		
Classwork grade:	20 ÷ 20	Pts.
Convert decimal above to % (multiply by 100) to find YOUR HW GRADE %		100%

Figure 10.2 Completed Grade Calculation Sheet

PART III

Building Relationships

CHAPTER 11

Building a Productive Classroom Environment

What Is It?

In Chapter 4: Promoting Mathematical Communication, we discussed ways to help students read, write, and speak mathematically. However, we find that running a successful classroom takes more than just teaching students the mechanics of mathematics. It also involves building positive relationships with students and establishing good student behavior.

In this chapter, we discuss some of the ways that we keep our classroom productive, efficient, and positive.

Why We Like It

We find that running a well-organized classroom sets a foundation for successful learning and helps communicate high expectations for our students. When a class runs smoothly, students can focus on our instruction. When we have a supportive relationship with students, they tend to be more willing to take chances. This sense of safety promotes learning.

Supporting Research

Many researchers have described the benefits of building a positive relationship with students. Students may feel that they can work on challenging tasks without being punished or criticized (Artzt, Armour-Thomas, & Curcio, 2008, p. 14). Positive teacher–student relationships can improve students' self-esteem (Gallagher, 2013, p. 14) and motivation (Farmer, 2018, p. 20). When teachers acknowledge students' needs and differences, students tend to be more engaged in learning (Rimm-Kaufman & Sandilos, n.d.). Students who believe that their teachers have high expectations for them are more likely to meet those expectations, which can result in higher academic achievement (Gallagher, 2013, p. 13).

Running an efficient classroom has many benefits. When teachers spend less time on administrative tasks in class, students are more likely to be focused on learning (Artzt et al., 2008, p. 15). Procedures and routines help students feel secure in the classroom, which, in turn, enables them to focus on content (Wong & Wong, 2004, pp. 170–171).

Common Core Connections

Effective classroom routines and positive teacher–student relationships can create an environment where students are supported to meet both the Standards for Mathematical Practice and the content-based standards.

Application

In our experience, cultivating a good relationship with our students and developing an effective classroom environment can't be done in 1 day. Many of the strategies we outline here take time to implement successfully. When done effectively, these strategies can help develop confidence, fairness, competence, and organization among students and ourselves.

MAKING A GOOD FIRST IMPRESSION

Researchers have found that people can form an impression of someone in as quickly as one-tenth of a second (Willis & Todorov, 2006, p. 592). Such factors as a person's clothing style, posture, or handshake can shape a person's impression (Rowh, 2012, p. 32). We find that the image that we project on the first day of school can set the tone for the rest of the year. Before the school year starts, we spend time reflecting on how we want students to perceive us. Here are some factors that we consider.

Greeting Students

We think carefully about how we greet students both on the first day and beyond. Welcoming students as they enter the room can make our class feel like a more inviting place. Raising our hand, making eye contact with students around the room, or clapping rhythmically can bring the class to order in a nonthreatening way. Starting our class with a smile and a friendly greeting like "Good morning!" can also show students that we care. Research shows that engaging students in positive dialogue (such as a greeting by name, handshake, or fist bump) results in less disruption and more academic engagement (Cook et al., 2018, pp. 152–154).

While making personal contact with students is important, we also try to be mindful of their beliefs and identities. Some students may feel uncomfortable shaking our hand for religious or cultural reasons. Others may feel excluded if we make assumptions about their gender. We recommend being as honest, accommodating, and considerate as possible. At the same time, we also find that regularly making an individual connection with students can help establish a positive tone in our classrooms.

Introductory Activity

We usually have an introductory activity (often called a Do Now or Warm-Up) ready for students when they walk into the classroom. As we said in Chapter 7: How to Plan Lessons, this activity sets the tone for mathematical learning and prepares students for the lesson. Having an introductory assignment on the first day (as well as every day after that!) also keeps students focused while we take care of quick administrative tasks. Having a Do Now on the board every day makes our class more consistent and predictable.

For the first day of school, we often use a "low floor, high ceiling" Do Now activity that helps alleviate "math anxiety" by enabling students to experience mathematical success on the first day (we define *math anxiety* in the Introduction of this book). We list some sample Do Now problems for the first day of school in the First-Day section of Chapter 4: Promoting Mathematical Communication.

Clothing

One of the most important factors that we consider is how we dress—not just on the first day but throughout the school year. What teachers wear can affect how students perceive them. Professional dress can enhance teachers' credibility and acceptance (Wong & Wong, 2004, p. 55). Teachers who dress more professionally may improve student learning (Sampson, 2016, p. 63) and be perceived as more organized and knowledgeable (Phillips & Smith, 1992, p. 20). Teachers who wear more casual attire may be perceived as friendlier and better able to motivate students (Phillips & Smith, 1992, p. 18). However, sloppy or excessively casual clothing may suggest carelessness or apathy toward students or teaching (Wong & Wong, 2004, p. 59).

We don't recommend a particular style of dress because we feel this should reflect each teacher's individual personality. (We recognize that the more formal dress that we prefer is not for everyone!) Instead, we suggest that you consider what image you want to project and how your students might perceive you, and then dress appropriately.

LEARNING NAMES

Learning how to pronounce students' names correctly sends a powerful message that we respect their individual and cultural identity (as people with unusual first names, we are particularly sensitive to this). Many names have personal significance to students and their families. Taking the time to say students' names correctly can be especially helpful for English Language Learners (ELLs) by making them feel welcome in their new school (Mitchell, 2016).

In contrast, getting students' names wrong (such as misspelling Larisa as "Larissa") or giving them nicknames without their consent (like calling Bobson "Bob") can show a lack of respect for their culture, erode their self-esteem, and negatively affect their social-emotional learning and academic performance (McLaughlin, 2016). Suggesting that a name is hard to say can be problematic since it implies that students are to blame for the teacher's unfamiliarity with the student's culture (Mitchell, 2016).

On the first day of school, we introduce ourselves to each student individually as they are working on our first-day activity and ask them to pronounce their name for us. We repeat the pronunciation as best we can, ask them to correct it, and write a phonetic pronunciation of the name in our seating chart if necessary.

To help students learn each other's names, we ask them to create name cards by folding a piece of paper into thirds to create a triangular prism, writing their names in large letters on two of the faces, and putting it on their desks for the first few days of class. Students can then see their classmates' names no matter where they sit in the room. We also make a point of saying students' names whenever they speak in class.

GETTING TO KNOW STUDENTS

On the first day of school, we ask students to tell us more about themselves. One simple way to get to know students better is to ask them to provide their contact information on blank index cards or forms like the one in Figure 11.1: Student Information Sheet.

In addition to getting basic contact information from students, we also like to ask questions that help us understand them better, such as the following:

- What about this class makes you *excited*? Explain.
- What about this class makes you *anxious*? Explain.
- Describe a time where you experienced success in a math class.
- Describe a time where you experienced frustration in a math class.

We like to select one or two of these questions and ask students to answer them in class on the first day. Teachers could also have students turn to a neighbor and share some of their responses as a way for them to get acquainted. (This works best for questions that we think students would feel comfortable discussing with strangers. Students may not want to share their fears and anxieties on the first day!) Teachers could also assign some of these questions for homework in the first few days of school.

We read through student responses to help us customize our instruction to meet their needs. We look for opportunities to personalize instruction for individual students as well as trends in the class. Here are some examples:

- If students express frustration with fractions, we may try to include additional support or alternate methods of instruction in lessons that use fractions.
- If students say they excelled at solving equations, we may incorporate more practice with equation-solving in our lessons to boost confidence.
- If students are ELLs, we may try to pair them with students who are more proficient in English but also speak the same language.

Even if we get students' contact information from our school, we ask students to provide it anyway because we find that the school's information is often inaccurate. We encourage students to provide the email address and mobile phone number of their parent or guardian so that we can communicate with them directly. We ask what language is spoken at home so we know what kind of assistance we may need when connecting with parents (we discuss communicating with them more in Chapter 12: Building Relationships with Parents).

CLASSROOM ORGANIZATION

A classroom's physical setting is an important element of our classroom environment. Here are some physical aspects of our room that we take into consideration.

Seating Arrangements

Teachers can arrange desks in different ways to match their instructional goals. Putting desks into rows might reduce off-task behavior but can make collaborative work more challenging (Simmons, Carpenter, Crenshaw, & Hinton, 2015, p. 62). Desks arranged in a circular pattern facing the front or center of the room can provide more intimacy and increase student contact (Falout, 2014, p. 183) but may not be appropriate if the class does a lot of group work.

We vary the seating arrangement in our classrooms depending on the lesson. Most of the time, we arrange our desks into pairs so that students have a partner but don't have as many opportunities to get off-task. When our students work in groups, we ask them to put two pairs of desks together to form a group of four. We arrange desks into rows only when we want them to work independently (such as when they take tests and quizzes) or if we want to reduce the possibility that students can get distracted. If necessary, teachers can have students practice moving their desks into different configurations as quickly and quietly as possible.

In addition, your school's building culture or available furniture also affects possible seating arrangements. One-piece school desks with attached chairs and slanted desks are difficult to arrange into pairs or groups. Desks whose tops are shaped like isosceles trapezoids give both students and teachers a great deal of flexibility in arranging desks into groups (Harvey & Kenyon, 2013, p. 9). Some schools may encourage or require a particular seating arrangement. Teachers who share a room may have limited options depending on their colleagues' preferences. As with the other strategies in this book, we recommend being flexible over the course of the school year.

Wall Space

Decorations on the wall, such as inspirational quotes or posters of different mathematicians from around the world, can make our room more culturally responsive. Highlighting mathematicians who are female, non-European, or active can help students identify with mathematics and make them feel that they can be a part of the field (Flores & Kimpton, 2012, p. 37). (See Chapter 2: Culturally Responsive Teaching for more ideas.)

However, a cluttered room can distract students from learning, so we try to minimize decorations to well-defined spaces while leaving space for student work, word walls, or anchor charts (we discuss word walls and anchor charts in Chapter 3: Teaching Math as a Language).

CLASSROOM RULES AND ROUTINES

To run a well-organized classroom, teachers need to establish the following:

- **Rules:** expectations of how students should behave
- **Procedures:** expectations of what teachers want done, such as what to do when students walk into the classroom, submit homework, or have a question
- **Routines:** what students do automatically (Wong & Wong, 2004, pp. 143, 170)

Rules

We find that classroom rules work best when they are short and simple so that they can be posted in the room for easy reference. Having no more than five rules also helps to make them easy to manage and understand. Good ideas for classroom rules can be found by consulting school and district policies, colleagues, and the internet (we list some online resources in the Technology Connections section of this chapter).

Some of our favorite rules include:

- Respect yourself and each other.
- Think before you speak or act.
- Listen to each other.
- Keep trying or use a different approach.

Since we like to make our students feel more comfortable about math, we don't overwhelm them on the first day of school with a long discussion about rules. We prefer to discuss rules as they come up in class. If students realize that their method of solving a problem isn't working and try another way, we may praise them for following our "Keep trying or use a different approach" rule.

Experts disagree on whether students should be involved in setting up classroom rules. Some encourage teachers to co-create rules with students, arguing that students who feel that they have more control over the classroom are more motivated to learn (Wilson, 2018, p. 128). Teachers could work with students on the first day of school to put rules in their language and mutually decide the consequences for breaking those rules (Shalaway, n.d.). Other educators believe that allowing students to decide class rules undermines teachers' authority. Students may incorrectly think that teachers lack the confidence or experience to know what is best for students (Linsin, 2014; Wong & Wong, 2004, p. 148).

In general, we find that allowing students to negotiate class rules uses valuable time on the first day that we would rather use to address other issues, such as boosting students' confidence to do math. Research indicates that allowing students to make decisions about their learning fosters their autonomy and improves their motivation (Reeve & Halusic, 2009, p. 147; Ryan & Deci, 2000, p. 58). However, we prefer to promote autonomy in other ways, such as allowing students to choose their own seats (although we reserve the right to change seats if necessary), to select an appropriate starting level for classwork, or to pick partners for group work.

Procedures and Routines

Establishing good classroom procedures and routines helps students focus on learning, makes classrooms more predictable, and reduces discipline problems (Wong & Wong, 2004, p. 170). Students with learning differences can especially benefit from predictability in the classroom (Brain Parade, 2015). Here are some of our most useful procedures and routines.

At the beginning of class, we have a Do Now or other introductory activity on the board so students can get to work immediately after entering the classroom. We have pencils and calculators on our desk if students need to borrow supplies for the lesson.

If students have a question or want to make a comment to the entire class, we ask that students raise their hands. This common technique prevents students from calling out and allows us to give enough time for students to think. (We talk more about wait time in Chapter 7: How to Plan Lessons and Chapter 4: Promoting Mathematical Communication.) However, we don't monitor learning simply by calling on students who raise their hands. We use other techniques, such as random selection (we discuss these options more in Chapter 18: Formative Assessment). If students want to use the bathroom, we ask that they use a different signal (such as holding up two fingers) so that we can silently respond to their request with a nod without interrupting the lesson or classwork. Of course, just because we ask that they use this signal doesn't mean they always do!

We follow Harry and Rosemary Wong's (2004, p. 174) three-step method to teach classroom procedures:

1. Explain (and model, if necessary) the procedure.
2. Rehearse the procedure under our supervision.
3. Reinforce the procedure by practicing (and reteaching, if necessary) the procedure.

COURSE DESCRIPTIONS

In the first few days of school, we often distribute a course description that introduces our class to both parents and students. Teachers can distribute these descriptions by giving them to students in class or mailing them home. We try to keep course descriptions as short as possible since we prefer not to overwhelm students with too much paperwork in the first few days of school. Our course descriptions (like the one in Figure 11.2: Course Description) typically include a list of supplies, expectations, and grading policy. If space permits, class rules can be included. Teachers can ask students and parents to acknowledge that they have read and understood the course descriptions by signing them, making them similar to class contracts. We discuss other ways to work with parents in Chapter 12: Building Relationships with Parents.

In our experience, sending course descriptions home has less value today than when we first started teaching since many students and parents expect to find

CHAPTER 11: BUILDING A PRODUCTIVE CLASSROOM ENVIRONMENT

information online. Nowadays, we find that posting basic class information on our class webpage is often more effective than distributing a piece of paper. We believe that these descriptions can work better for students or parents who are unfamiliar with the expectations for our course (such as a freshman class) or don't have reliable internet access. Written course descriptions can also be effective for ELLs and their parents if we can get the information translated. In addition, some schools may prefer that teachers send some type of introductory letter home. We recommend seeking suggestions from your colleagues and administrators.

We find that what matters most is providing basic information about our class at the beginning of the year in a format that students and parents can easily access. This information can be put online, on paper, or both.

SOLICITING STUDENT OPINION

When possible, we often ask students to periodically fill out evaluations to help us improve our teaching and classroom environment. Researchers have found mixed results on the usefulness of student surveys as summative assessments. Some find a positive association between student surveys and classroom observations (van der Lans, 2018, p. 359; White, 2013, p. 75), some find no association between student surveys and growth (White, 2013, p. 76), and others were unable to reach a clear conclusion (Bacher-Hicks, Chin, Kane, & Staiger, 2017, p. 27). We find that student surveys work best as formative assessments throughout the year so that we can adjust instruction for students based directly on their feedback.

If possible, we suggest keeping student surveys anonymous. Anonymous evaluations can ensure honest feedback by reassuring students that teachers won't use negative comments against them (Bill and Melinda Gates Foundation, 2012, p. 12). (We list some ways to run student surveys online in the Technology Connections section of this chapter.)

We try to keep our student surveys short and simple. We assign them as homework at the end of each marking period so that students have time to reflect individually. If time permits, student surveys could also be done in class.

When making student surveys, we prefer to gather qualitative instead of quantitative data. We find that students usually provide more meaningful and specific feedback on open-ended survey questions than on multiple-choice ones, which can make the survey feel more like a class test.

On student surveys, we like to ask no more than four questions, such as:

- Describe a time this marking period that you were successful. Why do you think you were successful?
- Describe a time this marking period that you were not successful. Why do you think you didn't do well?

- What do you think you should do differently next marking period? Explain. (Be specific.)
- What do you think I should do differently next marking period? Explain. (Be specific.)

To help students with learning differences or ELLs answer these questions, teachers may consider sharing a model response, paragraph outline, or sentence starter. For example, here is a possible outline for an answer to the question "Describe a time this marking period that you were successful":

TOPIC SENTENCE: I was successful this marking period when I _____.

- State details of your success (such as who was involved, what grade you got, when it happened).
- Explain why you think it happened (such as what you did differently this time compared to similar situations).

CONCLUSION: What do you think you could do to repeat this success in the future?

Of course, teachers don't have to limit these supports to ELLs or students with learning differences—all students may benefit from having access to them!

We look for general trends in individual responses. For example, after many of Bobson's students expressed concern in an end-of-semester survey that the lessons moved too quickly, he added more time for review in the following semester and paused longer between classwork problems.

TAKING NOTES

One routine that we try to establish at the beginning of the year is effective note-taking. We find that students who take good notes have several advantages:

- Notes provide students with a written record of what happened in class (Pauk, 2001, p. 235).
- Research indicates that people who don't take notes can forget most of the information within 2 weeks (Walmsley & Hickman, 2006, p. 615).
- As we said in Chapter 4: Promoting Mathematical Communication, when students write about math, they can improve their understanding (Curcio & Artzt, 2007, p. 262).
- We find that students can feel proud of well-organized notes, which often boost their self-confidence and motivation in class.

Here are some strategies that we use to help students take better notes and use them to study effectively.

Format of Student Notebooks and Handouts

In our classes, we encourage students to use a separate graph paper notebook for math. In our experience, maintaining a separate notebook makes their notes more organized and helps them more easily look up information from previous lessons. We prefer that students use graph paper because we find that it tends to improve the accuracy of their graphs and diagrams.

To make notebooks into effective study tools, students need to have questions or problems from the lesson in their notes. One option is for students to copy problems by hand. Research indicates that students may process information more deeply if they write in longhand (Mueller & Oppenheimer, 2014, p. 1166). Students with learning differences can especially benefit from getting better at writing by hand, but spending too much time on handwriting drains mental resources needed for higher-level aspects of writing, such as elaborating details or organizing ideas (Spear-Swerling, 2006).

Waiting for students to copy problems can take up a great deal of class time. Students with poor note-taking or handwriting skills may struggle to write legibly or spell correctly, eroding their motivation and distracting them from learning (Igo, Bruning, & Riccomini, 2009, p. 2; Igo, Riccomini, Bruning, & Pope, 2006, p. 98).

We strike a balance between fostering note-taking skills and minimizing student frustration by providing handouts with the date, aim or learning objective, and relevant information that we plan to discuss in the lesson. Students can tape, glue, or staple these handouts into their notebooks so that they have a chronological record of what we do in class. If the lesson's problems are short or if we don't have time to make a sheet, we ask students to copy the questions into their notebooks. We sometimes ask students to draw diagrams related to the problem if they are relatively simple to draw. Illustrating a concept can help students understand and process information (Fernandes, Wammes, & Meade, 2018, pp. 304–305). We choose this option if we feel that students won't get too frustrated drawing it, if they are expected to create a similar diagram on their own, or if we don't have the time to create the diagram ourselves.

To save paper, we sometimes make handouts that students can attach into their notebooks and show their work underneath, as shown in Figure 11.3: Brief Handout. To keep their work organized, students can divide their paper into four or six roughly equal boxes (which they can do more easily if they use graph paper) so that they can write the work for each problem in a box.

We make full-page handouts (like the one shown in Figure 11.4: Full-Page Handout) if we teach lessons involving concepts like graphing (so students don't have to draw a new set of coordinate axes for each problem), word problems (so they can highlight text), and complicated diagrams (so students can mark them up). We find that students who have trouble organizing their notes often benefit from receiving these handouts every day. We get paper with prepunched holes or punch holes in paper ourselves so that students can keep their handouts in a separate looseleaf binder.

Research about the benefits of electronic note-taking is inconclusive. Some researchers have found that students who write longhand notes do better on tests (Mueller & Oppenheimer, 2014, p. 1166) and are less likely to get distracted than students who use computers (Holstead, 2015). Other studies indicate that students who took notes on computer performed better (Fiorella & Mayer, 2017, p. 28; Morehead, Dunlosky, & Rawson, 2019, p. 20). In addition, students increasingly use digital technology for academic and personal use (Freeman, Higgins, & Horney, 2016, p. 283). Certain groups, such as students with learning differences, may benefit from using digital note-taking tools (Freeman et al., 2016, p. 306).

In our experience, neither handwritten nor digital notes have a clear advantage in a math class. Although many of our students feel more comfortable typing than writing longhand, we find that they struggle to type mathematical symbols and draw diagrams on a computer, both of which require different software with their own commands and interfaces. Also, since our state assessments are paper-based, our students need to be comfortable writing mathematics by hand. Teachers of students who take computer-based assessments may consider encouraging students to take notes electronically. Research indicates that students who take notes and assessments using the same method (by hand or computer) outperform students who use different methods (Barrett et al., 2014, p. 54). For now, we prefer handwritten notes, but our opinion may change as new technology develops in the future.

Fostering Good Note-taking Strategies

We often model on the board how to write the solutions to problems so that students learn how to structure their work. For procedures such as simplifying expressions or solving equations, we write each step of the solution on a separate line of text. We also encourage students to annotate their solutions with short explanatory comments. We model these comments ourselves by writing them on the board in a different color so students can distinguish them from the rest of their work. Figure 11.5: Annotated Work shows an example for simplifying radicals.

When someone makes a noteworthy observation in class, we gently remind students to put it in their notebooks by saying something like, "That was a great point! Let's write it down." As students are writing, we look around the room to see if they need more time. We may also pause to allow students enough time to think about what was said. To help ELLs, we make sure that important or complicated words are spelled out and printed on the board. If necessary, we ask students to repeat the comment or ask students to discuss it with a partner (using the Think-Pair-Share strategy that we described in Chapter 17: Cooperative Learning) so they can put it into their own words if possible.

As we teach a lesson, we keep our own record of class notes and post it online for students to use as a baseline to revise their notes. When we use a computer projector in our room, we save our presentation file, which includes the problems that we covered in class and their complete solutions. (Teachers who don't have an interactive projector can take pictures of board work and post them online.) We find that when we regularly post notes online, students often refine their notebooks by writing down material that they missed in class. Posting notes online also helps students who were absent from class, as well as ELLs and students with learning differences, who may want additional time to reflect on the lesson. Teachers can also print out notes for students to use as a reference in class or for students who don't have internet access at home.

Cornell Note-taking Method

To improve students' ability to organize their notes and reflect on lessons, teachers can use the Cornell system (sometimes called a *double-entry journal*), developed by Cornell University professor Walter Pauk over 50 years ago. Students divide their notebooks into two columns (a "cue column" about 2 inches long and a wider column called the "notes column"). In the cue column, students write questions based on the notes or record their thoughts. Students can rephrase theorems in their own words, explain steps in solving a problem, or add a diagram to clarify a definition. In the notes column, students record important points from the lesson, such as a definition, theorem, or word problem. Students can study by covering the notes column, answering the questions in the cue column, and reflecting on the content (Pauk, 2001, pp. 236–237).

Double-entry journals may be particularly helpful for ELLs, who can write in their native language to bridge the gap between a math problem and an explanation in English (Zhao & Lapuk, 2019, p. 291). They may also write explanations of relevant mathematical symbols in their own language.

Figure 11.6: Double-Entry Journal shows an example of a double-entry journal template for a Geometry class.

What Could Go Wrong

CLASSROOM TONE

We find that setting the wrong tone at the beginning of the school year can lead to future problems.

Sometimes, we've been too lenient with students by doing things like giving them work that is too easy, overlooking disruptive behavior, or giving students too much freedom. In our experience, not setting reasonable boundaries for students can create a chaotic atmosphere that disrupts learning.

At other times, we've been too strict by micromanaging student behavior, not giving students enough time to think after someone asks a question, banning students from using the bathroom, or not showing any part of our personality. Some teachers have even warned us not to smile until December because it could undermine our authority! We find that being too strict can discourage students from learning and give the impression that we are more concerned about maintaining order than teaching math and getting to know them as individuals.

The best advice that anyone has ever given us about teaching came from a veteran teacher who told us, "Act like you enjoy what you're doing." We find that showing enthusiasm for the material that we teach helps keep students more engaged and focused. Smiling often can put students more at ease and help them focus on their work. People who smile appear more sincere and sociable (Reis et al., 1990, p. 265)—characteristics that teachers need to build trust with students and foster learning.

While we try to be as energetic about our work as possible, we also recognize that we—like students or anyone else—sometimes have days when we're not at our best. At these times, we admit to our students—on that day or the next one—that we may not be as positive as we'd like to be but that we still have a job to do, which helps model resilience and humility.

MISHANDLING THE TEACHER–STUDENT RELATIONSHIP

Sometimes, we don't develop effective relationships with students. One way this can happen is when teachers maintain too much distance from students (which may lead them to conclude that we don't care about them or their learning). At other times, we may maintain too little distance from them. Spending too much class time talking about our and their personal lives can distract from math lessons. The key is maintaining a balance between the two.

Research indicates that teachers who develop a rapport with students improve class participation and learning (Frisby & Martin, 2010, pp. 155-156). We find that occasionally mentioning a few noteworthy details of our lives (such as our family or favorite hobbies)—without getting sidetracked from our lesson—helps students see us not just as sources of mathematical knowledge but also as individuals. Providing personal details can strengthen relationships (Smith, 2015, p. 36). In addition, we try to pay attention to how students are behaving in class. If they seem unusually quiet or distracted, we may speak to them privately and see if we can offer any help.

CHAPTER 11: BUILDING A PRODUCTIVE CLASSROOM ENVIRONMENT

In short, we find that having a friendly but professional relationship with students allows us to maintain our authority while supporting student motivation. Focusing on being consistent, fair, reasonable, and human at the beginning of the year helps make our classrooms be more productive and fun for the rest of the year!

TAKING NOTES

Many times, despite our best efforts, we have students who don't take useful notes. Often, they write down some of what we do in class because they have been trained in school to copy what is on the board, but they rarely use them to study. In these situations, we show students the value of good notes by frequently asking them to recall information from previous lessons. If we teach a lesson on proofs involving squares and rectangles, we ask students to refer to earlier lessons so that they can recall the properties of those quadrilaterals. Students can learn that referring to their notes is a tool of empowerment, not a sign of weakness (Wieman, 2011, p. 407). We model the process of recalling previously taught information by going through our lesson plans as students look up information. We find that showing students that we don't always remember everything helps humanize both us and the math we teach.

Student Handouts and Examples

Figure 11.1: Student Information Sheet

Figure 11.2: Course Description

Figure 11.3: Brief Handout

Figure 11.4: Full-Page Handout

Figure 11.5: Annotated Work

Figure 11.6: Double-Entry Journal

Technology Connections

CLASSROOM ENVIRONMENT

The Mathematicians Project has a list of diverse mathematicians online (http://awesome-table.com/-Kq4eNy0oVl-JUj1pK7I/view). Users can search by gender, ethnicity, year of birth or death, or mathematical field. The Women You Should Know project has posters of female STEM role models that can be downloaded for free (http://womenyoushouldknow.net/downloadable-stem-role-models-posters).

Dr. Harry K. Wong and Dr. Rosemary T. Wong, authors of the best-selling book *The First Days of School*, have resources related to establishing good classroom rules, routines, and procedures on their website (http://harrywong.com).

The ThoughtCo (http://thoughtco.com), Edutopia (www.edutopia.com), Yale University's Poorvu Center for Teaching and Learning (http://poorvucenter.yale.edu/FacultyResources/Managing-the-Classroom), and Scholastic (www.scholastic.com) have many useful resources on classroom management, including sample rules and routines. The articles can be found by using the search function on the sites.

In addition, Larry Ferlazzo's EdWeek blog (http://blogs.edweek.org/teachers/classroom_qa_with_larry_ferlazzo/2014/08/q_a_collections_best_ways_to_begin_end_the_school_year.html) has a collection of educator responses on the best ways to begin and end the school year.

STUDENT SURVEYS

Google Forms (http://forms.google.com) and SurveyMonkey (http://surveymonkey.com) enable teachers to run free anonymous student surveys.

NOTE-TAKING

Cornell University's Learning Strategies Center (http://lsc.cornell.edu/notes.html) has a description and sample template of the Cornell Note-taking System. WikiHow also has an article (https://www.wikihow.com/Take-Cornell-Notes) that describes the Cornell system.

Many web-based note-taking apps, such as OneNote (www.onenote.com), Evernote (www.evernote.com), and Zoho Notebook (http://www.zoho.com/notebook) work across several platforms (such as Windows and Mac computers, as well as Android and iOS devices). They also can include audio notes, video clips, and other files. Many of these apps are free, but some have features that are only accessible with a paid subscription. A comparison of online note-taking tools can be found at the Zapier website (https://zapier.com/blog/best-note-taking-apps).

Figures

PERSONAL INFORMATION

Math class code: _____ Period: _____

First: _____ Last: _____

Preferred Pronoun (e.g. he/she/they): _____

Guidance counselor: _____ Official class: _____ ID #: _____

Street address: _____

City: _____ Zip: _____

Parent/Guardian name: _____ Mobile phone: _____

Parent/Guardian email: _____ Other phone: _____

address: _____

Languages you speak:	Level of fluency in each language:	Languages your parents speak:	Level of fluency in each language:
1. _____	☐ High ☐ Medium ☐ Low	1. _____	☐ High ☐ Medium ☐ Low
2. _____	☐ High ☐ Medium ☐ Low	2. _____	☐ High ☐ Medium ☐ Low
3. _____	☐ High ☐ Medium ☐ Low	3. _____	☐ High ☐ Medium ☐ Low

ABOUT YOUR MATH COURSE LAST SEMESTER:

School you attended: Math course you took: Math teacher's name:

_____ _____ _____

Figure 11.1 Student Information Sheet

OTHER INFORMATION

What are you most excited about when you think about this class? Explain.

_____

_____

_____

What are you most afraid of when you think about this class? Explain.

_____

_____

_____

Where do you see yourself in five years? (If in college, explain what college you might be at and what your major is. If working, state what job you would like to have and why.)

_____

_____

_____

What else about yourself would you like to share?

_____

_____

_____

Figure 11.1 (Continued)

Bayside High School: Mathematics Department (2020–21)

INTRODUCTION TO MR. WONG'S ALGEBRA II CLASSES

WHAT IS ALGEBRA II?

Algebra II is an extension of the Algebra I course you took before. The course is divided into four main topics: algebra, exponential and logarithmic functions, trigonometry, and probability and statistics.

WHAT SHOULD I BRING TO CLASS EVERY DAY?

- Graphing calculator (See the list of recommended graphing calculators on the class website.)
- One graph paper notebook (marble or spiral, preferably 8.5" × 11")
- Homework or any other assignment that is due in class that day
- Folder: for tests, quizzes, review sheets, and homework
- Transparent tape, stapler, or glue (for attaching classwork sheets to your notebook)
- Recommended: Colored pens or pencils (any colors are fine), highlighter

HOW DO I KNOW WHAT ASSIGNMENTS ARE GIVEN AND WHEN THEY ARE DUE?

All assignments are posted on Google Classroom. To log into Google Classroom (http://classroom.google.com) or your Bayside HS email (http://gmail.com), use the following:

USERNAME: First letter of first name + Last name + Last four digits of student ID number @ baysidehighschool.org

PASSWORD: Nine-digit student ID number

Example: The email of Leonardo Fibonacci, whose student ID number is 1123581321, is lfibonacci1321@baysidehighschool.org .

HOW IS MY GRADE CALCULATED?

- Your grade is calculated according to Bayside High School's uniform grading policy.

Figure 11.2 Course Description

HOMEWORK:

- Homework is given every day except the day before and the day of a full-period test.
- Homework is usually a PDF file with 4–6 questions. We discuss and review homework at the beginning of class. It is collected and graded every day using the following rubric:

0%	50%	75%	95%	100%
Work is incoherent OR has no name.	About half of the work is missing or incorrect.	Some work is missing or incorrect OR questions are not written out in full (if handwritten).	All problems are correct or corrected, with one minor mistake.	All problems are complete and correct or corrected.

PARTICIPATION:

- Participation is measured by the quality and quantity of work done in class.
- The class notebook must be a graph paper notebook.

All grades are posted on Pupilpath, the online grade reporting system that Bayside High School uses. See the Bayside HS website for details on how to log in.

WHAT EXTRA CREDIT IS AVAILABLE?

- Extra credit assignments will be announced in class and posted on Google Classroom.
- You may put up the answer to a homework question and explain it to the class when we review homework. You will get one extra credit test point each time you put up a question, whether your work is correct or not (if it's incorrect, you must fix it as we discuss the question).

WHERE ELSE CAN I GET HELP?

- Our Google Classroom page has a list of websites and review books. See our Google Classroom webpage for details.
- Mr. Wong is available for help during periods 2 and 8 in the Math Office (room 204A)—no appointment needed.
- Get free tutoring from math teachers or Arista students during your lunch period. No appointment is necessary.

HOW CAN I CONTACT MR. WONG?

- Email is the best way to reach Mr. Wong.
- Please give your parent's or guardian's mobile phone number to the school so that you can correspond with Mr. Wong via text message.

Figure 11.2 (Continued)

ALGEBRA I DATE: _____

AIM #32: How do we use the quadratic formula to solve quadratic equations?

Solve for the variable in the following equations. Express the answer in simplest radical form. Show all work in your graph paper notebook. Attach this sheet to the front of your work.

QUADRATIC FORMULA: If $ax^2 + bx + c = 0$, then $x = \dfrac{-b \pm \sqrt{b^2 - 4ac}}{2a}$.

LEVEL 1

1. $x^2 + 9x + 5 = 0$
2. $x^2 + 7x + 11 = 0$
3. $2x^2 + 11x + 6 = 0$

LEVEL 2

4. $m^2 - 7m + 8 = 0$
5. $x^2 - 9x + 6 = 0$
6. $2q^2 + 3q - 7 = 0$

LEVEL 3

7. $r^2 + 10r + 21 = 0$
8. $n^2 + 6n + 9 = 0$
9. $2m^2 + 8m + 3 = 0$

LEVEL 4

10. $k^2 - 4k - 8 = 0$
11. $2x^2 + 4x - 7 = 0$
12. $3h^2 - 6h - 5 = 0$

Figure 11.3 Brief Handout

GEOMETRY (TERM 2) _____ **DATE:** _____

AIM #15: How do we solve problems involving the slopes of parallel lines?

DO NOW

The following diagram shows points K(3, 2), C(7, 3), H(2, 6), and D(6, 7).

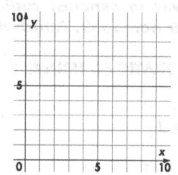

a. Graph points K, C, H, and D and lines $\overleftrightarrow{HD}$ and $\overleftrightarrow{KC}$.

b. What appears to be the relationship between $\overleftrightarrow{HD}$ and $\overleftrightarrow{KC}$?

c. Find the slopes of $\overleftrightarrow{HD}$ and $\overleftrightarrow{KC}$.

d. Your work above illustrates the following theorems:

   If two lines have the same _____ , then they are _____ .
   If two lines are _____ , then they have the same _____ .

LEVEL 1

Find the slope of a line parallel to the graph of each equation.

1. $y = 4x + 9$
2. $y = -3x - 6$
3. $y = 7$

4. $x = 1$
5. $5x + 6y = 11$
6. $2x - 5y = 8$

LEVEL 2

7. Determine if the line whose equation is $9x - 2y = 14$ is parallel to the line whose equation is $y = \frac{9}{2}x - 5$.

8. Determine if the line whose equation is $3x - 4y = 7$ is parallel to the line whose equation is $-6x + 8y = 14$.

Figure 11.4 Full-Page Handout

LEVEL 3

9. If the line whose equation is $3x + 2y = 8$ is parallel to the line whose equation is $y = ax - 5$, then find the value of a.

10. If the line whose equation is $2x + 7y = 15$ is parallel to the line whose equation is $y = ax + 15$, then find the value of a.

LEVEL 4

11. Write an equation in slope–intercept form of the line that passes through the point $(1, 3)$ and is parallel to the line $y = 2x - 1$.

12. Write an equation in slope–intercept form of the line that passes through the point $(-2, -1)$ and is parallel to the line $-6x + 3y = -3$.

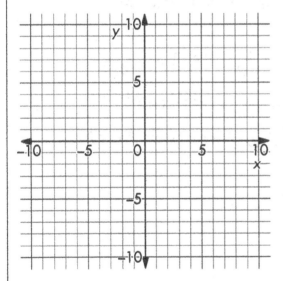
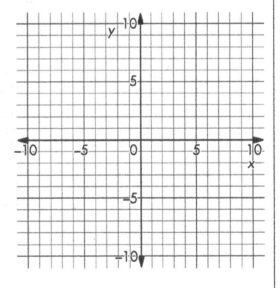

Figure 11.4 (Continued)

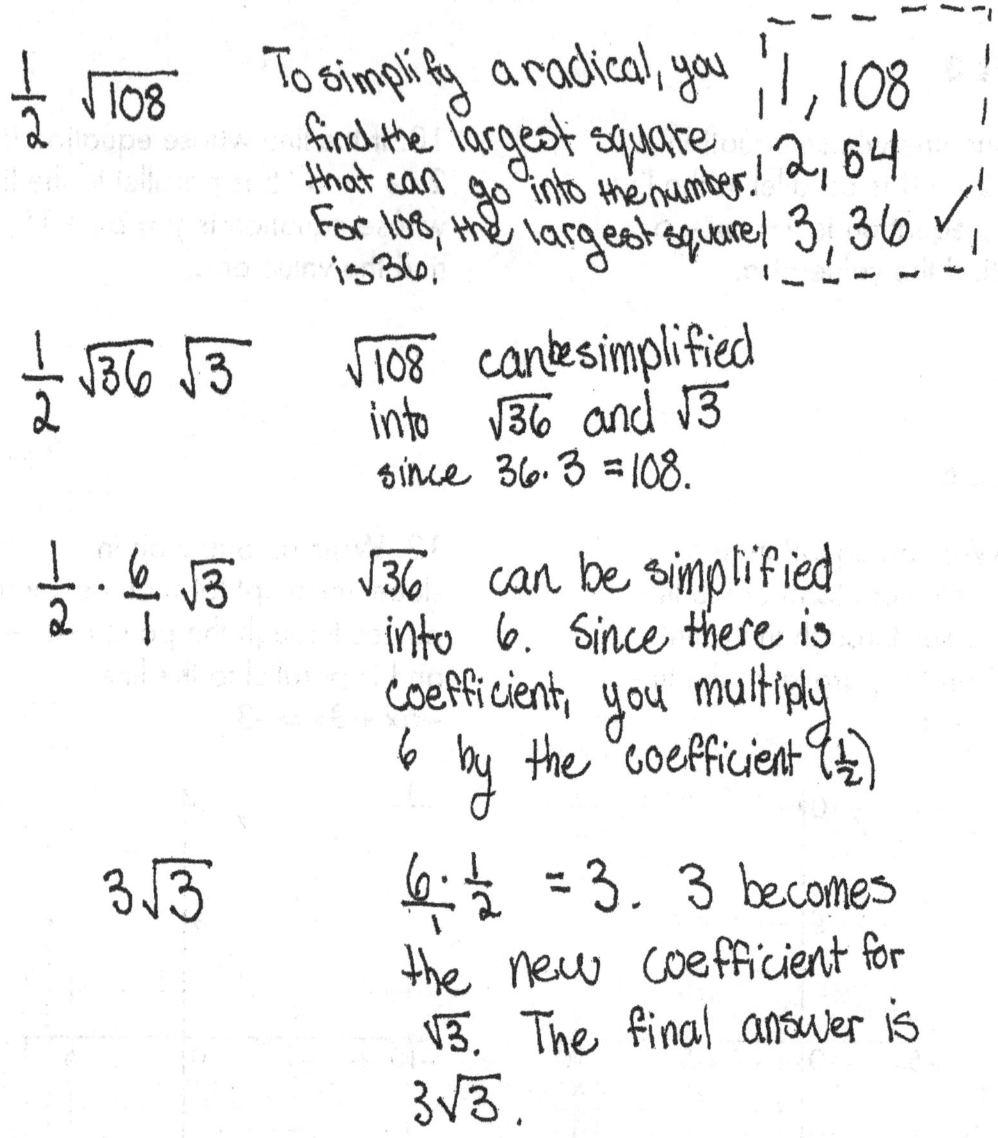

$\frac{1}{2}\sqrt{108}$ To simplify a radical, you find the largest square that can go into the number. For 108, the largest square is 36.

| 1, 108 |
| 2, 64 |
| 3, 36 ✓ |

$\frac{1}{2}\sqrt{36}\sqrt{3}$ $\sqrt{108}$ can be simplified into $\sqrt{36}$ and $\sqrt{3}$ since $36 \cdot 3 = 108$.

$\frac{1}{2} \cdot \frac{6}{1}\sqrt{3}$ $\sqrt{36}$ can be simplified into 6. Since there is coefficient, you multiply 6 by the coefficient ($\frac{1}{2}$)

$3\sqrt{3}$ $\frac{6}{1} \cdot \frac{1}{2} = 3$. 3 becomes the new coefficient for $\sqrt{3}$. The final answer is $3\sqrt{3}$.

Figure 11.5 Annotated Work

GEOMETRY   NAME: _____

AIM #42: How do we prove triangles congruent using ASA or SSS?

PROBLEM (Show all work in this column.)	YOUR COMMENTS (Record any tips, questions, or comments in this column.)
1. Given: $\angle M \cong \angle N$, P is the midpoint of $\overline{MN}$. Prove: $\triangle MOP \cong \triangle NOP$. [Triangle with vertex O at top, base M—P—N]	
2. Given: $\overline{AB} \perp \overline{BC}$, $\overline{DE} \perp \overline{EC}$, $\overline{AD}$ bisects $\overline{BCE}$. Prove: $\triangle ABC \cong \triangle DEC$. [Figure with points A, B, C, E, D]	

Figure 11.6 Double-Entry Journal

CHAPTER 12

Building Relationships with Parents

What Is It?

Many parents want to improve their children's mathematical education but face barriers that limit their ability to support their children's learning. Parents may work long hours, lack fluency in English, have incorrect or outdated information about their children's school, or feel uncomfortable about math. Building the right relationship with parents at the beginning of the school year can help make parents a valuable partner in the academic lives of their children.

The role of parents in their children's education can range from *involvement* (in which parents receive academic information and participate in meetings set and run by school personnel) to *engagement* (in which parents have more ownership and choice in their interaction and are considered partners in their children's academic lives) (Ferlazzo, 2009). In this chapter, we focus on ways to build a relationship with parents by fostering engagement, which we define to include all of the ways in which they influence children's overall actions. This includes not just the academic support that parents provide (such as monitoring their children's homework or getting tutors for them) but also the social and emotional support.

Since many of our students are raised or influenced by people who are not their biological parents, we use the term *parents* in this chapter to include relatives, guardians, and any other adults who help raise and support children.

Why We Like It

We find that building a productive relationship with parents improves the overall educational experience of our students. Parents feel validated when we recognize the many ways in which they support their children. This helps reduce both parents' and students' math anxiety, which in our experience is often a major impediment to students' academic success.

Supporting Research

Researchers differ on the exact relationship between parent involvement and student achievement (McNeal, 2014, p. 564). Many have found that increased parental engagement can boost children's self-esteem, motivation, and academic outcomes (Goodall & Montgomery, 2013, pp. 401–402). Students who believe that their parents value their education and have high expectations for them tend to be more confident and interested in academics (Fan & Williams, 2010, p. 69). A qualitative synthesis of meta-analyses found a strong positive relationship regardless of the definition of parent involvement, grade level, and ethnicity (Wilder, 2013, pp. 379, 392–393). However, if teachers reach out primarily to parents of students with academic or behavioral problems, increased communication between schools and parents may actually *decrease* student motivation (McNeal, 2012, p. 86; Sebastian, Moon, & Cunningham, 2016, p. 18). Since there is no universally accepted way to measure parental involvement, determining exactly how it affects student achievement is difficult (Fan & Chen, 2001, pp. 3–4).

Common Core Connections

Developing an effective relationship with parents can help students meet multiple standards.

Application

Here are some of the strategies that we use to improve parent engagement.

COMMUNICATING WITH PARENTS

At the beginning of the school year, we try to establish an ongoing dialogue with parents. We start by making sure that we have their correct contact information. In our experience, schools often don't have the most current contact information (including email address and mobile phone number) for parents. We ask students to provide this information on the first day of school when they fill out our student information sheets (we discuss them more in Chapter 11: Building a Productive Classroom Environment).

We reach out to as many parents as we can in the first few days of school. Of course, this depends on the available time and the number of students we have in our classes. If contacting every parent isn't feasible, we make sure to contact the parents of students who might require more attention (based on our initial assessments). We find that making an introductory phone call to these parents *before* any serious problems arise sets a better tone for the parent-teacher relationship throughout the year.

When possible, we prefer to make our first contact with parents by phone or in person. We find that in most cases, having a real-time initial conversation is more personal, which helps minimize the possibility of misunderstanding. We then stay in touch with parents using various methods:

- **Texts for short messages.** Examples would be reminders of upcoming assessments or simple questions for which parents can write a quick response.

- **Emails for more detailed messages.** We find that emails are also helpful for reaching out to parents of English Language Learners (ELLs) when we can't find a translator. In our experience, parents of ELLs can read through emails more carefully and get them translated if necessary. We also send emails if we have difficulty reaching parents during the day or if we need a written record of our parent outreach. Many parents prefer to use texting and email because smartphones allow them to communicate quickly and easily (Thompson, Mazer, & Flood Grady, 2015, p. 197; Zieger & Tan, 2012, p. 36).

- **Phone calls for conversations about more sensitive issues.** These might require a lengthier or more complex discussion.

- **In-person meetings for the most sensitive or difficult conversations.** In these situations, facial expressions and other nonverbal communication are especially helpful in conveying a particular message.

When we communicate with parents, we often use the outline in Table 12.1.

Figure 12.1: Parent Communication Script is a template that teachers can use to help them remember what to say. For more experienced teachers, Figure 12.2: Parent Communication Log is a shorter version of a parent-communication log.

ADDRESSING PARENTS' MATH ANXIETY

Many parents say to us, "I've always been bad at math," or, "I was never good at math in school." They often share a story about how much they struggled with math in school or how they can't help their child because they don't understand the material. The widespread use of Common Core–aligned math has also heightened parents' math anxiety since many of them don't understand the methods that their children are learning (Bay-Williams, Duffett, & Griffith, 2016, p. 5).

Table 12.1 Parent Communication Outline

Step	Examples
1. **INTRODUCTION**: We say who we are and how we're related to their child.	"Hi, I'm Mr. Wong, Joanna's Algebra I teacher."
2. **PURPOSE**: We explain why we're calling. Even if we're calling to describe a problem, we phrase our purpose in a nonconfrontational way.	"I wanted to give you an update on how she's doing in our math class." "I'd like to introduce myself and tell you a little bit about this course and how it's different from what he took before."
3. **PRAISE**: We always try to make a specific positive comment about the child. Sometimes, we also ask general questions about the student as a way to start the conversation in a less negative way.	"She asks a lot of questions in class and is always helping her classmates when she can." "What would you like him to accomplish this year in math?"
4. **CONCERNS**: If we have any concerns, we then describe them along with our expectations. We also explain how the child's behavior hinders learning.	"Although she's been trying very hard in class, she failed our last quiz. I don't want her to get too discouraged and give up studying for our test next week." "I've noticed that he rarely writes anything down in his notebook. Unfortunately, in this topic, students have to explain their thinking using specific vocabulary."
5. **ACTION**: We explain what we've done so far to address the situation or offer ideas about what we'd like to do. If possible, we draw on our experience to point out what students in the past have done to succeed.	"She told me that she gets very nervous during tests. In my experience, this happens when students don't get enough practice. To build her self-confidence, I'd like her to go to tutoring during her free period and do additional problems for practice." "In my experience, students who take good notes get higher scores on our state test because they remember the vocabulary better."

Table 12.1 *(Continued)*

Step	Examples
6. **SUPPORT**: We ask for the parent's help or suggestions for what else we could do. We emphasize that we would like to work together in the child's best interests. We also outline any next steps that we can take.	"What was your child's experience with math like in the past? When did he like it or do particularly well? What do you think was the reason for his success?" "Could you encourage her to go for tutoring a few times a week, especially when she's having difficulty? I find that getting help a few times a week is typically a manageable commitment for many students." "Would you mind checking with your child periodically to make sure that he is taking notes regularly? His notes should have the date, aim, and classwork written out." "What else can I do to help?"
7. **QUESTIONS**: See what questions or concerns the parent has. Often, parents share information that helps put the child's challenge in context. If appropriate, offer to share information with other relevant school personnel.	"What other questions or concerns do you have?" "May I share this information with her guidance counselor?"
8. **THANKS**: Thank the parent for working with us.	"Thank you for everything you've done to help her in math." "Thank you for your support."

Acknowledging parents' discomfort toward math and giving them concrete ways to address it can help both parents and students. Research shows that children learn less when parents who have negative feelings about math help their children with their homework (Maloney, Ramirez, Gunderson, Levine, & Beilock, 2015, p. 1485). In addition, students with more positive attitudes toward math tend to have strong social support and encouragement from parents (Mohr-Schroeder et al., 2017, p. 219; Rice, Barth, Guadagno, Smith, & McCallum, 2013, p. 1047).

Parents who make their children feel more discouraged about math can reinforce students' negative feelings about the subject (Usher, 2009, pp. 301–302).

To address parents' math anxiety, we provide reassurance that they shouldn't feel pressured to help their children with their homework. Many parents may not remember the content in middle and high school math. Expecting them to learn the material in order to help their children places an unfair burden on parents. Many teenagers no longer want to ask their parents for help (McNeal, 2012, p. 88). Furthermore, we often teach methods (such as the area method of division) that differ substantially from the traditional algorithms that parents learned in school. Parents who help their children can easily get frustrated with our methods, even if these methods can foster deeper conceptual understanding. Parents who help can actually do more harm than good if they criticize these techniques without realizing their benefits.

Instead of giving direct help to their children, we encourage parents to discuss schoolwork with them in a more general way. We ask parents to look at their children's notebooks—not to see if the work is accurate, but rather to make sure that they took notes and completed the homework. We also encourage parents to talk to their children about what they learned in math class, how difficult it was, and how learning it can help in the future. Research indicates that children whose parents talk about the importance of education have higher academic achievement (McNeal, 2014, p. 575).

When we communicate with parents, we often share common struggles with math and how to overcome them. We talk about our experiences of difficulty with math and what we did (or wish we had done) to improve our understanding. For example, when Larisa was a student, she never had to study for math until she entered college. There, the courses covered so much material that she started experiencing panic attacks before tests. She found that studying her notes, forming a study group, practicing many problems, and using various test-taking strategies (like doing the easiest problems first) helped her gain more confidence and overcome her test anxiety. Sharing these stories can serve as a model for parents, who can then turn their own anxiety into a positive motivation for their children. Parents can also foster a growth mindset in their children by pointing out that their mathematical intelligence is not fixed or inborn but can increase with more effort and studying (Allen & Schnell, 2016, p. 400; Dweck, 2007). We talk more about the importance of promoting a growth mindset in Chapter 4: Promoting Mathematical Communication.

In addition, we encourage parents to support their children by enforcing rules about homework (such as when and where it's done), which research indicates can improve academic achievement more effectively than giving direct help (Wilder, 2013, p. 386). Parents can set aside a quiet place and time at home for children to study.

PARENT-TEACHER CONFERENCES

Parent-teacher conferences are an important opportunity for many parents to learn about their child's progress. Here are some strategies that we use to make them worthwhile for everyone involved.

Before parent-teacher conferences, we think about what we want to say about each student. We sometimes jot down brief notes to ourselves as a reminder, such as, "Works well with other students in class but doesn't do homework." If possible, we print progress reports for parents to review (most electronic gradebooks give teachers the ability to print individual reports for each student—see the Technology Connections section of this chapter for more details). We also make copies of the introductory course description that we send out at the beginning of the year for any parents who didn't get it or would like another copy (see Figure 11.2: Course Description for a sample).

When we meet with parents, we follow the general outline that we described in the Communicating with Parents section earlier in this chapter: we greet them, point out positive comments, ask for their comments or suggestions, address any concerns that we have, outline possible remedies, and thank them for their support.

We encourage students to come with their parents. Many times, we allow students to take the lead in guiding the conversation. We find that doing so helps students take ownership over their work. They can explain in their own words why they do well in certain areas and struggle in others. To help students outline important points before the meeting, teachers can share progress reports with students and ask them to think about how they would describe their academic performance. During the meeting, we go through students' individual progress reports and point to noteworthy parts so that students can explain areas of both success and concern.

If parents have difficulty understanding English, we try to secure a translator from our school. Asking students to translate can create potentially awkward situations in which they are explaining academic or behavioral concerns to their parents, so we only recommend this as a last resort if nobody else is available. In those situations, we try to use nonverbal communication, such as gestures and facial expressions, to convey the meaning behind what we say. A translation app can also be used if no other option is available. We describe some options in the Technology Connections section of this chapter.

HOME VISITS

A growing number of schools are using home visits to build trust between families and educators and reduce implicit biases. Implicit biases can lower student achievement by creating a self-fulfilling prophecy in which teachers have low expectations of students, who then match those expectations, reinforcing those negative beliefs

(McKnight, Venkateswaran, Laird, Robles, & Shalev, 2017, p. 8). By conducting home visits, educators can see students as individuals with unique characteristics (a process called *individuation*), elicit positive emotions, promote collaboration, and reduce implicit biases (McKnight et al., 2017, p. 25). Research indicates that home visits can improve student attendance and achievement (Sheldon & Jung, 2015, p. 20; 2018, p. 13).

Before conducting a home visit, teachers should research culturally appropriate etiquette for meeting families (Johnson, 2014, p. 378). Visits should always be voluntary and prearranged so that parents are prepared for the visit (McKnight et al., 2017, p. 25). Reaching out to all parents or a cross-section of parents can ease tension by reassuring them that they are not being singled out.

During the visit, teachers can discuss topics not directly related to school. School-related conversations can focus on students' strengths, such as, "What are her strengths at home?" and "What can we do to build on these strengths?" (McKibben, 2016). Unobtrusively gathering information about the home environment can help teachers get a better overall understanding of students. After a home visit, teachers should record details of the visit and thank families for allowing them to enter the student's home (Johnson, 2014, p. 378).

Although home visits can have many benefits, they require a significant amount of time and resources, especially if teachers visit on their own. We suggest that teachers check with their school or district to determine policies regarding home visits. Going with a guidance counselor, translator, or another teacher can make the experience less stressful for school personnel. The Parent-Teacher Home Visit Project also recommends that teachers be trained and compensated for visits outside the school day (McKnight et al., 2017, p. 29)—something that can obviously be done only with school or district support.

Although we aren't able to conduct home visits on a regular basis, we try to incorporate many of the ideas behind home visits in all aspects of our dealings with parents.

WORKING WITH PARENTS OF CULTURALLY DIVERSE STUDENTS

When working with diverse students, teachers need to consider both the assets that parents contribute and the challenges they face.

Parents of culturally diverse students support their children's education in many ways. Compared to parents born in the United States, immigrant parents are more likely to value the importance of education and believe that their children will graduate from college (Raleigh & Kao, 2010, pp. 1085, 1089). They often provide encouragement and economic support for their children's academic success (Cherng & Ho, 2018, p. 88). Many families belong to religious organizations that provide tutoring, afterschool, and summer educational programs (Latunde, 2017a,

p. 255). In addition, parents' previous life experiences and current sacrifices may serve as a motivation for all students to stay focused on school. Teachers can encourage parents who are comfortable sharing their stories to do so with their children or the entire class.

Cultural differences may help explain why some parents may not respond to outreach in expected ways. Immigrant parents across all class levels are generally less likely to participate in school activities. They often lack familiarity with the US educational system or lack the language skills to communicate comfortably with school personnel (Cherng & Ho, 2018, p. 102). Some Latino parents (as well as others) may believe they can best support their children's education by teaching the moral and ethical values that can boost their children's self-confidence (Civil & Menéndez, 2010, p. 2; Ramos, 2014, p. 5). These parents and others who defer to teachers as the experts on teaching content can unfortunately be interpreted as being uninvolved or apathetic about their children's education. Some African American families accustomed to the welcoming social and cultural network of churches might find schools to be intimidating and unwelcoming (Latunde, 2017b).

WORKING WITH PARENTS OF STUDENTS WITH LEARNING DIFFERENCES

Parents of students with learning differences often feel that schools marginalize them since their children have a history of academic and behavioral difficulties (Adkins-Sharif, 2017). Often, these parents are made to feel like they are to blame for their children's poor academic performance or discipline-related issues.

Instead, being more sensitive to the frustrations of parents can reduce the possibility of judging them or making them feel defensive (Davern, 1996). Allowing parents to express their concerns, focusing on what's best for the child, and inviting them to work together to devise the right course of action can help turn parents into true partners in their children's education (Adkins-Sharif, 2017).

When we reach out to *all* parents, but especially those of students from different cultures or with learning differences, we try to do the following:

- Learn about what they hope for their children and what positive experiences their children may have had in school. We find that parents are often the best sources of advice for how we can better educate their children.

- Acknowledge the many ways in which they support their children's education, such as communicating the value of education or modeling appropriate behavior (Latunde, 2017a, p. 254).

- Provide regular information about upcoming events and regularly solicit feedback to facilitate a meaningful two-way conversation and make schools less intimidating.

- Provide more information about learning strategies and accommodations, which can be particularly helpful for parents of children with learning differences (Latunde, 2017a, p. 268). We explain to parents what modifications we make in our instruction or on tests for their children.

What Could Go Wrong

Many times, we get frustrated with students. Unfortunately, we've sometimes taken out those frustrations on their parents. We find that calling or writing to parents simply to complain about their children's behavior—no matter how infuriating, rude, or disrespectful—usually backfires. Many parents, instead of supporting us, become defensive and feel that we are attacking them as well as their children. These conversations can quickly turn into shouting matches that can often lead to further problems. Instead, we recommend taking a few minutes before communicating with parents to reflect on whether the behavior is worth the phone call, email, or text. We also try to tone down our anger by imagining how we would feel if we got such a message from a teacher—and as parents ourselves, we have! Many times, we share experiences of similar parenting challenges that we've faced when dealing with our own children.

As we said earlier, not being sensitive to parents' unique circumstances can also lead to problems. Asking parents to purchase an expensive graphing calculator or pay for a private tutor may not be reasonable for low-income families and can foster resentment. Instead, we try to provide more affordable solutions, such as lending school calculators or providing information about free tutoring from school, libraries, or other community resources.

At other times, parents get frustrated with us. They may tell us that we're not teaching properly or they may insist on having their children do math the way that *they* learned it (as both parents and math teachers, we've been on both sides of this as well). Sending communications to parents that explain what we're doing in class and why we're using certain methods can help alleviate their concerns.

In all of these situations, we try to keep two things in mind:

- Focus on what we can do together to help the student instead of quibbling over the details of who said what to whom.
- Treat parents as partners in their children's education—after all, they've known our students longer than we have, so they have many more years of experience dealing with their children.

Finally, we sometimes get overwhelmed with trying to keep in touch with parents. Trying to stay in constant communication with every parent about every positive or negative event in class can quickly spiral out of control. Trying to make ourselves

CHAPTER 12: BUILDING RELATIONSHIPS WITH PARENTS

constantly available, such as by giving parents a personal email address or cell phone number, may only makes things worse.

To maintain our sanity, we try to maintain a separation between our personal and professional lives and set limitations for when parents can contact us and when they can expect a response. We tell students and parents that they can expect responses only during certain hours. We never respond to emails late at night and rarely make phone calls at night or on weekends unless we make arrangements with parents beforehand. In addition, we always use a school or district email address, never a personal one, when emailing parents. Similarly, we never give out our personal phone number to parents.

Student Handouts and Examples

Figure 12.1: Parent Communication Script
Figure 12.2: Parent Communication Log

Technology Connections

Many schools or districts have online gradebooks that enable parents to keep track of their students' grades. Often, these gradebooks have tools that notify parents when grades have been uploaded and can produce individualized grade reports, which can be useful for regular updates and meetings with parents. We recommend checking with your school or district to see what is available. One free option for teachers whose school or district doesn't have an online gradebook is ThinkWave (www.thinkwave.com).

In addition, many schools have content management systems that allow teachers to create class pages for students and parents. Other options include the free Google Classroom (http://classroom.google.com), Edublogs (www.edublogs.org), WordPress (www.wordpress.com), and Weebly for Educators (http://education.weebly.com).

Websites like TeacherVision (www.teachervision.com), ThoughtCo (www.thoughtco.com), and Edutopia (www.edutopia.org) have many articles and other resources on parent-teacher conferences and keeping in touch with parents. Free translation apps like Google Translate and Microsoft Translate are available on Android (http://play.google.com/store) and iPhone (http://itunes.apple.com) devices.

Free apps like Remind (www.remind.com), Class Dojo (www.classdojo.com), TalkingPoints (http://talkingpts.org), and Edmodo (www.edmodo.com) enable teachers and parents to communicate via mobile text messages in several languages.

The Parent-Teacher Home Visit Project (www.pthvp.org) has resources and research related to home visits on its website. The ¡Colorín Colorado! website (http://www.colorincolorado.org/article/making-your-first-ell-home-visit-guide-classroom-teachers) has a guide for visits to the homes of ELL students.

Figures

Parent Communication

Date	Time	Class	Student	Parent	Method
					☐ Email ☐ Text ☐ Phone ☐ In-Person

Step	Notes
1. INTRODUCTION	
2. PURPOSE	
3. PRAISE	
4. CONCERNS	
5. ACTION	
6. SUPPORT	
7. QUESTIONS	
8. THANKS	

Figure 12.1 Parent Communication Script

Parent Communication

Date	Time	Class	Student	Parent	Method	Notes
					☐ Email ☐ Phone ☐ Text ☐ In-Person	
					☐ Email ☐ Phone ☐ Text ☐ In-Person	
					☐ Email ☐ Phone ☐ Text ☐ In-Person	
					☐ Email ☐ Phone ☐ Text ☐ In-Person	
					☐ Email ☐ Phone ☐ Text ☐ In-Person	

Figure 12.2 Parent Communication Log

CHAPTER 13

Collaborating with Other Teachers

What Is It?

In the United States, many teachers work in isolation—compartmentalized into classrooms, lonely, and placed in situations where working with administrators or other teachers is difficult. This is not a healthy situation for anyone, and is even more unhealthy for new teachers. The "egg crate" metaphor, first articulated by sociologist Dan Lortie (1975, p. 14), has also been used to describe how some school leaders and educational reformers view teachers—easily replaceable or interchangeable (Schleifer, Rinehart, & Yanisch, 2017, p. 5).

By working together, teachers may be able to feel less isolated and improve their instruction. Researchers have defined *teacher collaboration* to include a wide variety of practices, including co-planning activities or tests, mentoring, observing peers, and discussing common problems (Schleifer et al., 2017, p. 7; Vangrieken, Dochy, Raes, & Kyndt, 2015, pp. 26-27). In this chapter, we define teacher collaboration to mean any effort in which partners work together to perform a task (Vangrieken et al., 2015, p. 23).

Why We Like It

Collaboration has been an integral part of our teaching. We work with over 30 math teachers in our school and over 1,000 STEM educators in the Math for America professional learning community. In our experience, working with other teachers enables us to share workloads, improve our teaching with new ideas, and reduce our stress level. In fact, we developed all of the ideas in this book through collaboration with our colleagues!

Supporting Research

Researchers have found both positive and negative consequences for teacher collaboration (Vangrieken et al., 2015, pp. 27-29). When teachers work together, student achievement often improves (Kraft & Papay, 2014, pp. 488, 494; Ronfeldt, Farmer, McQueen, & Grissom, 2015, p. 507), especially for students of color and low-income students (Moller, Mickelson, Stearns, Banerjee, & Bottia, 2013, p. 188). Collaboration can improve teachers' motivation, reduce their feelings of isolation, and encourage teachers to collaborate more as their performance improves (Vangrieken et al., 2015, pp. 35-36). However, collaboration can also reduce the time that teachers have for individual tasks like grading (Bae, 2017, p. 23) and lead to the formation of cliques, which can increase tensions among teachers (Bovbjerg, 2006, p. 249).

We believe the benefits of collaboration far outweigh the disadvantages when utilizing the characteristics described in the Application section of this chapter.

Common Core Connections

In our experience, the Common Core standards cannot be understood and interpreted clearly without collaboration with other educators.

Application

Researchers have found several characteristics of successful teacher collaboration:

- **Trust:** Teachers who trust their colleagues and are valued as professionals are more likely to collaborate (Jao & McDougall, 2016, p. 560; Leana & Pil, 2014).

- **Shared goals:** Having shared goals and strengthening collaboration appear to reinforce each other (Brook, Sawyer, & Rimm-Kaufman, 2007, p. 215). Teachers at schools with similar educational goals are more likely to work together, which can in turn help their goals to become more similar.

- **Choice:** Teachers tend to work together more often in an informal setting, so allowing teachers to form their own partnerships can lead to more effective collaboration (Brook et al., 2007, pp. 238-239).

- **Time:** To collaborate effectively, teachers need time to build trust and reach a consensus on goals (Jao & McDougall, 2016, p. 560; Schleifer et al., 2017, p. 41).

- **Support:** Administrators can support collaboration by giving teachers time to collaborate (Schleifer et al., 2017, p. 41) and freedom to innovate (Bae, 2017, p. 32). Some research indicates that principals who provide strong instructional guidance tend to collaborate more (Goddard, Goddard, Sook Kim, & Miller, 2015, p. 524). Other research has found that teachers are more likely to seek help from their colleagues than their principal (Leana & Pil, 2014). We find the instructional guidance that principals give is often less important than the school climate that they create and the freedom that they give teachers to work together.

Here are some of our favorite methods that incorporate many of these five criteria of successful teacher collaboration.

DISCUSSING VALUES

Having common values is critical for any collaboration. The underlying assumptions that teachers have about math, students, learning, and teaching strongly affect how they teach (Artzt, Armour-Thomas, & Curcio, 2008, p. 20). To develop common values among teachers, we start by discussing what we believe about how students learn and how we can create an effective learning environment. Here are some of the questions that we discuss with colleagues while collaborating:

- **How do we motivate students to learn math?** We discuss motivation more in Chapter 1: Motivational Strategies.

- **How do we interact with students of different ethnicities?** We discuss self-reflection and other ways to connect with all students in Chapter 2: Culturally Responsive Teaching.

- **How do we write effective lessons to promote student learning?** We discuss this more in Chapter 6: How to Plan Units and Chapter 7: How to Plan Lessons.

- **How do students work with each other to communicate mathematical ideas?** We discuss this more in Chapter 3: Teaching Math as a Language, Chapter 4: Promoting Mathematical Communication, and Chapter 17: Cooperative Learning.

- **What rules, procedures, and routines will make our classroom operate smoothly?** We discuss this more in Chapter 11: Building a Productive Classroom Environment.

- **How do we effectively assess student learning?** We discuss this more in Chapter 9: How to Plan Tests and Quizzes, Chapter 16: Project-Based Learning, and Chapter 18: Formative Assessment.

- **How do we differentiate instruction for our students?** We discuss this more in Chapter 14: Differentiating Instruction and Chapter 15: Differentiating for Students with Unique Needs.

- **How should we use technology to facilitate student learning?** We discuss this more in Chapter 19: Using Technology.
- **What should students do outside of class to promote their learning?** We discuss this more in Chapter 8: How to Plan Homework.
- **How should we grade students?** We discuss this more in Chapter 10: How to Develop an Effective Grading Policy.
- **What role do parents have in their children's education?** We discuss this more in Chapter 12: Building Relationships with Parents.

Sometimes, though, we can't always reach a consensus on values. We discuss this more in the What Could Go Wrong section of this chapter.

PLANNING WITH OTHER MATH TEACHERS

One simple way that we collaborate with our colleagues is by sharing materials (such as lesson plans, homework assignments, and assessments). We find that putting our files on shared online drives allows us to access each other's material easily without having to email each other (we describe some of our favorite sharing tools in the Technology Connections section of this chapter).

When possible, we also talk to others who are teaching the same course to maintain a common pacing (we discuss pacing and unit plans more in Chapter 6: How to Plan Units). Getting feedback on pacing is especially useful when we follow a curriculum written by someone else since the suggested pacing can be unrealistic for our students.

Another way to facilitate collaboration is to have common planning time, which is a period of time during the school day specifically set aside for teachers to write lessons, reach out to parents, and discuss concerns about students. Administrators can schedule common planning time for teachers who want to plan together. This can be effective when teachers have a shared vision and clearly defined goals (Cook & Faulkner, 2010, p. 7). We often use it to analyze and reflect on our lessons for that day. This time is also a valuable way for us to receive emotional support from colleagues. Getting sympathy for a bad lesson or sharing joy for an effective one helps keep us grounded!

When we don't have common planning time, we have to resort to other measures. We discuss some of them in the What Could Go Wrong section of this chapter.

INTERDISCIPLINARY COLLABORATION

Working with colleagues in other departments can also be effective. One simple way that we work with teachers in other departments is to share information about students. Many times, we discover that students who struggle in our classes make a connection with other teachers. In these situations, we talk to the student or the teacher so we can learn from them. Conversely, we often talk to teachers of students who excel in our class to determine ways that we can bring them to an even higher level. In our experience, these conversations with other teachers give us a different perspective on students and help us adjust our instruction. Colleagues may share strategies that work with students in other classes or share information about students' lives that can help put their performance in context.

We sometimes consult with teachers in other departments who have more expertise in certain skills. Social studies teachers have asked us for advice on the best ways for students to create and interpret graphs. We have asked science teachers for ways to teach science-related topics like periodic functions or exponential growth. Statistical analysis can be another rich area for interdisciplinary collaboration (we discuss this more in Chapter 5: Making Mathematical Connections). If time permits, teachers can even invite colleagues to come into their classes to teach certain topics.

Teachers can also plan units or lessons with colleagues in other departments. An integrated curriculum can give students meaningful opportunities to connect learning in school with the outside world (Senn, McMurtrie, & Coleman, 2019, p. 1) and encourage teachers to make curricula more relevant to students (Stolle & Frambaugh-Kritzer, 2014, pp. 67–68). Students can examine changes to a biome in a biology class (by looking at the changes in the environment), in a social studies class (by examining how social and economic changes over time affected the environment), and in a math class (by analyzing data related to environmental changes over time).

Writing lesson plans with colleagues in other departments can be challenging in large schools where students take classes from many different teachers. Interdisciplinary lesson planning is often easier where most students take the same classes (i.e., in small schools or middle schools). In a large high school like ours, we often have students from several grades in the same class, so it can be difficult to make connections to what they are learning in other subjects. Trying to coordinate lessons across different courses can also be tricky. We feel that coordinating terminology and solution methods is easier and ultimately more important than trying to make sure that the same concept is being taught simultaneously in several classes. Chemistry teachers and math teachers can use the same methods to solve proportions. History teachers and math teachers can use the same vocabulary when interpreting tables and graphs.

OBSERVING OTHER TEACHERS

Another way that teachers can work together is to observe each other in the classroom. Observing other teachers can raise awareness of important aspects of both their instructional practice (the mathematical tasks, learning environment, and discourse) and their cognitions (what they believe, know, and want to achieve) (Artzt et al., 2008, pp. 17, 39). In addition, colleagues who are willing to take risks may be encouraged to allow others into their classrooms, which can promote a sense of experimentation and shared learning (Gonzalez, 2013).

We often observe our colleagues to get ideas about how to improve specific aspects of our own instruction. While teaching a lesson, we are often so busy that we may not see everything that's going on in our classroom. We find that having another pair of eyes in the room can help us better understand the effectiveness of our teaching.

Schools or districts can facilitate peer observations by allowing teachers to observe each other and (if necessary) get substitutes to cover the observer's regular classes.

Before observing other teachers, we try to discuss the goals and overall plan of the lesson, as well as the specific areas of concern to be observed. Here are some of the questions that we try to answer during an observation. These are based on a template used in the math teacher training program at Queens College (Artzt et al., 2008, pp. 178–182):

- How did the teacher motivate students to make conjectures about the math?
- How appropriate was the sequencing and difficulty level of the mathematical tasks?
- How did the teacher respect and value students' thinking?
- How did the teacher manage classroom rules and handle administrative routines?
- How did the teacher communicate with students in a nonjudgmental manner, monitor their understanding, and respond appropriately?
- How did the teacher encourage students to communicate with each other and justify their thinking so that they could reach mathematical conclusions?
- How did the teacher show a desire to help students develop a conceptual and procedural understanding?
- How did the teacher assess students' prior knowledge and interests, plan suitable tasks that address their potential misconceptions, and make them active participants in the lesson?
- To what extent did the teacher feel that the lesson was successful? What modifications did the teacher suggest to improve the lesson?

During the lesson, the observer writes down anything that is related to the pre-agreed areas of concern. If necessary, the observer may talk to students to get a better understanding of what they are doing.

After the lesson, both the observer and the teacher individually reflect on the lesson, using the questions that were previously discussed. Both teachers then meet to share their thoughts on the lesson and ways that it could be improved.

We find that peer observations, when done voluntarily, can be an honest, nonjudgmental source of information about our teaching. Our pre- and post-lesson meetings feel less like evaluations and more like informal conversations. Peer observations can also build trust between colleagues and encourage further collaboration.

CO-TEACHING

Co-teaching, sometimes called *collaborative team teaching*, is a teaching setup in which two or more professionals work together to educate a diverse group of students in the same physical space (Cook & Friend, 1995, p. 3). Common examples include a general education teacher working with a special education or ELL teacher. Co-teaching has many advantages: it can reduce the student–teacher ratio, provide additional support for students with learning differences, and improve overall instruction (Keeley, 2015, p. 2).

Typically, co-teaching involves one or more of the following techniques:

- **One teach/One assist:** In this whole-class setting, one teacher leads while the other circulates around the room, assisting students as necessary.
- **Station teaching:** Students circulate among two stations set up around the classroom, receiving instructional content from a teacher at each station (or independently or with partners, if a third station is set up).
- **Alternative teaching:** One teacher works with a small group of students while the other teacher works with the rest of the class.
- **Parallel teaching:** The class is divided into two heterogeneous, roughly equal groups and each teacher simultaneously works with one group.
- **Team teaching:** In this whole-group setting, both teachers share instruction. They take turns leading the discussion, or one talks while the other demonstrates (Cook & Friend, 1995, pp. 5–8).

While the one teach/one assist model is often the easiest to implement, research indicates that many students prefer parallel teaching or team teaching, perhaps because these methods split the class into smaller groups, providing more individualized instruction (Keeley, 2015, p. 12). When we co-teach classes, we rarely use the alternative teaching or parallel teaching model since we find that two simultaneous conversations in the same room can be noisy and difficult to manage. Furthermore, using the alternative teaching model runs the risk of stigmatizing those being separated from the rest of the class, especially if they are consistently the students with learning differences or behavioral issues (Hunt, 2010, p. 160).

As a result, we tend to use the alternative teaching or parallel teaching models sparingly. If several students missed the previous day's lesson, we often pull them aside at the beginning of class and quickly summarize what they missed. In situations like these, we also invite anyone who wants a recap of the lesson (whether they were in class or not) to join the small group. Of course, if we find that most of the class joins the "pull-out" group, then we adjust our instruction and either re-teach the lesson or separate those students who *are* ready to move on and get them started on the next lesson!

Most of the time, we use a combination of the one teach/one assist and team teaching methods. Co-teachers often use the one teach/one assist model when the general education teacher is stronger in mathematical content and the special education teacher is stronger in learning strategies (Hunt, 2010, p. 157). We find that the team teaching model can often be more engaging for students since they are interacting with two teachers at once. However, we find that a team teaching lesson works well only when the co-teachers know and trust each other well since it obviously requires a great deal of interplay between them. We often insert other teaching methods, such as the station teaching method, into our lessons as well.

We typically use a mix of co-teaching techniques throughout the year, depending on each lesson's goals, our students' needs, and each co-teacher's preferences. We may even use more than one technique in the same period. Here is how Bobson and his co-teacher typically structure their collaborative class:

1. Bobson's co-teacher guides the class homework discussion at the beginning of class while he takes attendance and walks around the room to answer specific questions.
2. Bobson starts the lesson while his co-teacher circulates around the room. Sometimes, she interjects by clarifying a point or demonstrating a technique.
3. Depending on the lesson, he and his co-teacher then alternate guiding the class through practice or supervising an activity (such as station work).
4. Both teachers manage student behavior, monitor student learning, and adjust instruction as necessary.
5. Both Bobson and his co-teacher bring all students together at the end of the period to summarize the lesson.

As shown in the previous example, we try to alternate classroom roles with our co-teachers as much as possible. When both teachers circulate around the room at different times, each one can assess student understanding and have a voice in the classroom (Hunt, 2010, p. 158).

To summarize, here are our tips for making co-teaching a rewarding experience for both teachers and students:

- **Agree on instructional beliefs.** Sharing beliefs (such as the roles of teachers and the proper method of student learning) is critical since teachers' instructional beliefs guide their practice (Cook & Friend, 1995, p. 11).
- **Agree on rules and procedures.** Having the same procedures prevents students from playing teachers off each other to get the answer that they want (Cook & Friend, 1995, p. 11).
- **Work as partners.** We try to treat our co-teachers as partners, actively involving them in all aspects of planning, administrative tasks, classroom management, and lesson implementation. Often, our co-teachers have more experience working with particular groups of students than we do. (Both of us are general education teachers.) For example, Larisa's co-teacher, a special education teacher, often suggests good ways to accommodate students' needs and scaffold lessons with appropriate supports. ELL teachers can gather resources to differentiate instruction, devise appropriate language development goals, and help modify instruction to match the needs of ELLs. Often, suggestions from these co-teachers help us improve instruction for *all* students.
- **Find time to plan.** Setting aside time with co-teachers not only gives us time to plan lessons but also to work out any pedagogical differences.
- **Demonstrate equality.** We try to show our students and each other that both co-teachers are equals and should be treated as such. This includes putting both teachers' names on all handouts, alternating roles in class, and referring to "our" students instead of "your" students or "my" students.
- **Build trust.** Taking the steps outlined above can foster trust between co-teachers, which is critical for any co-teaching relationship to work.

Unfortunately, things don't always go as well as we'd like. We describe some of these situations and what teachers can do in the What Could Go Wrong section of this chapter.

We describe other strategies for working with ELLs and students with learning differences in Chapter 15: Differentiating for Students with Unique Needs.

MENTORING

Mentoring, where an experienced teacher provides guidance and support to a newer teacher, can provide valuable professional development for both mentors and mentees. Mentoring can range from a formal relationship set up by a school or district with scheduled meetings to an informal collaboration between colleagues. Research shows that mentoring has mixed results on student achievement but can reduce teacher turnover (Schleifer et al., 2017, pp. 26–27) and help teachers feel more satisfied about their jobs (Weisling & Gardiner, 2018, p. 64).

Formal mentoring relationships work best when mentors and mentees have clearly defined roles and interact with each other both inside and outside of their class (Weisling & Gardiner, 2018, pp. 65–67). Informal mentoring relationships rely heavily on social relationships, so they tend to thrive in positive, open, and stable environments (Du & Wang, 2017, p. 16). All mentoring relationships, whether formal or informal, succeed when they are based on trust so that colleagues can open up to each other and share their vulnerabilities (Spooner-Lane, 2016, p. 16).

When we mentor other teachers, we benefit by reflecting on our own practice. When we work with mentees, we ask ourselves many of the questions that we list in the Observing Other Teachers section in this chapter. We try to treat our mentees as co-teachers, using many of the strategies that we describe in the Co-Teaching section in this chapter.

In addition, even though we are both veteran teachers, we often receive guidance from our colleagues when we lack expertise in a certain area, such as when we use new technology or teach a course for the first time. We find that such informal mentoring strengthens relationships between colleagues by creating a collaborative and supportive environment in which teachers share their knowledge with others.

LESSON STUDY

Lesson study is an ongoing process in which teachers collectively plan a lesson, implement and observe it, and reflect on it in a debriefing session (Groth, 2011, p. 446). Although the term *lesson study* refers to a specific professional development cycle that originated in Japan, we use the term more broadly to refer to any effort in which teachers, administrators, and sometimes outside experts work together to plan, teach, observe, and refine lessons.

Lesson study has many benefits. Teachers who engage in lesson study report that they put more thought into their planning, use more tasks that elicit student thinking, and monitor student learning more closely (Post & Varoz, 2008, p. 477). Since teachers work collectively to devise norms for discussion, reflect on their work, and develop shared values, lesson study can strengthen existing collaboration (Perry & Lewis, 2009, p. 383–384). Research suggests that students may even get some modest academic benefit from instruction produced by a lesson study group (Cheung & Wong, 2014, p. 144; Godfrey, Seleznyov, Anders, Wollaston, & Barrera-Pedemonte, 2018, p. 334).

Lesson study typically consists of the following steps:

- Teachers work in groups to write a lesson and determine a focus for lesson observations.
- One teacher implements the lesson while the other teachers observe it, either in person or via videotape if attending in person is impractical.
- Teachers meet to reflect on what went well in the lesson and what could be improved.
- Teachers repeat the cycle of implementation with observation and reflection (Post & Varoz, 2008, pp. 473–474).

To make lesson study more effective, researchers recommend encouraging a wide variety of professionals (including outsiders) to participate, distributing leadership among teachers, and supporting lesson study groups with funding and other resources (Perry & Lewis, 2009, pp. 383–385).

We feel that lesson study can help build a strong and productive professional community. However, like any of the other strategies that we describe, lesson study requires a sizable commitment since participants must find the time to plan a lesson together, observe each other, and reflect together. We suggest that lesson study can only be implemented effectively if supported by the school's administration through common planning time, funding, and other resources.

Even if a full-fledged lesson study is impossible, we believe that *any* type of collaboration incorporating any of the key elements of lesson study—planning lessons together, observing other teachers, and reflecting afterward—is usually more effective than working alone.

PROFESSIONAL LEARNING COMMUNITY

Another way for teachers to work together is to be part of a professional learning community (PLC). Defining a PLC precisely can be tricky since the term has been used to refer to almost any group of individuals that work together on an education-related issue, such as a grade-level teaching team, school faculty meeting, or national professional organization (DuFour & Reeves, 2016, p. 69). We use the term *professional learning community* to mean a group of educators who meet regularly to identify their professional learning needs, reflect on the effectiveness of their work, and develop resources to improve their instruction (Blitz & Schulman, 2016, p. 1).

In effective PLCs, teachers work collaboratively to do the following:

- Develop a set of pedagogical beliefs (Vangrieken et al., 2015, p. 23).
- Engage in an ongoing, reflective conversation in which teachers analyze their practice (DuFour, 2004).
- Determine what content students will learn and what schools will do when they don't. If students receive a poor report card grade, their teachers and guidance counselors may talk to them, notify their parents, and set up tutoring sessions (DuFour, 2004).
- Make meaningful decisions about how their learning community is run, such as what curricula to use and what resources schools can provide (Vescio, Ross, & Adams, 2008, p. 85).
- Promote continuous teacher learning driven by the needs and interests of *teachers*, not administrators (Vescio et al., 2008, p. 86).
- Share ideas from their personal practice through activities like peer coaching and classroom observations (Brown, Horn, & King, 2018, p. 54).

Many PLCs are run within schools. In addition, other organizations, such as school districts, regional cooperatives, and local professional organizations, enable teachers from different schools to collaborate. Math for America enables recipients of its STEM educator fellowships to create and facilitate professional development workshops. In addition, local affiliates of the National Council of Teachers of Mathematics often have professional development opportunities for educators (we list some of the available online resources from these organizations in the Technology Connections section of this chapter).

Teachers can also build their own PLCs online by posting on social media platforms like Twitter or Facebook, following other educators online, responding to posts from other people, and creating websites where they share their own thoughts on education. We list several useful technology resources in the Technology Connections section of this chapter.

In our experience, working in PLCs often gives us valuable information about how to improve our teaching practice. Collaborating in professional development workshops has enabled us to work with teachers in schools whose student populations are very different from ours. Learning about best practices from their experiences has helped us become more culturally responsive to our students. We also find that collaboration that is created and run by teachers tends to be deeper and more meaningful since participants tend to be more motivated and focused.

What Could Go Wrong

Here are some of the things that can go wrong when we try to collaborate with other teachers:

LACK OF TRUST

Sometimes, teachers have to work with colleagues who have different pedagogical beliefs or teaching styles. A teacher who wants to emphasize conceptual understanding may want students to work through lengthy discovery activities, while another teacher may prefer that students spend the time mastering procedural skills. Other problems can arise when collaborating teachers don't share information with each other due to lack of time or motivation. This can reduce the other teacher to an "assistant" who walks into the classroom unsure of what is supposed to happen.

In our experience, taking the time to talk to colleagues about their values often helps us build trust. Talking about the questions that we list in the Discussing Values section earlier in this chapter can help reach a consensus. No matter how heated and personal these conversations may get, we try to treat our colleagues as professionals whose opinions deserve to be respected.

REINFORCING NEGATIVE STEREOTYPES

Collaboration can sometimes reinforce negative stereotypes. One study that examined teacher collaboration and math achievement of Hispanic students found a negative relationship for ELLs but a positive relationship for children of immigrants (Bottia, Valentino, Moller, Mickelson, & Stearns, 2016, p. 524). The researchers hypothesized that teachers who perceived ELL students to be poor math students reinforced those stereotypes when they talked with their colleagues, while teachers who perceived native-born children of immigrants as hard working reinforced those stereotypes after collaboration (Bottia et al., 2016, p. 525).

To combat these stereotypes, we try to remember that *all* students are capable of learning. Teachers who adopt and maintain this kind of a growth mindset (which we discuss more in Chapter 4: Promoting Mathematical Communication) for every student can also help ensure that we don't hinder their academic achievement.

LACK OF COLLEAGUES

Sometimes, a teacher may not be able to collaborate with anyone else in their school, perhaps because nobody else teaches the course or wants to share ideas.

In these situations, one option is to self-reflect, using the observation process (outlined in the Observing Other Teachers section of this chapter) of a pre-lesson reflection, note-taking during the lesson, and post-lesson reflection. Obviously, making observations during the lesson can be difficult, so they may be limited to brief notes like "this question confused students and I had to rephrase."

Another possibility is to have students who are not in the class (such as former students of ours or students recommended by another teacher) to act as observers. Since these students aren't in the class, they won't have to worry about missing a lesson (as long as they have a free period, such as lunch or study hall, while we teach) and they can observe more objectively than if they had to comment on what their classmates were doing. Teachers could use students in the class, as long as they can be trusted to observe objectively and can make up the lesson later. Teachers could also ask their students to fill out an anonymous survey during the lesson, afterwards, or both, using some of the questions in the Observing Other Teachers section of this chapter.

In addition, teachers can get feedback from colleagues in local professional organizations. Teachers can also use social media to find and learn from others. Although obviously not the same as collaborating with people who can observe your classroom, such feedback can often be insightful since colleagues in other schools or districts bring different perspectives that can help us improve our instruction.

LACK OF TIME

Perhaps the most serious obstacle to collaboration is lack of time. Often, this results from scheduling issues, such as co-teachers who don't have common prep time.

Unfortunately, when we don't have common prep time with our collaborating teachers, we often have to make sacrifices in order to make the partnership work. This can mean meeting outside of school hours or communicating by phone or email. Online file-sharing and collaboration tools like the ones we list in the Technology Connections section of this chapter can help.

Technology Connections

Many resources for peer observations can be found on websites like ASCD (www.ascd.org), NCTM (www.nctm.org), and Edutopia (www.edutopia.org). Although the peer observation guidelines used by Queens College's math teacher training program are not available online, a similar template appears on the US State Department's online resource for English language teachers (http://americanenglish.state.gov/files/ae/resource_files/peer_observation_handout.pdf).

The websites listed above also have many resources available for working with other math teachers and teachers of other subjects. Additional resources on interdisciplinary collaboration can be found on the WNET Education (http://www.thirteen.org/edonline/concept2class/interdisciplinary/index.html) and College

CHAPTER 13: COLLABORATING WITH OTHER TEACHERS

Board (http://secure-media.collegeboard.org/apc/AP-Interdisciplinary-Teaching-and-Learning-Toolkit.pdf) websites.

Many school districts have resources for mentors and mentees. We recommend checking with them to see what information is available. For example, the New York State Education Department has links to different mentoring programs (http://www.highered.nysed.gov/tcert/resteachers/models.html). In addition, ASCD's *Educational Leadership* journal had a special July 2011 focus issue on mentoring (http://www.ascd.org/publications/newsletters/education-update/jul11/vol53/num07/toc.aspx).

Resources for lesson study are available from the Lesson Study Group at Mills College (http://lessonresearch.net), the Chicago Lesson Study Group (www.lessonstudygroup.net), and the University of Wisconsin-La Crosse (http://www.uwlax.edu/sotl/lsp).

Many schools and districts have professional learning communities. In addition, local affiliates of the National Council of Teachers of Mathematics (NCTM) provide resources for members. NCTM's complete directory of local affiliates (http://www.nctm.org/Affiliates/Directory) is available online. Math for America (www.mathforamerica.org) has a network of teacher-led professional development for recipients of its fellowships in several areas, including New York City; Boston; Washington, DC; San Diego; and Los Angeles.

Several online tools exist to help teachers collaborate. Google Drive (http://drive.google.com), Dropbox (http://dropbox.com), and SugarSync (http://sugarsync.com) allow teachers to share documents with each other. Online word processors like Google Docs (http://docs.google.com) enable teachers to share and edit lesson plans online simultaneously.

In addition, teachers can get advice on Twitter (www.twitter.com) from the Math Twitter Blogosphere (http://twitter.com/hashtag/MTBoS) and I Teach Math (http://twitter.com/hashtag/ITeachMath) communities. There are math educators around the world who follow Twitter posts with the #MTBoS or #ITeachMath hashtags. Resources for how to follow hashtags are available on the http://Dummies.com website (http://www.dummies.com/business/marketing/social-media-marketing/how-to-track-twitter-hashtags).

Teachers can also blog using any of the countless website tools available, including Edublogs (www.edublogs.org), WordPress (www.wordpress.com), and Weebly for Educators (http://education.weebly.com). The Edmodo website and app (www.edmodo.com) also enable teachers to access a community of educators.

Edutopia (http://www.edutopia.org/blog/new-teachers-becoming-connected-educators-lisa-dabbs) and TeachThought (http://www.teachthought.com/education/8-ideas-10-guides-and-17-tools-for-a-better-professional-learning-network) have guides for teachers who want to become more connected online.

PART IV

Enhancing Lessons

CHAPTER 14

Differentiating Instruction

What Is It?

Differentiation, or *differentiated instruction*, is a model in which teachers modify aspects of instruction (such as student grouping or learning time) to support students' different learning methods, learning rates, talents, and interests (van Geel et al., 2018, p. 52). In differentiated classrooms, teachers try to challenge all students and provide several possible paths for meeting educational objectives (Little, Hauser, & Corbishley, 2009, p. 36; Tomlinson, 2014, p. 16). Differentiated instruction is designed to keep students moving forward regardless of their levels of readiness (Taylor, 2015, p. 17).

Why We Like It

Differentiation can make our classes more engaging. In order to differentiate, we have to learn students' needs and interests. Treating students more like individuals instead of interchangeable parts helps them connect with *us*. By differentiating, we can also connect more with *them*. This is particularly important since US classrooms are becoming increasingly diverse. Non-white students now make up a majority of public school classrooms, 10% of public school students are English Language Learners (ELLs), 14% receive special education services, and 7% are in gifted and talented programs (US Department of Education, 2015, 2017, 2018a, 2018b). This strategy can facilitate a sense of trust, which creates a more positive classroom environment (we discuss this topic in Chapter 11: Building a Productive Classroom Environment) and improves student motivation (which we discuss more in Chapter 1: Motivating Students).

Furthermore, when we adjust our instruction to match students' needs and interests, they tend to be more focused on the lesson and their learning improves. We can't customize *every* lesson to accommodate *every* individual student. However, *any* amount of differentiation that we can fit into our instruction makes us better teachers.

Finally, differentiation sends the message that *all* students can learn and can have access to an excellent education. As a result, our students are more likely to believe that their mathematical learning is not fixed but can improve with high-quality instruction and the right amount of effort (we discuss this idea of a growth mindset more in Chapter 4: Promoting Mathematical Communication).

Supporting Research

Differentiated instruction is grounded in three research-based concepts:

1. **Multiple intelligences.** *Intelligences* are the capacities that all individuals have to process a certain type of information: linguistic, musical, logical-mathematical, spatial, bodily-kinesthetic, interpersonal, intrapersonal, and naturalist (Gardner, 2006, pp. 6, 8–19). Since students learn in a variety of ways, teachers can improve their instruction by using delivery methods that appeal to different intelligences (Kapusnick & Hauslein, 2001, p. 156; Morgan, 2013, p. 35). Multiple intelligences differ from *learning styles*, which are modes of instruction or study that theoretically are most effective for students. Research shows little evidence that matching instruction to a particular learning style will improve instruction (Goodwin & Hein, 2017; Pashler, McDaniel, Rohrer, & Bjork, 2008, p. 117; Willingham, Hughes, & Dobolyi, 2015, p. 267). In other words, learning styles suggest that each student learns best in *one* way, while multiple intelligences mean that every student learns in many *different* ways.

2. **Zone of proximal development.** The *zone of proximal development*, proposed by Soviet psychologist Lev Vygotsky (1978, p. 86), is defined as the difference between what learners can do independently and what they can do with assistance. Teachers can work with students with diverse skills by stretching them beyond their comfort level but not to the point of frustration (Kapusnick & Hauslein, 2001, p. 157). Some have criticized Vygotsky's theory for being rhetorical and not intended for theoretical development (Valsiner, 1998, p. 69; Valsiner & van der Veer, 1993, p. 43). However, other researchers have found evidence to support Vygotsky's claims in various educational settings (Freund, 1990, p. 124; Verenikina, 2010, pp. 8–10).

3. **Scaffolding.** *Scaffolding* is the set of activities that support learning as students move through their zones of proximal development (Wood, Bruner, & Ross, 1976, p. 90). Scaffolding can be removed as students gain confidence and competence in being independent learners (Moschkovich, 2015, p. 1068).

Differentiation can incorporate a wide variety of instructional strategies and is thus difficult for researchers to analyze precisely. However, differentiation seems to have a small overall positive effect on academic performance, especially when it is done in a supportive context, such as a computer-assisted environment or a broader school reform (Deunk, Smale-Jacobse, de Boer, Doolaard, & Bosker, 2018, p. 35). Differentiation appears to be significantly related to teacher self-efficacy (De Neve, Devos, & Tuytens, 2015, p. 37; Dixon, Yssel, McConnell, & Hardin, 2014, p. 13; Suprayogi, Valcke, & Godwin, 2017, p. 297).

Research validates several practices that are central to differentiation (Huebner, 2010), including effective classroom management (Rimm-Kaufman & Sandilos, n.d.) and adjusting instruction to match students' needs and interests (Parsons et al., 2018, p. 27).

Common Core Connections

The Common Core standards do not specify how to differentiate instruction for any particular group of students. However, they are based on the idea that all students must have the opportunity to learn and meet the same high standards in order to access the knowledge required after they graduate. As a result, the standards should be interpreted so that all students can participate in mathematical activities with appropriate accommodations (NGA & CCSSO, 2010, p. 4).

For example, several standards, such as drawing geometric shapes with given conditions (7.G.A.2), finding the approximate solutions to the equation $f(x) = g(x)$ (A-REI.D.11), and estimating areas under the normal curve (S-ID.A.4), ask students to solve problems using different methods or to represent mathematical ideas in different ways (NGA & CCSSO, 2010, pp. 50, 66, 81). Other standards ask students to compare properties represented in different ways (8.F.A.2), to model with mathematics (MP.4), or to use appropriate tools strategically (MP.5) (NGA & CCSSO, 2010, pp. 55, 7), which allow students to differentiate by process using multiple mathematical methods.

Application

Carol Ann Tomlinson (2014, p. 20) defines four areas in which teachers can differentiate instruction—content, process, product, and affect. In this section, we describe some of our favorite techniques for each of them.

DIFFERENTIATION BY CONTENT

Teachers can differentiate *content*—what students will learn or how students access the information needed to learn (Tomlinson, 2014, p. 82).

Tiered Activities

We often use *tiered activities* (sometimes called *parallel tasks*), in which we divide work into levels by complexity so that students with different levels of understanding on a topic can work simultaneously on it (Pierce & Adams, 2005, p. 145; Small & Lin, 2010, p. 11; Tomlinson, 2014, p. 133). We sequence work to move students through their zone of proximal development. In other words, the tasks are designed to be accessible enough so that students can use their past knowledge to understand it, but challenging enough so that students can extend their learning (Artzt et al., 2008, p. 13).

Typically, our tiered activities are divided into different levels of complexity, as shown in Table 14.1: Levels of Complexity for Tiered Lessons. We typically use four or five levels since our students are used to seeing these scales on the state tests and AP exams that they take.

We fit our tiered lessons into the familiar framework of a whole-group introductory discussion, guided independent practice, and a whole-group summary. Figure 14.1: Tiered Lesson—Literal Equations and Figure 14.2: Tiered Lesson—Midpoint show outlines of tiered lessons from Algebra I and Geometry. Both lessons have Do Now questions that are accessible to students based on their prior knowledge. The class discussion of the Do Now activity leads into the lesson. Classwork problems range from the easiest questions in Level 1 to the hardest questions in Level 5. The lessons conclude with a summary question that students can answer individually as an exit ticket or with the class in a whole-group discussion.

We provide students with classwork sheets that contain all of the problems in the lesson. By providing all levels to everyone, students can monitor themselves, know

Table 14.1 Levels of Complexity for Tiered Lessons

Level	Description
1	Requires direct recall of a fact or method or requires a one-step procedure
2	Requires an additional step beyond a Level 1 problem
3	Requires several steps to be combined, usually two or more theorems or facts without explicit guidance *(approaches appropriate standard)*
4	Requires reasoning through complex problems or applies solution method to a new context *(meets appropriate standard)*
5	*(optional)* Requires additional synthesis, interpretation, or analysis beyond a Level 4 problem *(exceeds appropriate standard)*

what they need to do to improve to the next level, and work through the lesson at their own pace (Wong & Bukalov, 2013, p. 56). We find that most of our students prefer to follow along with us while we discuss the lesson and do classwork examples. After they become comfortable with the new material, many choose to continue without our guidance and often work with a classmate. As we circulate around the room, we can quickly assess student progress by seeing the level at which they are working.

We discuss how we organize classwork at the beginning of the year by sharing Table 14.1: Levels of Complexity for Tiered Lessons with students. We clearly label the levels on classwork sheets so that students understand our expectations and everyone has a common language for describing their level of understanding for each lesson.

To create a tiered lesson, we use the following steps:

1. **Determine the goals and skills for the lesson.** We typically take these from our unit plan (which we describe more in Chapter 6: How to Plan Units).

2. **Determine the Level 4 problems that are appropriate for those goals.** Since Level 4 problems meet the standards for the lesson, we use them as our end goal. We refer to sources such as textbooks, websites, and end-of-year assessment questions to help us.

3. **Adjust the Level 4 problems as necessary to match the skills and readiness of our students.** Matching the levels to our students' current skills and readiness is critical to their social-emotional learning. We use the levels to divide work into increments that are small enough so that students can process and retain information. Doing so helps students feel that progress is more attainable, which will decrease their stress level and improve the likelihood that they will remember what they learn (Hammond, 2015, p. 130; Sousa, 2017, p. 50). If the Level 4 problems are too easy for our students, we may label them Level 3 so that students can get challenging work. If the Level 4 problems are too difficult, we may label them Level 5 or put them off into a future lesson to prevent students from getting too discouraged.

4. **Identify three or four skills necessary to complete Level 4 problems.** These skills help us determine the difficulty of Level 1 (one skill required), Level 2 (two skills required), and Level 3 (three or more skills required) problems. For example, Figure 14.2: Tiered Lesson—Midpoint leads to coordinate geometry proofs involving quadrilaterals. To prepare students for these Level 4 proofs, Level 1 contains straightforward applications of the midpoint formula, Level 2 contains a slightly more complicated application, and Level 3 contains simpler proofs in which two segments bisect each other.

5. **Put an appropriate number of problems for each level.** We typically put between two and four questions in each level, depending on the difficulty of the questions and the amount of time we have in class. This usually gives most students enough time and practice to reach Level 3 or 4. Students who don't reach Level 4 by the end of class often need additional guidance or instruction (we discuss this more in the "What Could Go Wrong" section of this chapter).

6. **If we have students that we believe can complete Level 4 problems and want an additional challenge, we try to create Level 5 problems.** Figure 14.1: Tiered Lesson—Literal Equations contains a Level 5 real-world application that involves solving a literal equation. In Figure 14.2: Tiered Lesson Midpoint, Level 5 asks students to use the midpoint formula to complete a proof of a theorem—an extension of the coordinate geometry proofs typically seen on our state exams. We often use textbooks for advanced courses or questions from math competitions as inspirations for Level 5 questions.

Tiered activities tend to work best for topics that require relatively little direct instruction so that students have enough time to work through the levels. Lessons that require a great deal of direct instruction, such as an introduction to an unfamiliar topic, don't work well with tiered activities. In these situations, we rely on other techniques, such as concept attainment (described in Chapter 3: Teaching Math as a Language), open-ended questions (described in Chapter 4: Promoting Mathematical Communication), and project-based learning (described in Chapter 16: Project-Based Learning).

Tiered activities can also work well with standards-based grading (which we discuss more in Chapter 10: How to Develop an Effective Grading Policy). Classwork levels can correspond to the levels of understanding in a standards-based grading rubric.

Tiered activities often require a great deal of time and energy to organize. We have to think carefully about putting problems in a logical order. Relying on textbooks and other sources that order questions by increasing difficulty (even if they don't explicitly group problems by level) can make this task easier. Nevertheless, we believe that the many benefits of tiering—identifying and communicating the skills required to meet the lesson's goals—make the effort worthwhile. In addition, once we prepare tiers for a lesson, we can reuse them every year.

When we tier lessons, putting too few questions at each level won't give students enough practice or expose them to a sufficient number of challenging questions. Putting too many questions can discourage students by making them feel that they aren't making any progress. We try to put enough questions at each level so that most students can reasonably finish about three levels in one lesson.

As students are working, we monitor their progress constantly to see where they encounter difficulty or move quickly. If many students are getting stuck on the same problem, we stop the class and have a whole-group discussion about common misconceptions. We find that writing good tiered lessons requires years of refining to determine the right amount of difficulty and number of questions for each level. As we said in Chapter 7: How to Plan Lessons, reflecting on a lesson after we teach it and making brief notes, such as "add more problems with negative numbers" or "move #8 to Level 3—too hard," can prevent us from making the same mistakes in the future.

Presentation Style

Teachers can also differentiate content by varying their *presentation style*, the way in which new information is presented to students. They could watch a video, read a textbook, or view a website. In addition, students can use manipulatives (which we discuss in Chapter 15: Differentiating for Students with Unique Needs) and technology (such as graphing calculators).

We recommend allowing students to work in pairs or small groups when possible. This helps to promote *active learning* (in which students *do* something to enhance their understanding, such as answer questions, find and explain a pattern, or choose an appropriate mathematical model for a data set) instead of *passive learning* (in which students merely absorb information, such as listen to a lecture, watch a video, or read a textbook). Research indicates that active learning helps students monitor their thinking and promotes retrieval practice (Markant, Ruggeri, Gureckis, & Xu, 2016, pp. 145–146). Although students can obviously answer questions on their own, we find that students are more likely to interact with the material when they work with other students. To keep everyone accountable, we may randomly select a student from each group to answer a question or distribute small whiteboards or electronic clickers so that students can answer a question simultaneously. We may also use activities in which students teach each other parts of the lesson. We discuss these and similar techniques more in Chapter 17: Cooperative Learning.

Teachers can provide questions that students can answer while reading a book or watching a presentation. For a lesson on literal equations (shown in Figure 14.1: Tiered Lessons—Literal Equations), students can answer questions like:

- What is the definition of a literal equation?
- List at least two similarities and differences between solving a literal equation and solving other types of equations.

Ability Grouping

Many teachers identify differentiation with *ability grouping*, an instructional practice in which teachers group students into homogeneous groups based on their prior performance or initial readiness level. Unlike *tracking*, which is typically done by schools between classes, ability grouping is done by teachers within classes (Loveless, 2013, p. 13).

Ability grouping is controversial. Proponents argue that it helps teachers to modify the pacing and content of instruction. A recent meta-analysis indicates that students benefited from being grouped by ability within the classroom (Steenbergen-Hu, Makel, & Olszewski-Kubilius, 2016, p. 41). Critics argue that it increases achievement gaps and reduces the self-esteem of students labeled "low-ability." Some research indicates that students grouped by ability have lower test scores than students in heterogeneous grouping, and the achievement gap increases with more exposure to ability grouping (Buttaro & Catsambis, 2019).

In our experience, consistently separating all students by their level of readiness stigmatizes students who receive less-challenging work, which quickly decreases their motivation. However, we find, as others (Johnson, 2011; Steenbergen-Hu et al., 2016, p. 41) do, that such grouping can occasionally benefit students who are very far behind or ahead of the rest of the class. In these situations, we suggest using reteaching assignments or curriculum compacting, respectively, which we describe in more detail below.

Reteaching Assignments

Some of our students are so far behind the rest of the class that they struggle to retain any new information. They often score below 20% on tests and quizzes and cannot do problems independently. We look at their grades from previous years and talk to them privately to understand their prior knowledge and the reasons for their struggle.

To help these students succeed, we give them *reteaching assignments*, which reteach material covered in previous courses that we don't have time to cover in our curriculum. Reteaching assignments include instruction that students can read or view on their own with minimal intervention, such as videos, websites, or textbook pages. If we have several students in a class who need reteaching assignments, we encourage them to work together so they can support each other and so that we can answer their questions more easily. The reteaching assignments replace the regular assignments that we give other students.

To help motivate students to work on these reteaching assignments, we give them properly scaffolded work that matches their current level of readiness (Usher & Kober, 2012a, p. 4). If we have time, we meet with these students individually to develop their own reteaching plan. We provide a list of six or seven options (such as a packet of worksheets or an online assignment) and ask them to choose three. Together with the students (and their parents, if possible), we develop an individualized plan for the assignment. Putting this plan in writing clarifies expectations for everyone and helps hold students (and us) accountable. We discuss other ways to engage students in Chapter 1: Motivating Students.

Of course, giving reteaching assignments isn't as effective as working with students one-on-one outside of class or having small-group instruction with a tutor. Unfortunately, we often don't have enough time or resources to make other options possible.

Some students are generally able to keep up with our presentation of new material but may have difficulty with a particular concept from a previous course. In these cases, we offer the opportunity to complete a reteaching assignment only for that topic. Although we allow students to work on these assignments in class, we point out to them that they would then miss out on lessons and potentially fall behind. As a result, we encourage them to take time outside of class to complete reteaching assignments. Constantly assessing students and giving them reteaching assignments if they don't demonstrate mastery takes too much time and effort to be sustainable. Instead, we allow *all* students to decide on their own whether they need a reteaching assignment for a particular topic.

Creating reteaching assignments can be challenging since we don't have the time to make them for *every* topic in every course we teach. Instead, we design assignments only for the most important prerequisite skills. For example, when we teach Geometry, we use reteaching assignments for algebraic skills such as combining like terms and solving linear equations, which students should have learned in Algebra I. Eighth grade teachers may prepare reteaching assignments on operations with fractions, which students should have learned in seventh grade.

In addition, we rarely write problems from scratch for reteaching assignments. Instead, we put together packets of pages from textbooks, worksheets from our own or our colleagues' lessons from previous courses, or assignments from tutoring and practice websites (we list resources in the Technology Connections section). Many websites include online tutorials that allow students to learn on their own, although we monitor their progress in class.

Another challenge that we encounter with reteaching assignments is that some students are so far behind that they never reach grade-level material. Many times, these students have experienced years of frustration in math, so the idea of trying to learn several years of material in a few months can heighten their math anxiety. In addition, the pressure that they get from parents, guidance counselors, or schools to move on to the next course or graduate even if they haven't shown mastery can discourage even the most motivated student. In these situations, we focus on the growth that they make in our class. We monitor and praise their progress constantly and encourage them to become more self-directed learners. We find that students who make significant progress can build their self-confidence and motivation, which will help them in the future.

To assess students who complete reteaching assignments, we often grade student work on the assignments and count it as a test or quiz. To make grading more manageable, we grade handwritten work on a simplified scale, such as two points for a correct solution, one point for a partially correct solution, and zero points for a missing or completely incorrect solution. Using websites that automatically grade student work can save a lot of time. Since we often allow students to complete reteaching assignments outside of class, they may cheat and ask a friend or use an app to do the work. Pointing out to them that reteaching assignments are designed to help them catch up to where they should be may discourage student cheating. We discuss how we help students think about ethics in Chapter 9: How to Plan Tests and Quizzes.

Another option for grading students who complete reteaching assignments is to give them in-class tests or quizzes. Often, we create these assessments by taking questions from the packets, textbooks, or websites that we used to create the reteaching material. In-class assessments can keep students more accountable for their work. Sometimes, we have enough time to give both a reteaching assignment and an in-class test or quiz.

Curriculum Compacting

Curriculum compacting is an instructional technique in which students with prior knowledge can skip work they already know and instead learn more challenging content. Research indicates that students who used a compacted curriculum scored significantly higher on math tests (Reis & Renzulli, 1992). This strategy can benefit any student who shows strength or high levels of interest in a particular topic. Differentiating for these learners can help them become more self-directed and engage them at their level (Gregory & Chapman, 2007, p. 77).

Researchers Sally Reis and Joseph Renzulli (1992) divide curriculum compacting into three parts:

1. **Identify instructional goals that represent new content.** Teachers can use the instructional goals from resources like unit plans (which we discuss in Chapter 6: How to Plan Units) and curriculum guides. Comparing learning objectives from students' current courses to those from previous courses can help teachers identify what students should have learned before.

2. **Identify students who have already mastered instructional goals to be taught.** Although teachers can use pretests, we think that this would

overwhelm students and teachers with a barrage of testing. Instead, we suggest occasionally incorporating questions from future lessons into more informal assessments like homework questions, Do Now assignments, or exit tickets. These questions should be clearly labeled as previews of an upcoming topic so that students wouldn't feel that they are being tested on something that they haven't learned yet. Presenting these questions as extra credit or a challenge can even motivate some students to try to learn some of the material on their own, especially if they know that demonstrating mastery of a topic will enable them to learn more advanced material.

3. **Provide acceleration and enrichment options.** This can include project-based learning (which we discuss in Chapter 16: Project-Based Learning) or group work (which we discuss more in Chapter 17: Cooperative Learning). This work would replace the regular classroom assignments for that topic.

Figure 14.3: Curriculum Compacting—Coordinate Geometry shows an example of an enrichment project that can be used in place of a unit on coordinate geometry proofs. In Part 1, students create their own coordinate geometry proofs with numerical coordinates (these questions are the main goal of the unit, so they can serve as a starting point for this project). In Part 2, students are guided through the process of writing coordinate geometry proofs with variables. In Part 3, students use coordinate geometry to prove theorems.

Ideally, the enrichment projects in a compacted curriculum would be done with minimal teacher guidance. Students assigned to these projects could work together in class.

We like the idea of differentiating for students whose levels of readiness are significantly different from those of their classmates. However, keeping track of which students are working on which projects can become overwhelming for many teachers. We believe that differentiating work with tiered activities (which we described earlier in this chapter) can be more effective. It requires less work (since teachers don't have to create large projects) and allows students to immediately self-select an appropriate starting point for work for every lesson. Furthermore, tiered activities can avoid many of the stigmas that can arise from giving different work to students in class (which we discuss more in the What Could Go Wrong section).

DIFFERENTIATION BY PROCESS

Teachers can also differentiate *process*—the activities through which students make sense of the content (Tomlinson, 2014, p. 82). Before moving on to the next part of a lesson, students need time to digest the material and assess their understanding (McCarthy, 2015).

Multiple Mathematical Methods

We often ask students to learn mathematical content by using multiple mathematical methods. This technique can strengthen their reasoning by helping them find more efficient solutions and make connections across various topics (Lim, Kim, Stallings, & Son, 2015, p. 290; Pólya, 1945, p. 64; Zhang et al., 2015, p. 139).

Here are examples of problems that can be solved in multiple ways:

- Calculate $\frac{1}{5} \div \frac{3}{4}$ using a number line or circle model.
- Using a coordinate geometry or paragraph proof, prove that the altitude drawn to the base of an isosceles triangle is also a median.
- Solve the following system of equations by substitution, elimination, or graphing:

$$4x + 3y = 29$$
$$2x - 3y = 1$$

- Determine whether the quantities represented in the accompanying table are in a proportional relationship by graphing or finding equivalent ratios.

Table 14.2 Ratio Word Problem Table

Miles	0	1	2	3	4
Cost	$3	$5	$7	$8	$9

We use this technique after we have taught several methods as a way to synthesize what students have learned. It can take as little as 10–15 minutes or can be a full-period activity, depending on the topic and the amount of time we have. Chapter 5: Making Mathematical Connections describes other ways that we relate mathematical ideas to each other.

One way to implement this strategy is for students to work in small groups to solve the same problem in different ways. They can write the solution in their notebooks or on separate pieces of paper so others can see their work. Students can choose a method that they prefer as long as each group can produce solutions that use all of the methods we discussed (another possibility is to ask students to select a method they *don't* like so they can see if it works better for them). They then share their work with others in their group and compare to determine the advantages and disadvantages of each. We then ask groups to share their findings with the class in a whole-group discussion. We discuss group work more in Chapter 17: Cooperative Learning.

Multiple Intelligences

When possible, we try to appeal to students' multiple intelligences by using physical, visual, musical, and other representations of mathematical ideas. We don't do this in order to cater to *one* particular "style" of learning. Instead, we introduce *several* ways of learning. Knowing how to solve a problem in different ways can improve student understanding (Gardner, 2013).

Here are some ways in which teachers can appeal to multiple intelligences when solving problems:

- Show a geometric sequence kinesthetically with stacks of counters, linguistically through word problems like Zeno's dichotomy paradox, or spatially with pictures that follow a geometric pattern.
- Learn the quadratic formula spatially with faces to represent variables, linguistically with a mnemonic or story, or musically with a song ("3 ways," n.d.).
- Determine whether a function is even or odd kinesthetically with arms or spatially with graphs.

We especially like to represent mathematical ideas by drawing. Drawing requires students to process information semantically, visually, and kinesthetically. As a result, they learn a concept three times in different ways, which can improve their memory (Fernandes et al., 2018, pp. 304–305; Terada, 2019; Wammes et al., 2016, p. 1773).

DIFFERENTIATION BY PRODUCT

Teachers can differentiate *product*—how students show what they know. Here are some ways that students can demonstrate their knowledge.

Tiered Tests

When we tier our lessons, we often differentiate our tests as well. On a typical differentiated exam, most of the test is the same for all students, but we include one or two constructed-response sections in which students can choose tiered questions. Typically, we tier the two or three most difficult questions, which appear at the end of our tests. Tiering the hardest questions offers the most variety for differentiation. We describe other aspects of testing in Chapter 9: How to Plan Tests and Quizzes.

A tiered question is a group of questions on the same topic with different levels of difficulty. Each question in a group has a different point value corresponding to its level of difficulty. Students select one question from each group to answer.

Figure 14.4: Tiered Test Questions shows two tiered questions from a test. For each question, students can choose from an easy two-point, a more challenging four-point, or an advanced six-point item. Since the question is worth four points, students who choose the easiest question can't earn full credit, while students who choose the hardest question can earn full or extra credit even if they don't get it completely correct. Although our tiered classwork often has four or five levels, we find that when we offer such diversity on tests, students take too much time choosing questions. We simplify the process by offering only three levels of difficulty per question.

Just as we allow students to select an appropriate level of difficulty in classwork, we also let them choose what level to answer. We find that students often experiment with answering different levels of a question.

To encourage students to focus on the level that matches their knowledge, we only grade one part for each question. We find that students often experiment with answering different levels of a question. We suggest two options for grading differentiated work:

- **Grade only one level for each question but allow students to choose what level they want graded for each question.** We like this option because it strikes a reasonable balance between giving students autonomy and holding them responsible for completing a task. We tell students to circle the work for the part that they want us to grade. If they don't indicate this, we either grade the work corresponding to the part that is worth the fewest points or grade the part that is most complete. If possible, we give feedback on work that students didn't want us to grade. For example, if students correctly answered a more challenging problem but asked us to grade an easier one instead, we may write a comment like "This was correct!" or have a private conversation. Encouraging them to work on harder problems during class can also build their confidence on tests.

- **Grade all student work shown at all levels.** This method has the advantage of encouraging students to try different levels. However, we find that students wind up trying so many parts that they often don't complete questions. Teachers could choose to add up all points that students earn from all levels, but such work could be difficult to grade.

Before the test, we explain how our tiered tests are structured so students know what to expect and how they will be graded. We sometimes incorporate tiered questions into classwork or homework so they can develop an appropriate strategy during tests. We also try to make nontiered sections of tests either slightly shorter or easier to give students enough time to try different problems in tiered sections of tests. Finally, we remind students during tiered tests or quizzes to choose an appropriate level for each tiered question.

While we would like to offer tiering on every test and quiz we make, we find that the extra time and effort required makes that goal virtually impossible. We find that offering tiered testing even once or twice per semester can offer benefits. Tiering reduces the chances that struggling students will get no credit while holding advanced students accountable with harder questions. Teachers can reserve tiering for a test on a particularly difficult unit or for a change of pace from the usual routine.

Portfolios

A *portfolio* is a collection of student work that demonstrates student understanding of specific skills or ideas. By providing initial samples of work and periodically adding samples over a given time period, students can monitor and reflect on their progress (Gregory & Chapman, 2007, p. 63).

Students can select pieces to add to their portfolios using guidelines that reflect both mathematical and social-emotional growth, such as an example of work that they:

- are proud to show others (such as pieces that show growth)
- are doing right now
- understand well
- struggled to complete (Gregory & Chapman, 2007, p. 64; Maxwell & Lassak, 2008, p. 406)

To get a representative sample of student work throughout the year, we ask students to include one or two examples of each criterion per unit or marking period (this can vary according to the course and our students). With each piece, students write a brief explanation of what criteria it satisfies and why it was chosen. Although we encourage students to select pieces themselves, we sometimes suggest ones that we think fit certain criteria.

Students can maintain physical folders that we keep in our classrooms or take pictures of their samples to include in a digital folder. We periodically remind students to check their portfolios during the year to add new material or replace existing samples with better examples.

Review Materials

Students can create *review materials*—brief summaries of key concepts from a unit. We start with a list of key concepts for the course (which we take from our curriculum, unit plan, or textbook) and organize them by units in the order in which we teach them. Students then use this list to create review materials.

We use several different formats for review material:

- **Review sheet.** Students can make a review sheet for each unit. To encourage students to emphasize important ideas, we limit the amount of material that students can include. This length varies, depending on the unit's complexity. Typically, we restrict it to no more than one page per unit. By the end of the school year, students will then have a five- or six-page summary of the entire course. To discourage students from handing in someone else's typed work as their own, teachers can ask them to submit handwritten review sheets (an excerpt of one is shown in Figure 14.5: Review Sheet). Sometimes, we provide more structure in the review sheet by organizing information into tables (as shown in Figure 14.6: Fill-in Review Sheet) or giving sentences with blanks that students can complete. This variation has the advantage of ensuring that all students have the same material.

- **Vocabulary chart.** A vocabulary chart consists of terms, definitions, examples, and illustrations (we discuss vocabulary charts more in Chapter 3: Teaching Math as a Language). This technique works particularly well for topics or courses like Geometry that have many terms or formulas.

- **Review booklet.** Students can also create a review booklet, which is a longer, more detailed version of a review sheet for a unit or topic (see Figure 14.7: Review Booklet for an example). They receive an outline of the key concepts of the unit (which we derive from our curriculum outline or unit plan) and a list of specific examples or pictures that they can use. We ask students to imagine that they are writing a chapter of a review book, so we encourage them to be clear and creative.

- **Presentation.** Students who are comfortable using technology can create a presentation or video for the unit.

These options for review material are not mutually exclusive. We sometimes differentiate this assignment by giving students some or all of the options and allowing them to choose (we discuss the role of autonomy in student motivation in Chapter 1: Motivating Students).

Grading Alternate Assessments with Rubrics

When we grade students on open-ended tasks like portfolios and review assignments, we use *rubrics*, which contain criteria for student work, including descriptions of levels of performance quality for the criteria (Brookhart, 2013, p. 4). In contrast, when we grade test and quiz questions, we use scoring guidelines, which we describe in Chapter 9: How to Plan Tests and Quizzes.

Most of the rubrics that we use are *analytic rubrics*, which describe work on each criterion individually. By separating criteria, students and teachers can more easily see what elements of their work need more attention (Brookhart, 2013, p. 6). Our rubrics typically have criteria for explanations of mathematical terms, examples with solutions, and formatting, as shown in Table 14.3: Rubric. When we give an assignment, we review the criteria with students so that they know what we expect.

Table 14.3 Rubric

PTS.	Explanations	Examples	Formatting
4	All relevant mathematical concepts (terms, formulas, theorems) are briefly and clearly explained.	All relevant examples with complete and correct solutions are included.	Review sheet is legible and fits in allotted space for each unit.
3	The explanations of most relevant mathematical concepts are briefly and clearly explained. A few have minor errors.	Most examples with their solutions are included. (A few important concepts lack examples or a few solutions have minor errors.)	Review sheet is slightly illegible or uses slightly more space than allowed.
2	Only about half of the explanations are briefly and clearly explained.	Only about half of the relevant examples or their solutions are included. OR All relevant examples but no relevant solutions are included.	About half of the review sheet is illegible. OR The review sheet uses a great deal more space than allowed.
1	Only a few of the explanations are briefly and clearly explained.	Only a few of the relevant examples or their solutions are included.	Review sheet is generally illegible.
0	No explanations are included.	No relevant examples are included.	Review sheet is completely illegible.

DIFFERENTIATION BY AFFECT

Teachers can differentiate by *affect*—the classroom conditions and interactions that set the tone and expectations for learning. Effective teachers create a classroom environment in which all students feel safe and ready to learn and establish a positive working relationship with students (Tomlinson, 2014, p. 82).

We find that treating students as individuals, learning their interests, giving them some autonomy in their learning, and even seemingly small things like greeting them by name can, over time, improve student motivation and academic achievement. As we teach a lesson, we constantly assess how our students react, looking for any signs that indicate their level of understanding and their emotional readiness to learn more.

We discuss these ideas more in Chapter 1: Motivating Students, Chapter 11: Building a Productive Classroom Environment, and Chapter 18: Formative Assessment.

What Could Go Wrong

One of the most serious challenges we face is the fear that differentiation requires a great deal of additional time or effort. In fact, most effective teachers already modify at least *some* of their instruction *some* of the time (Tomlinson, 2014, p. 14). Differentiation is less a *checklist* of procedures and more of a *philosophy* that helps educators plan strategically (Gregory & Chapman, 2007, p. 2).

Here are some simple differentiation strategies that should take little additional time and effort and can be easily incorporated into almost any lesson:

- **Clearly indicate difficulty level.** Most math lessons contain problems that are sequenced from easiest to hardest. We often group similar problems together into three to five levels and clearly label them so that students can more easily know the degree of difficulty for their work.

- **Create an alternate assignment.** Use a website like DeltaMath or a textbook review sheet to create an alternate assessment, which could be counted as a quiz or extra credit. Many websites can grade student work automatically, saving teachers valuable time.

- **Allow students to write their own word problem using a context that reflects their interests, family, or culture.** We discuss this more in Chapter 4: Promoting Mathematical Communication.

- **Give students a variety of ways to summarize a lesson.** At the end of class, let students summarize, using tools such as responding to a prompt in a short paragraph or drawing a picture that explains key concepts. We discuss lesson summaries more in Chapter 3: Teaching Math as a Language.

As with many other strategies we describe in this book, we find that *any* differentiation is usually better than none. Most of the time, our students are eager to get assignments that they feel are more appropriate for them. In the end, differentiating effectively will help us align our instruction more with students' talents, challenges, levels of readiness, and interests.

Student Handouts and Examples

Figure 14.1: Tiered Lesson—Literal Equations
Figure 14.2: Tiered Lesson—Midpoint
Figure 14.3: Curriculum Compacting—Coordinate Geometry
Figure 14.4: Tiered Test Questions
Figure 14.5: Review Sheet
Figure 14.6: Fill-In Review Sheet
Figure 14.7: Review Book

Technology Connections

The website of the IRIS Center at Vanderbilt University's Peabody College (http://iris.peabody.vanderbilt.edu) has many resources on differentiation. Howard Gardner's website (http://howardgardner.com) has articles and other information on multiple intelligences.

Websites with more information on cooperative learning include the Jigsaw Classroom website (www.jigsaw.org) and the International Association for the Study of Cooperation in Education (http://www.iasce.net/home/resources).

The National Center for Research on Gifted Education (http://ncrge.uconn.edu) has information on ways to help advanced students. The ¡Colorín Colorado! website (www.colorincolorado.org) has many resources for teachers of ELLs. The National Association of Special Education Teachers (www.naset.org) and Learning Disabilities Online (http://www.ldonline.org/educators) websites have resources for teachers of students with learning differences.

Several sites can help teachers create rubrics. Rcampus has a rubric gallery (http://www.rcampus.com/rubricshellc.cfm?mode=gallery&sms=publicrub&sid=21&) with thousands of math rubrics. Quick Rubric (www.quickrubric.com) and TeAchnology (http://www.teach-nology.com/web_tools/rubrics) allow teachers to create and save their own rubric online.

Students can use Google Sites (http://sites.google.com), Kidblog (http://kidblog.org), or Evernote (http://evernote.com) to create websites that can serve as online portfolios.

To create customized reteaching assignments, we like to use sites like the free DeltaMath (http://deltamath.com).

Figures

Algebra I

Aim #28: How do we solve a literal equation?

Goals:
- Compare and contrast a literal equation with an equation containing only one variable.
- Investigate and apply the procedure for solving a literal equation.
- Solve a literal equation for an indicated variable (the subject).
- Apply the solution of literal equations to real-world problems.

Standards: (A-CED.A.4) Rearrange formulas to highlight a quantity of interest, using the same reasoning as in solving equations.
(A-REI.B.3) Solve linear equations and inequalities in one variable, including equations with coefficients represented by letters.

Do Now

Solve each equation for x.

a. $3x + 7 = 19$ *Ans:* $x = \dfrac{19-7}{3} = 4$

b. $2x + 4 = 18$ *Ans:* $x = \dfrac{18-4}{2} = 7$

c. $3x + 5 = y$ *Ans:* $x = \dfrac{y-5}{3}$

How is your work to all four problems from the Do Now similar?
   Answer: Subtracted the constant and divided by the coefficient of x.
   DEFINITION: A **literal equation** is an equation with two or more variables.

Level 1

1. Solve $a + 5 = c$ for a.   Answer: $a = c - 5$
2. Solve $P = 4s$ for s.   Answer: $s = \dfrac{P}{4}$
3. Solve $x - m = 4$ for x.   Answer: $x = 4 + m$
4. Solve $\dfrac{x}{y} = 5$ for x.   Answer: $x = 5y$

Figure 14.1 Tiered Lesson—Literal Equations

Level 2

5. Solve $y = mx + b$ for m.   Answer: $m = \dfrac{y - b}{x}$

6. Solve $\dfrac{ax}{b} = c$ for x.   Answer: $x = \dfrac{bc}{a}$

7. Solve $V = lwh$ for h.   Answer: $h = \dfrac{V}{lw}$

8. Solve $m = \dfrac{a + b}{2}$ for b.   Answer: $b = 2m - a$

Level 3

9. Solve $28 = t(r + 4)$ for t.   Answer: $t = \dfrac{28}{r + 4}$

10. Solve $a(q - 8) = 23$ for q.   Answer: $q = \dfrac{23}{a} + 8$

11. Solve $aq - mq = d$ for q.   Answer: $q = \dfrac{d}{a - m}$

Level 4

12. Solve $pq = w + pt$ for p.   Answer: $p = \dfrac{w}{q - t}$

13. Solve $ax - ry = x + w$ for x.   Answer: $x = \dfrac{w + ry}{a}$

14. Solve $ax - by = xz + y$ for y.   Answer: $y = \dfrac{ax - xz}{b + 1}$

Level 5

15. The formula to convert temperature in degrees Celsius (C) to degrees Fahrenheit (F) is $F = \dfrac{9}{5}C + 32$.

   a. Solve the conversion formula for C.
      Answer: $C = \dfrac{5}{9}(F - 32)$

   b. The high temperature last year was 86° Fahrenheit. Use your answer to part a to convert this to Celsius.
      Answer: 30°C

Summary

How is solving literal equations similar to solving an equation with just one variable? How is it different?

Figure 14.1 (Continued)

Geometry

Aim #3.27: How do we use the midpoint formula?

Goals:
- Derive the formulas for the midpoint of a line segment, given the coordinates of its endpoints.
- Find the coordinates of the midpoint of any line segment, given the coordinates of its endpoints.
- Find the coordinates of one endpoint of a line segment given the coordinates of the midpoint and its other endpoint.
- Write coordinate geometry proofs by finding the midpoint of line segments.

Standards: (G-GPE.B.4) Use coordinates to prove simple geometric theorems algebraically. For example, prove or disprove that a figure defined by four given points in the coordinate plane is a rectangle; prove or disprove that the point $(1, \sqrt{3})$ lies on the circle centered at the origin and containing the point $(0, 2)$.

Do Now

1. $\overline{EF}$ has endpoints $E(4, 1)$ and $F(8, 9)$.
   a. Graph $\overline{EF}$ on the coordinate plane.
   b. Let M be the midpoint of $\overline{EF}$. Find the x-coordinate of M. How is it related to the x-coordinates of points E and F?
      Answer: 6 is the average of 4 and 8.
   c. Find the y-coordinate of M. How is it related to the y-coordinates of points E and F?
      Answer: 5 is the average of 1 and 9.
   d. Graph M. How can you find the midpoint of M without graphing the line segment? (HINT: Look at the relationship between the coordinates of M and the coordinates of E and F.)

2. Fill in the blanks:
   The midpoint of the segment connecting two points (x_1, y_1) and (x_2, y_2) is $\left(\dfrac{x_1+x_2}{2}, \dfrac{y_1+y_2}{2}\right)$.

Figure 14.2 Tiered Lesson—Midpoint

(Elicit: It is the average of the coordinates of the endpoints. Students don't have to use the formula—they can graph the points and find the midpoint graphically.)

PRONUNCIATION: Students should say "x-sub-1," not "x-1." Emphasize and repeat as necessary.

Level 1

3. Find the midpoint of the segment whose endpoints are:

   a. (1, 6) and (5, 0) b. (−2, −1) and (4, −3) c. (−7.2, 2) and (4.5, 7.5)
   (3, 3)   (1, −2)   (−1.35, 4.75)

Level 2

4. If one endpoint of $\overline{AB}$ has coordinates (6, 10) and M(7, −2) is the midpoint of $\overline{AB}$, find the coordinates of the other endpoint.
   Answer: (8, −14)

5. If one endpoint of $\overline{CD}$ has coordinates (5, −1) and M(−1, 3) is the midpoint of $\overline{CD}$, find the coordinates of the other endpoint.
   Answer: (−7, 7)

Level 3

6. Given points A(−2, 1), B(2, 6), C(8, 3), and D(4, −2).
   a. Show that $\overline{AC}$ bisects $\overline{BD}$.
      $\overline{AC}$ passes through (3, 2), the midpoint of $\overline{BD}$.
   b. Show that $\overline{BD}$ bisects $\overline{AC}$.
      $\overline{BD}$ passes through (3, 2), the midpoint of $\overline{AC}$.
   c. Write an appropriate conclusion that summarizes your findings from parts a and b.
      If two segments have the same midpoint, then they bisect each other.

Level 4

7. Quadrilateral EFGH has vertices E(8, −2), F(17, −2), G(16, −6), and H(7, −6).

Figure 14.2 (Continued)

a. Show that the diagonals of EFGH bisect each other.

Write appropriate work to show that they have the same midpoint, (12, −4). Then write: If two segments have the same midpoint, then they bisect each other.

b. What kind of quadrilateral is EFGH? Use your work from part *a* to explain.

Parallelogram: If the diagonals of a quadrilateral bisect each other, then it is a parallelogram.

8. Show that the diagonals of the quadrilateral whose vertices are A(3, 5), B(6, 4), C(7, 8), and D(4, 12) do not bisect each other.

Midpoint of $\overline{AC}$ = (5, 6.5), midpoint of $\overline{BD}$ = (5, 8). The diagonals don't have the same midpoint, so they don't bisect each other.

Level 5

9. △ABC has vertices A(0, 0), B(a, 0), and C(b, c).

a. Sketch a graph of △ABC on the coordinate plane.

b. Let M be the midpoint of $\overline{AC}$ and N be the midpoint of $\overline{BC}$. Find the coordinates of M and N. (HINT: Your coordinates should be in terms of *a*, *b*, and *c*.)

$$M\left(\frac{b}{2}, \frac{c}{2}\right), N\left(\frac{a+b}{2}, \frac{c}{2}\right)$$

c. Find the lengths MN and AB. How do they compare?

$$MN = \frac{a+b}{2} - \frac{b}{2} = \frac{a}{2}, AB = a$$

d. State in words the theorem that you have just proven.

The length of the line segment that joins the midpoints of two sides of a triangle is equal to half the length of the third side.

Summary

How is the midpoint of a line segment related to the average?

The midpoint's coordinates are the average of the x- and y-coordinates.

Figure 14.2 (Continued)

Geometry

NAME: _____

Enrichment Project: Coordinate Geometry Proofs

- Coordinate geometry provides a powerful tool for geometric proofs. This assignment will guide you through the proofs of two theorems related to the midpoints of polygons.
- Attach this sheet to the front of your assignment. Write your name on the top of this paper in the space provided. All pages with your work must be stapled together.
- Complete as much of this assignment as you can. Show all work, including appropriate formula substitutions, diagrams, graphs, etc., on separate pieces of graph paper. Partial credit will be given.

Part 1

1. Write and solve one example of any three of the coordinate geometry proofs listed below (use the examples from the Geometry Regents Exam on the class website as models):
   a. Prove that △ABC is isosceles.
   b. Prove that quadrilateral DEFG is a parallelogram.
   c. Prove that quadrilateral HJKL is a rectangle.
   d. Prove that quadrilateral MNPQ is a trapezoid.
   e. Prove that quadrilateral RSTU is a rhombus.
   f. Prove that quadrilateral WXYZ is a square.

For each example that you write, you must do the following:

- Choose appropriate coordinates for your polygons. All vertex coordinates must be integers. Do not use variables.
- Use correct mathematical language.
- Create an answer key that shows all correct work, including a graph.
- Do *not* just copy a question from a textbook or website—you must create your own.
- The vertices for each polygon must be in at least two quadrants.

Figure 14.3 Curriculum Compacting—Coordinate Geometry

Part 2

2. △EFG has vertices $E(0, 0)$, $F(x_1, y_1)$, $G(x_2, y_2)$. (Do *not* substitute numbers for x_1, y_1, x_2, or y_2.)

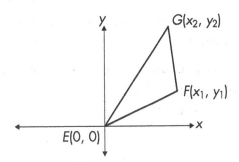

   a. H is the midpoint of $\overline{FG}$ and I is the midpoint of $\overline{EG}$. Use the midpoint formula to calculate the coordinates of H and I.

   b. Use the distance formula to show that $HI = \frac{1}{2}EF$.

   c. Find the slopes of $\overline{HI}$ and $\overline{EF}$. Based on this calculation, what is the relationship between $\overline{HI}$ and $\overline{EF}$?

   d. Your work has proven the following theorem (fill in the blanks): the line segment joining the _____ of two sides of any _____ is half the length of and _____ to the third side of the _____ .

3. Quadrilateral ABCD has vertices $A(0, 0)$, $B(p, 0)$, $C(q, r)$, and $D(s, t)$. Let p, q, r, s, and t represent real numbers. (To simplify calculations, we can place the quadrilateral so that one side of the quadrilateral lies on the x-axis and one vertex is at the origin.)

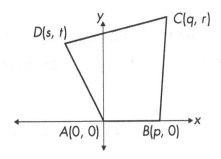

Figure 14.3 (Continued)

a. Let M be the midpoint of $\overline{AB}$, N be the midpoint of $\overline{BC}$, O be the midpoint of $\overline{CD}$, and P be the midpoint of $\overline{AD}$. Find the coordinates of M, N, O, and P. (Your answer will be expressed in terms of p, q, r, s, and t, not numbers.)

b. Express the slope of each side of MNOP as a fraction in simplest form.

c. Based on your answer to part b, what kind of quadrilateral is MNOP? Justify your answer.

d. Your work has proven the following theorem (fill in the blanks): the line segments joining the _____ of consecutive sides of any _____ form a _____ .

Part 3

4. Using the examples in Part 2 as models, prove any three of the following theorems. Your proofs must use variables as coordinates. To simplify your work, you can put one vertex of your polygon at the origin and another vertex on the x-axis.

   a. The midpoint of the hypotenuse of a right triangle is equidistant from the vertices.

   b. The length of the median of a trapezoid is one-half the sum of the lengths of the bases.

   c. If a quadrilateral is a parallelogram, then its diagonals bisect each other.

   d. If a quadrilateral is a rhombus, then its diagonals are perpendicular.

   e. If a quadrilateral is a rectangle, then its diagonals are congruent.

   f. If a quadrilateral is a square, then its diagonals are congruent and perpendicular.

   g. If a parallelogram has congruent diagonals, then it is a rectangle.

   h. If the diagonals of a rectangle are perpendicular, then the rectangle is a square.

Figure 14.3 (Continued)

CHAPTER 14: DIFFERENTIATING INSTRUCTION

PART IV (8 points total). Answer one part for each question. Circle the letter of the part to be graded. If no letter is circled, only the part with the lowest value will be graded. For all questions in this part, a correct answer with no work shown will receive no credit.

14	a. Solve for x: $5x + 1 = 11$. (2 points)	b. Solve for x: $-6x + 3x + 7 = -14$. (4 points)	c. If $6x - 1 - x = 19 + x$ and $2x + 1 = y$, what is the value of y? (6 points)
15	a. Express $(m + 4)(m - 5)$ in standard form. (2 points)	b. Express $(2y + 5)(3 - y) + y^2 - 18$ in standard form. (4 points)	c. Does $m^3 + n^3 = (m + n)(m^2 - mn + m^2)$ for all values of m and n? Justify your answer. (6 points)

Figure 14.4 Tiered Test Questions

Categorical/Qualitative Data - Data that describes a subject
> No meaningful calculations can be made from categorical data.

Bar Graph

Students' Eye Colors

(Bar graph showing Brown: 15, Green: 10, Blue: 5)

Comparing Bar Graphs
> Can graph categorical variables given that they are all measured in the same units
> Compare the heights of the bars

Frequency Table

Students' Eye Colors

Eye Color	Frequency
Brown	15
Green	10
Blue	5

Conditional Relative Frequency - Ratio of a frequency in the body of the table to the total

Two Way Table

What is your favorite fruit?

	Apples	Oranges	Total
Kids	38	12	50
Adults	26	24	50
Total	64	36	100

Marginal Frequency - Values in the "total" row and column

Joint Frequency - Values in the "body" of the table

Figure 14.5 Review Sheet

Geometry: Angle Measurement Relationships in Circles

Complete the table using the pictures and terms below. Then fill in the formulas for each based on the diagrams.

Position of vertex	Type of angle	Diagram	Measurement formula	Formula in words
Center of circle				
On circle				
Inside circle				
Outside circle				

Figure 14.6 Fill-In Review Sheet

Algebra I: Review Book on Functions

You are about to write your own chapters of your math textbook about functions. To help you organize your book, include the following parts:

PART 1: INTRODUCTION: WHAT IS A FUNCTION?

- State the formal mathematical definitions of *relation* and *function*.
- Explain in your own words what a function is. Include your own examples.
- Make sure your text answers the following questions:
- Provide one real-life example (*not mentioned in class or in the textbook*) of a relation that is a function. Explain your answer.
- Provide one real-life example (*not mentioned in class or in the textbook*) of a relation that is *not* a function. Explain your answer.

PART 2: DOMAIN AND RANGE OF FUNCTIONS

- Define the words *domain* and *range*.
- Explain how to find the domain and range of a function, both algebraically and graphically. Include appropriate examples.

PART 3: FAMILIES OF FUNCTIONS

- Write the equation and graph of a linear, quadratic, square root, cube root, exponential, and piecewise function.
- On each graph, label and state the following characteristics: minimum (absolute and relative), maximum (absolute and relative), intercepts, end behavior.

Figure 14.7 Review Booklet

PART 4: TRANSFORMATIONS OF FUNCTIONS

- Select a function $f(x)$ from one of the families in Part 3. Write its equation and graph.
- Write the equation and graph of $f(x)$ after the following transformations:
  - $f(x + 2)$, $f(x - 2)$, $f(x) + 2$, $f(x) - 2$.
- Describe any changes in the function's characteristics (minimum [absolute and relative], maximum [absolute and relative], intercepts, end behavior) after the transformations.

PART 5: SUMMARY

- Create a summary of your chapter book.

Grading (25 Points Total)

- *(5 pts.)* completeness and accuracy of your calculations
- *(5 pts.)* completeness and accuracy of your graphs
- *(5 pts.)* completeness and accuracy of your written explanations
- *(5 pts.)* organization
- *(5 pts.)* creativity

Other Directions

- A paper copy of this book is due by the date and time listed on the class webpage.
- Creativity counts. Feel free to make your work look like an actual book!

Figure 14.7 (Continued)

CHAPTER 15

Differentiating for Students with Unique Needs

What Is It?

In Chapter 14: Differentiating Instruction, we discussed techniques for differentiating instruction for all students. However, we find that some students have unique needs that require particular attention:

- **English Language Learners (ELLs)** are students whose first language is not English and who are learning English.
- **Students with learning differences** are students who have a physical, cognitive, or psychological condition that adversely affects their academic performance (Willis, 2007, p. 12). These children are often called *students with disabilities*, *special education students*, or *students with special needs*. Since many argue that these terms highlight deficiencies, we use the term *learning differences* to emphasize the idea that these students simply learn differently than others (Learning Disabilities vs. Differences, n.d.).
- **Advanced students** work at a level significantly above students of the same age. Many school districts call them *gifted* or *gifted and talented students*, but we prefer the term *advanced students* since we believe that all children, not just advanced ones, have gifts and talents.

We use the term *students with unique needs* to refer to ELLs, students with learning differences, and advanced students. In this chapter, we focus on these three groups for several reasons:

- ELLs, students with learning differences, and advanced students make up a growing portion of the student population nationwide. According to the US Department of Education, 10% of public school students are ELLs, 14% receive special education services, and 7% are in gifted and talented programs (US Department of Education, 2015, 2017, 2018a, 2018b).

- These groups are closely related and have considerable overlap. Some ELLs also have conditions that are categorized as learning differences (Miller, 2016, p. 59). Some researchers use the term *exceptional children* to refer to both students with learning differences and advanced students (Thomson, 2012, p. 159). *Twice-exceptional children* show both an unusual ability and a type of disability (Baldwin, Omdal, & Pereles, 2015, p. 3).

- The percentage of ELLs and students with learning differences has grown in the last 30 years, and their academic performance continues to lag behind the rest of the population. In the 2017 National Assessment of Educational Progress, Grade 8 ELLs and students with learning differences scored between about 40 and 60 points below their peers. Grade 12 students showed a similar gap (National Assessment of Educational Progress, 2017).

While the varied needs of these students may be "different" from those of the majority of our students, they also bring many different assets into the classroom.

Why We Like It

We find that when we work with students with unique needs, we often use strategies that are not only effective for them but for *all* children. In other words, good teaching for ELLs, students with learning differences, and advanced students can be good teaching for everybody!

Supporting Research

Research indicates that effective teachers of students with unique needs do the following:

- Believe that all students can succeed and do not let stereotypes negatively affect academic expectations (Shifrer, 2013, p. 469; Woodcock & Hitches, 2017, p. 312).

- Identify students' strengths and use them to overcome obstacles to student learning (Armstrong, 2012).
- Provide work that is challenging yet appropriate for students' level of readiness (Mun et al., 2016, p. 5; Rogers, 2007, p. 383).
- Build relationships with and among students and attend to their social-emotional needs (Gándara & Santibañez, 2016).
- Understand and appreciate how culture can affect academic achievement (Samson & Collins, 2012, p. 10).

Common Core Connections

The Common Core standards do not specify intervention methods or materials necessary to support students with unique needs. To help them meet the goals spelled out in the standards, teachers can encourage students to represent information and express ideas in multiple ways (6.EE.C.9, 7.RP.A.2, 8.EE.B.5, 8.F.A.2, F-IF.C.8, G-CO.D.12), use technology and manipulatives to promote student thinking (7.G.A.2, A-REI-D.11, F-BF.B.3), and emphasize understanding of problems before attempting to solve (6.RP.A.1, 7.NS.A.1, 8.EE.C.8a, A-APR.A.1, F-IF.A.1, G-SRT.C.6) (NGA & CCSSO, 2010, pp. 42, 44, 48, 50, 54, 55, 64, 66, 69, 70, 76, 77).

Application

In this section, we first discuss some of the strengths and challenges of students with unique needs. We then share some of our favorite strategies for working with them.

STRENGTHS AND CHALLENGES OF STUDENTS WITH UNIQUE NEEDS

As part of our effort to get to know our students better, we focus on finding out more about their assets and challenges. Using an introductory survey at the beginning of the year (which we describe more in Chapter 11: Building a Productive Classroom Environment), we ask students to answer questions like the following:

- What do you like to do for fun?
- In what activities or hobbies do you excel?
- What is your favorite way of expressing yourself?
- What about school do you enjoy? What about school makes you uncomfortable?
- What do you enjoy doing in class the most? What do you enjoy doing in class the least? (Baldwin et al., 2015, pp. 6–7)

In addition to any information that students give us, we observe them during the year to identify their strengths and areas of difficulty. We especially look for moments when they appear happy, focused, anxious, nervous, or distracted. If possible, we make a mental note of these situations or jot them down to give us a better picture of each student.

Here are some strengths and challenges that we see for ELLs, students with learning differences, and advanced students.

English Language Learners

ELLs bring strengths to classrooms in ways that may not be immediately visible. They often have learned different techniques for solving problems or have interests and life experiences that can be assets to our math instruction. The left side of Table 15.1: US and Latin American Prime Factorization Methods shows the factor tree method learned by many American students. Students in many Latin American countries learn a systematic prime factorization method (shown in the middle of the figure) in which they start with the least prime number factor first and then divide by successive primes (Perkins & Flores, 2002, p. 350). When we saw this, we decided to modify our instruction to accommodate both groups. The factorization on the right shows a factor tree that combines both strategies. Using a more systematic method takes much of the mystery out of algorithms like this, which can help students when performing more complex tasks like simplifying radicals and factoring expressions.

Many ELLs have literacy and math skills in other languages that can be applied to developing these skills in English (August & Shanahan, 2006, p. 5). In many

Table 15.1 US and Latin American Prime Factorization Methods

US method	Latin American method	Modified US method
(factor tree: 120 → 6, 20; 6 → 2, 3; 20 → 10, 2; 10 → 2, 5)	120 | 2 60 | 2 30 | 2 15 | 3 5 | 5 1	(factor tree: 120 → 2, 60; 60 → 2, 30; 30 → 2, 15; 15 → 3, 5)
$120 = 2 \cdot 2 \cdot 2 \cdot 3 \cdot 5$	$120 = 2 \cdot 2 \cdot 2 \cdot 3 \cdot 5$	$120 = 2 \cdot 2 \cdot 2 \cdot 3 \cdot 5$

East Asian languages, the words for numbers reflect their base-ten structure (43 in Chinese can be translated as "four ten three"). ELLs who are familiar with English grammar and language acquisition terms can often apply them in a math class. When we compare a mathematical operation like addition or multiplication to a "verb" that links two or more "subjects," we find that ELLs are often more likely to understand our analogy than native English speakers!

Like students with learning differences, ELLs often need more time and support to process information, especially when they have to perform complex tasks like reading word problems or writing explanations. It is important to remember that though ELLs may sometimes have similar needs to students with learning differences, they are based on language, not cognitive, issues. In addition, ELLs who endured trauma before coming here might be more likely to experience anxiety, exhibit aggressiveness, or have less energy for thinking. Since these responses are often dictated by the body's involuntary reaction to stress, trauma survivors may have less control of their behavior (Wright, 2017, pp. 143-144). A supportive, positive classroom environment can make these students feel safer. Experts recommend that teachers make an extra effort to engage trauma survivors regularly in warm one-on-one discussions. Since trauma survivors may have issues trusting people, offering structured opportunities for group work (such as the activities in Chapter 17: Cooperative Learning) and modeling strategies for working together can give them opportunities to practice connecting with their peers (Wright, 2017, pp. 145-147).

Students with Learning Differences

Research on students with learning differences has found evidence to support *neurodiversity*, the theory that neurological differences are natural variations that can have positive benefits (Rentenbach, Prislovsky, & Gabriel, 2017, p. 59). For example, individuals with autism tend to focus better on small details within more complex patterns (Armstrong, 2012). Advocates of neurodiversity argue that students with learning differences need help and accommodation, not a cure (Robison, 2013).

Many times, students with learning differences often ask us to clarify what we teach or explain a mathematical concept in a different way. While these questions could be viewed as an inconvenience, we find that addressing them can benefit all students. By not being afraid to say that they don't understand something, they can serve as role models for the rest of the class. Teaching students with learning differences has also motivated us to represent mathematical ideas in different ways, which can appeal to students' multiple intelligences (we discuss multiple representations more in Chapter 5: Making Mathematical Connections and Chapter 14: Differentiating Instruction).

Since learning differences cover a wide range of conditions, including emotional, intellectual, and physical differences, challenges can vary considerably. In our experience, students with learning differences often need more time to process information, so planning lessons with effective accommodations can be difficult. Like advanced students, students with learning differences can struggle to study and take notes. Unlike advanced students, though, they sometimes lack the memory skills to retain information easily (Miller, 2016, p. 64).

Advanced Students

Since advanced students generally tend to remember what they understand, they often need less review and can learn new information more quickly. They can handle more abstract tasks and tend to be comfortable working independently (Winebrenner, 2000). They may have excellent mental math skills, often see shortcuts that are missed by other students, and make mathematical connections independently. They can serve as a bridge between us and their classmates since advanced students can often explain our lessons to their peers better than we can!

Unfortunately, many advanced students rely heavily on their mental math skills, so they don't always see the value of expressing their thoughts in writing. After a while, they may not develop good note-taking or study skills, which can hinder them by the time they reach higher-level courses. Advanced students may be less experienced in struggling productively by grappling with a difficult concept and persevering with limited guidance until they understand it. As a result, they may get easily frustrated and give up when they encounter challenging material. Praising them for being "smart" can make things worse by reinforcing the idea that their understanding is fixed. Instead, advanced students can benefit from positive feedback that recognizes their effort and reinforces a growth mindset (which we discuss more in Chapter 4: Promoting Mathematical Communication).

TECHNIQUES TO SUPPORT STUDENTS WITH UNIQUE NEEDS

Since we often have ELLs, students with learning differences, and advanced students in the same class, devising appropriate lessons for each individual student every day is practically impossible. Fortunately, we find that the same strategies can be helpful for all three groups. In this section, we describe some of our favorite modifications.

Adjusting Instruction to Match Students' Strengths and Needs

When planning lessons for students with unique needs, we try to determine what they already know. We do this by looking at their prior grades, using our experience with previous students, giving them questions from upcoming topics in classwork or homework (to reassure students, we point out that these questions haven't been covered in class yet), or asking them in class what they remember about a topic.

We then make as many adjustments as we can reasonably fit into our unit and lesson plans (which we discuss in Chapter 6: How to Plan Units and Chapter 7: How to Plan Lessons). For each topic, we think about whether we should modify the depth or level of difficulty to match the needs of our students. These adjustments depend on the makeup of our class and the available time that we have. For example, when we teach solving quadratic equations, we may build in extra time for prerequisite skills like simplifying radicals if we find that most of our students have difficulty with that topic. If our students quickly grasp our lesson on solving two-step equations with integral coefficients, then we give them more challenging problems with fractions or decimals.

Students with learning differences and ELLs often need more time to process information (as we stated before, though, those needs are based on different reasons). During a lesson, this can mean giving students more wait time after a question or comment made during an in-class discussion (we discuss this more in Chapter 4: Promoting Mathematical Communication and Chapter 7: How to Plan Lessons), allowing students to look up words whose meaning is unclear to them, or providing additional scaffolding during discussions or classwork. Giving students more time to think also affects our unit planning. As a last resort, we may even cut out less important topics (such as topics that are not heavily emphasized in end-of-year exams, other units in the course, or in future classes) in order to accommodate the extra time that we give students.

Since advanced students frequently need less time to process information, we sometimes allow them to move more quickly through topics—a technique known as *curriculum compacting*. While this strategy can be effective, as we point out in Chapter 14: Differentiating Instruction, it can also be difficult to manage.

We find that we can manage pacing adjustments most effectively when we make them for the *entire* class as opposed to a small group of students. As a result, we usually try not to move a group of students ahead or behind the rest of the class.

Besides modifying the pacing, we also vary the types of activities that we do to match students' strengths, as shown by the following examples:

- Students with ADHD are often more restless than their peers, so we may incorporate more high-energy strategies such as games into lessons.
- Since children with autism are frequently very good at paying attention to detail, we may design activities in which students find the differences in pictures or definitions that appear similar.

- Since some students with learning differences may have outstanding artistic abilities, we include drawing in many of our lessons. For example, students can illustrate a mathematical concept or a lesson summary (Armstrong, 2012).

Ensuring Productive Struggle

We design work that enables students to engage in *productive struggle*, in which they actively work with little guidance on a task that is just beyond their abilities. When students struggle productively, they are not experiencing pointless frustration or practicing something that has just been demonstrated (Hiebert & Grouws, 2007, p. 387).

Productive struggle requires teachers and students to adopt a growth mindset, which we discuss in Chapter 4: Promoting Mathematical Communication. If students see struggle as a necessary step toward eventual mastery, they are more likely to persist in academic endeavors (Johnson, 2018). In contrast, if students believe that they can't improve their learning, they can quickly get discouraged and lose their motivation to work.

Sometimes, we allow students to work on alternate assignments that more closely match their levels of readiness. We may allow them to choose from a variety of options so that they can self-select the assignment that works best for them. For example, when discussing how to write the equation of a line given certain conditions, students can choose from the following options:

- Create an anchor chart that has a labeled diagram defining and illustrating the relevant vocabulary for writing the equation of a line, such as slope, x-intercept, y-intercept, and zero. ELLs can include terms and definitions in both English and their native language.
- Work through a packet that reteaches material learned in previous years, such as plotting points on a coordinate plane and determining the slope of a line from its graph.
- Create a review book or make a video that explains how to write the equation of a line given various conditions.
- Work on a sheet that has problems divided into levels of difficulty so that students can choose problems that are challenging for them.

We discuss anchor charts more in Chapter 4: Promoting Mathematical Communication and reteaching material, review books, and tiered lessons more in Chapter 14: Differentiating Instruction.

To ensure that students don't get discouraged while being challenged in their work, we pay close attention to how they react while working. If they get distracted, question the value of the assignment ("Why are we doing this?"), or express feelings of discouragement or frustration ("I don't get it!"), we adjust by giving them a hint or asking them what part of our assignment needs clarification. We find that students who believe that we are mindful of their feelings as well as their academic progress tend to persist more often through difficult work. Students with learning differences can be good role models for the rest of the class since they frequently are more willing to ask for help.

Frayer Models

We frequently use techniques that help ELLs and students with learning differences (as well as other students) develop their vocabulary skills. One method that we use to develop vocabulary skills is the *Frayer model*, devised by researchers Frayer, Frederick, and Klausmeier (1969). As shown in Figure 15.1: Frayer Model (Blank), the Frayer model is a graphic organizer in which the vocabulary term appears in the center, surrounded by its definition, a diagram, examples, and nonexamples. Our own experience confirms research indicating that students who visualize and illustrate math vocabulary showed a better conceptual understanding (Bruun, Diaz, & Dykes, 2015, p. 536).

To ensure that students understand our expectations, we first complete a few Frayer models together in class. Figure 15.2: Frayer Model—Perpendicular Bisector shows an example of a completed graphic organizer. Students can work individually or in small groups, referring to their notes as necessary. Once students become familiar with the concept, we assign Frayer models as homework for relevant vocabulary terms.

Since each course can have dozens of vocabulary terms, students need a way to keep Frayer models organized. We encourage students to write them on index cards. If possible, we print blank Frayer models on hole-punched index cards and give them to students. Students can then use key rings or index card binders to keep them together.

Since ELLs are often familiar with Frayer models, we sometimes rely on these students to model for others how to complete the graphic organizers. ELLs can complete a Frayer model either in English in another language if they know the term in that language. Since students have different levels of mathematical and language fluency, we allow students to choose the language in which they write Frayer models.

Concept Maps

We sometimes ask students to create a *concept map*. Similar to a concept map that we use to organize lessons in a unit plan (which we discuss in Chapter 6: How to

Plan Units), concept maps enable students to summarize all of the concepts that relate to a term or compare different problem-solving methods.

Figure 15.3: Concept Map shows different representations of a ratio. In this example, students write the definition and write in the appropriate words to explain the relationship between numbers. They can also see how the graph, table of values, words, and pictures are related to each other (we discuss multiple representations of a concept more in Chapter 5: Making Mathematical Connections). Showing pictures and verbal representations of the ratio 1:2 can also help students understand that the ratio of a group of 1 to a group of 2 is equivalent to a relationship of 1 out of 3.

Concept maps can easily be translated into different languages for ELLs. In addition, we adjust the level of support that we give advanced students or students with learning differences by varying the amount of scaffolding that we give them. For example, advanced students can be asked to create a concept map that includes a definition, different representations, and related terms, while other students may have other prompts and completed examples.

We typically assign concept maps toward the end of a unit. At first, we complete them together in class until students understand what we expect. Students can then complete concept maps in class or as homework. Students can use these concept maps throughout the year. Teachers can even allow students who need additional support to use concept maps on tests and quizzes as a reference. Teachers could give extra credit to students who don't use the concept maps to encourage them to memorize the information.

Simplifying Language

Another technique for improving students' language skills is to have them rewrite mathematical problems into simpler language. This strategy can be helpful for native English speakers, who can rewrite problems using high-frequency English words from everyday language. Since ELLs may not be as familiar with these simpler English words, they may need more direct instruction on this vocabulary and may need to translate them in their native language (Miller, 2016, p. 60).

Here is an example of an algebra word problem whose language is difficult for many students to understand:

> Marcos is designing a rectangular garden in his backyard. He wants the length of the garden to be 20 feet more than twice its width. Marcos also wants to create a walking path around each side of the garden that is 5 feet wide. Determine an algebraic expression in terms of w, the width of the garden in feet, that represents the perimeter of the rectangle surrounding the walking path.

We start by asking students to read through the problem several times, highlighting or underlining important as well as unknown terms. We find that students often have difficulty interpreting the phrases "20 feet more than twice its width," "determine an algebraic expression," "in terms of w," "perimeter," and "surrounding." Then we ask students to work in pairs or small groups to rephrase them in their own words. If necessary, students can use their notes, glossaries or dictionaries, or other resources to help them.

We allow students to use informal language as long as it's not mathematically incorrect or misleading, such as:

- "20 feet more than twice its width" means $20 + 2w$ (where w is the garden width in feet).
- "Determine" means write.
- "Algebraic expression" means "use letters and numbers."
- "In terms of w" means "use w somewhere."
- "Perimeter" means "total distance around."
- "Surrounding" means "on the outside."

By eliminating language that is not required for the solution (such as "in his backyard"), students can then extract the essential parts of the problem into a simplified version:

A rectangle has a width of w and a length of $20 + 2w$. Also, add 5 to each side. Use w and numbers to write the total distance around the rectangle.

When we rewrite mathematical language, we find that ELLs can be very helpful. By necessity, ELLs must first look up unfamiliar words and reinterpret them into another language before they can even begin to solve the problem. Teachers can ask ELLs to share their experiences in simplifying language with the rest of the class. ELLs can also model patience, persistence, and their own best practices for their classmates.

We also employ the four-step thinking process of understanding the text, creating a plan, solving, and checking the solution to help students re-express mathematical language. We describe this process more in Chapter 4: Promoting Mathematical Communication.

Manipulatives

One way to help students with unique needs represent mathematical ideas in multiple ways is to have them use manipulatives like algebra tiles or pattern blocks to concretely represent abstract mathematical concepts. Research indicates that play-oriented movements like manipulating objects can improve cognition (Bouck & Park, 2018, p. 97; Jensen, 2005). Here are some ways that manipulatives can benefit these students:

- ELLs don't need to be fluent in English to understand mathematical ideas with manipulatives (Borgioli, 2008, p. 188). For example, by constructing a trinomial like $x^2 + 6x + 9$ with algebra tiles, they can see that it is a square with side length $x + 3$. They can use nonverbal communication (such as hand gestures) or phrases and sentences in any language to conclude that $x^2 + 6x + 9 = (x + 3)^2$.

- Students with learning differences can experience math in a more tactile way. To see the relationship between the area of a parallelogram and the area of a triangle, students can move pattern blocks or cutouts of congruent triangles to see that the area of a triangle is half the area of a parallelogram with the same base and height.

- Advanced students can answer more sophisticated questions by explaining how manipulatives show abstract mathematical ideas. Students can construct the centroid of a triangle by folding squares of parchment paper. By analyzing the steps in the construction, they can map out a justification of it.

Independent Study

Many times, the differentiation strategies that we outline in Chapter 14: Differentiating Instruction (such as curriculum compacting or tiered assignments) and this chapter may not meet the academic or personal needs of all students.

Some students can't work at the same pace as most of their peers, such as the following:

- advanced students for whom even a highly accelerated assignment is not challenging enough
- students whose circumstances (such as health issues, a job, or family responsibilities) prevent them from attending class regularly
- students who have a strong interest in a subject not taught at their school

In these situations, students can engage in an *independent study*, in which they work with teachers to design a long-term project that allows them to explore their interests more deeply with little direct supervision. We discuss projects more in Chapter 16: Project-Based Learning.

Social-emotional Learning

To build effective relationships with all students who have unique needs, we work to create a classroom environment in which they feel safe and welcome. As we've said throughout this book, attending to students' social-emotional learning can reduce their math anxiety and make them more productive learners.

When planning lessons for different students with unique needs, we try to imagine ourselves in a situation where we don't know the culture's dominant language or we have a condition that is deemed problematic by society. Even if we successfully implement all of the strategies we list above, we can't anticipate every possible issue that can arise. Establishing meaningful relationships with our students is the best way to reach them both academically and emotionally. Here are some ways that we make those connections:

- **Treat all students respectfully.** This includes greeting them by name every day, getting to know their interests and concerns, assuming that they are competent, and giving them opportunities to contribute meaningfully to class discussions. Through cooperative learning activities like Think-Pair-Share and jigsaws (which we discuss in Chapter 17: Cooperative Learning), students can read, write, speak, and listen to others. These activities allow them to make sense of what we teach and develop their language skills with peers in a low-risk environment (Borgioli, 2008, p. 188; Miller, 2016, p. 64; Rentenbach et al., 2017, p. 60).

- **Establish a familiar routine.** Routines can include having an introductory assignment for students to do at the beginning of class or passing up homework for collection after we review. When we transition from one part of the lesson to the next, we clearly indicate (with a verbal or visual cue) what we plan to do. For example, if we have a summary exercise at the end of class, we may say, "We're approaching the end of class, so let's wrap up our lesson by filling out the exit ticket that I'm passing out right now." Predictable routines can be especially helpful for students with learning differences and ELLs, who can focus on the current task instead of worrying about what is going on around them (Cornelius, 2015; Rentenbach et al., 2017, p. 62). Predictability doesn't mean that every class must have exactly the same format. It does mean that we make our goals and intentions clear to students at all times. We discuss rules and routines more in Chapter 11: Building a Productive Classroom Environment.

- **Use clear, concise language when communicating with students.** Simplifying language doesn't mean "dumbing down" our ideas. Instead, it means minimizing the use of slang expressions ("find the odd man out" instead of "determine which one doesn't belong with the others") or obscure cultural references ("follow Ram Dass's advice" instead of "let's focus on what we're doing right now"), which can confuse and intimidate students (Cornelius, 2015; Robertson, n.d.). Speaking too softly or speaking in a monotone loses students' attention and makes conveying important information difficult (Yale Poorvu Center, n.d.).

- **Give students the freedom to move around the classroom.** Movement can be particularly helpful for students with learning differences. Students with ADHD may think best when their bodies are active, so they could walk around while discussing ideas (Rentenbach et al., 2017, p. 61).

What Could Go Wrong

When working with students who have learning differences or with ELLs, we can succumb to stereotypical thinking and make errors in judgment. Sometimes, we mistakenly believe that these students are less capable of higher-order thinking and that they would be better off memorizing facts and learning formulas. We find that students can tell when teachers are "talking down" to them and often respond by losing motivation and focus. As we said in Chapter 3: Teaching Math as a Language, teaching math as a "bag of tricks" often leads students to make more mistakes if they don't fully understand them in context.

Developing meaningful relationships with students enables us to discover aspects of their lives that may not appear in a transcript or individualized education plan. This helps us devise better instruction for them. For example, one of Larisa's students had difficulty solving word problems. After several conversations, she found that he learned mathematical words in English since he was born in the United States and attended American schools. However, his parents and friends spoke to him only in Spanish. Larisa found that when he was allowed to use Spanish-language and English-language versions of test questions (which she could easily provide since many of our state exams are translated into different languages), he was able to combine his fluency in Spanish with his knowledge of mathematical terms in English.

While advanced students can often be a great benefit to struggling students, we find that using them too often as tutors or giving them more practice at the same level can slow their learning and demoralize them. It's important to balance the benefits that are often gained when advanced students help their peers with other times when advanced students receive more challenging individual work. Often, we incorrectly assume that advanced students will always be productive and complete their work on time and with little assistance. In fact, they can be just as disinterested as any other students—usually because they're not receiving sufficiently challenging work (Winebrenner, 2000).

Another mistake we make with some advanced students is to misread their lack of focus as an indication that they don't have a higher level of skills. We find that the best way to avoid this is to observe students more closely and keep an open mind. For example, one of Bobson's students appeared very disorganized in class. She rarely showed coherent work on constructed-response questions and couldn't maintain an organized notebook. Over time, he noticed that the student did very well on multiple-choice questions. At first, Bobson assumed that she was copying from another student or guessing. After having some private conversations with the student in which he asked her to explain how she got her answers, he realized that she actually had an above-average sense of estimation and often found unusual shortcuts. While this allowed her to correctly eliminate multiple-choice distractors, it also proved a hindrance when she was required to explain her reasoning. As a result, Bobson worked with her to improve her note-taking skills so that she could do better on free-response questions. He also called on her more often in class so she could explain her shortcuts to her peers. Finally, he recommended that she join our school's math team, where she wouldn't need to show work and could solve more challenging questions.

Many times, parents of students with unique needs don't know what additional resources schools can provide to help their children learn. Often, these parents believe having a 504 plan or being placed in an ELL class will stigmatize students or put them into lower-level classes. Some parents may even refuse to have their child tested. In these situations, we try to reassure students and their parents that these accommodations are designed to put children on a more equal footing with other students. We ask other school personnel, such as a guidance counselor or social worker, for help in working with parents. Finally, we try to provide whatever reasonable modifications we can in class, such as allowing them to work with students who speak the same language or providing more appropriate work.

Student Handouts and Examples

Figure 15.1: Frayer Model (Blank)

Figure 15.2: Frayer Model—Perpendicular Bisector

Figure 15.3: Concept Map

Technology Connections

For parents and teachers of advanced students, the National Center for Research on Gifted Education (http://ncrge.uconn.edu), the National Association for Gifted Children (www.nagc.org), and the National Society for the Gifted and Talented (www.nsgt.org) have research articles and other resources.

The websites of math competitions are often excellent sources of more demanding problems. These include MathCounts (www.mathcounts.org), the American Mathematics Competition (http://www.maa.org/math-competitions), the Math League (www.mathleague.org), and the New York City Interscholastic Mathematics League (www.nyciml.org).

For teachers of students with learning differences, Vanderbilt University's Center for Teaching (http://cft.vanderbilt.edu/guides-sub-pages/creating-accessible-learning-environments), the Center on Online Learning and Students with Disabilities (http://www.centerononlinelearning.res.ku.edu), the National Association of Special Education Teachers (www.naset.org), Learning Disabilities Online (www.ldonline.org), and TeacherVision (http://www.teachervision.com/teaching-strategies/special-needs) have lessons, strategies, and other relevant information. Do2Learn (http://do2learn.com) has worksheets, songs, games, behavior management plans, and other resources for students with learning differences.

For teachers of English Language Learners, ¡Colorín Colorado! (www.colorincolorado.org), and Edutopia (http://www.edutopia.org/topic/english-language-learners) have information and activities. Many Things (http://manythings.org) has quizzes, expressions, videos, word games, and other activities for ELLs. The Mathematics for English Language Learners project (www.mell.org) has lesson plans and other instructional tools. The Understanding Language website from the Stanford Graduate School of Education (http://ell.stanford.edu/teaching_resources/math) has a list of general principles, guidelines for developing new materials, and templates for language-focused activities. In addition to Google Translate (http://translate.google.com), many school districts and states provide glossaries and other print resources for English Language Learners to use on standardized tests. For example, New York State provides bilingual mathematical glossaries in over 30 languages (http://steinhardt.nyu.edu/metrocenter/resources/glossaries).

For independent studies, Kathy Schrock's website (http://www.schrockguide.net/assessment-and-rubrics.html) has many rubrics that are suitable for student research projects.

The ThoughtCo (http://www.thoughtco.com/the-frayer-model-for-math-2312085) and Teacher Toolkit (http://www.theteachertoolkit.com/index.php/tool/frayer-model) websites have articles on using the Frayer model in a math class.

LINKS Learning (http://www.linkslearning.k12.wa.us/Teachers/1_Math/2_Curriculum_Planning/2_Math_Concept_Maps/index.html) has interactive concept maps for number sense, measurement, geometry, algebra, and probability and statistics. Achieve the Core has an interactive coherence map at http://achievethecore.org/coherence-map that shows the connections between Common Core K–12 math standards.

Figures

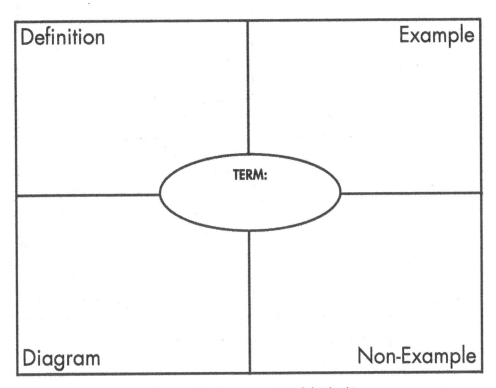

Figure 15.1 Frayer Model (Blank)

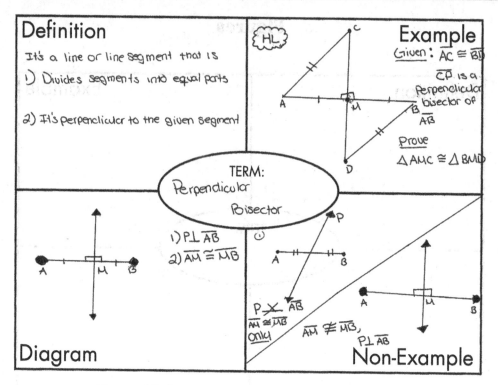

Figure 15.2 Frayer Model—Perpendicular Bisector

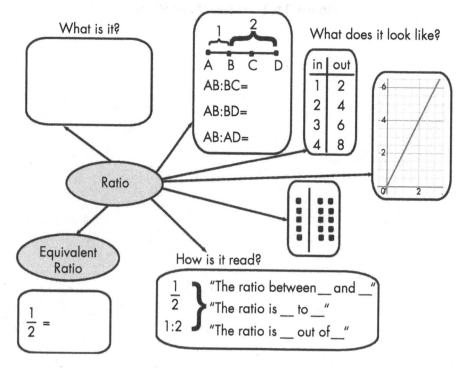

Figure 15.3 Concept Map

CHAPTER 16

Project-Based Learning

What Is It?

Students have to learn more than facts and computations. They need to know how to learn independently and apply that knowledge in future situations. Only providing students with direct instruction may not foster a deeper understanding necessary to solve complex problems. One way to strengthen student understanding is to use *project-based learning*, in which students work for an extended period of time on a real-world, open-ended question and demonstrate their knowledge by creating a public product or presentation ("What Is PBL?," n.d.).

Many times, we use open-ended questions in our lessons (as discussed in Chapter 7: How to Plan Lessons) and homework (as discussed in Chapter 8: How to Plan Homework). However, some questions are so complex that they can't be covered adequately in one lesson or homework assignment. In this chapter, we focus on ways to address these larger questions through projects.

Why We Like It

We like projects because they give students an opportunity to show their knowledge outside of a traditional test or quiz. Projects allow students to showcase talents and skills that they don't often use in class. We find that many of our students enjoy a break from our typical routine. They can also work at their own pace on a topic that interests them. Projects enable students to learn important life skills, such as planning their time and developing research skills (Gregory & Chapman, 2007, p. 146).

In addition, solving open-ended problems—the heart of project-based learning—is an important part of mathematical thinking. Well-chosen problems can help students solidify what they already know and extend their knowledge (NCTM, 2000, p. 52). Learning how to solve problems efficiently can strengthen mathematical skills and improve people's ability to make decisions. In fact, Harvard psychologist Howard Gardner (2006, p. 6) defined intelligence as the ability to solve problems or create products that are important to a community. We believe that with proper scaffolding and differentiation, project-based learning can be an effective strategy for teaching *all* students.

Supporting Research

Project-based learning stems from the work of researchers like John Dewey (1938, p. 20) and Jean Piaget (1952, p. 7), who developed constructivism, the theory that students actively create knowledge through experience. In addition to project-based learning, other pedagogical strategies that are based on constructivism include:

- **Problem-based learning.** Students solve authentic open-ended problems (Barrows, 1996, pp. 5–6; Barrows & Tamblyn, 1980, p. 18; Delisle, 1997, p. 2).
- **Discovery learning.** Students discover new knowledge either without teacher assistance (called *unassisted* or *pure* discovery learning) or with teacher assistance (called *assisted* discovery learning) (Marzano, 2011).
- **Performance-based learning.** Students complete a task by constructing an original response or performing an activity (Darling-Hammond & Adamson, 2010, p. 7; Hibbard et al., 1996).

Constructivist strategies encourage students to learn by answering nonroutine questions. Unlike *routine* exercises (which can be easily solved because the solution method is already known), *nonroutine* questions present a challenge since their solutions are not immediately clear (de Freitas & Zolkower, 2011, p. 230). Often, nonroutine questions are open-ended—they may have many possible answers, starting points, or solution methods. Although nonroutine and open-ended questions are technically not the same (a nonroutine question could theoretically be closed-ended, and open-ended questions could be routine), we find that in practice almost all nonroutine problems have at least some open-ended component, and most open-ended questions are nonroutine. Thus, we use the two terms interchangeably.

Research suggests learning that emphasizes open-ended problems and projects may be more effective than direct instruction in improving academic outcomes (Condliffe, Visher, Bangser, Drohojowska, & Saco, 2017, p. 49; Laitsch, 2007), especially for promoting long-term retention of knowledge (Strobel & van Barneveld, 2009, p. 55) and designing solutions to complex problems (Walker & Leary, 2009, p. 25). The effectiveness of these learning strategies may vary by discipline. Some studies indicate that using open-ended problems and projects may have a positive impact in science (Carrabba & Farmer, 2018, p. 170; Drake & Long, 2009, pp. 11–12) and social studies (Parker et al., 2011, pp. 555–556; Summers & Dickinson, 2012, p. 98) classes. A 2017 meta-analysis did not find a similarly significant impact in math classes but noted that teachers may feel that project-based learning is difficult to implement in a math class (Condliffe et al., 2017, pp. 37–39).

Common Core Connections

The Common Core's Standards for Mathematical Practice list several necessary skills for solving open-ended problems, including making sense of them (MP.1), reasoning abstractly and quantitatively (MP.2), using appropriate tools strategically (MP.5), and making use of structure (MP.7) (NGA & CCSSO, 2010, pp. 6–8). These skills are particularly useful for projects.

Application

Before beginning larger assignments that are more typically defined as projects, we start with helping students become more familiar with the basic idea of open-ended questions.

OPEN-ENDED CLASSWORK PROBLEMS

To help students get more comfortable with open-ended questions, we often incorporate them into our lessons before assigning larger projects. These questions can be used in class to introduce or summarize a lesson, as we describe in Chapter 7: How to Plan Lessons. We find that many students may have difficulty understanding the context of the real-world applications typically found in projects. As a result, they may fail to see the math that underlies the problem. For example, the following word problem attempts to elicit the definition of a circle—the set of points that is the same distance (radius) from a fixed point (center):

> A dog's 6-foot chain is tied to a stake in the middle of a backyard. Describe the shape of the area in which the dog can walk.

This question may be problematic because students may not know the definition of a stake and may not have pets or a backyard, so visualizing the context can be difficult. Including a picture may be confusing (they may have trouble seeing that the boundary of the area is formed when the dog pulls the chain taut) or distracting (they may focus on the dog's appearance).

Asking students to find a mathematical pattern or make a mathematical connection can often be more accessible for many students than using a real-world context.

Another version of this problem that removes the real-world context would be the following:

> On the coordinate plane, draw all the points that are 6 units away from the origin.

If students are given or shown a coordinate plane, they only need to know the location of the origin and to measure a distance of 6 units to start answering the question.

Figure 16.1: Discovering Pi shows a series of nonroutine questions that can be used to introduce π. Students measure the circumferences and diameters of different-sized circles and discover that the ratio of the circumference to the diameter is slightly larger than 3. We then conclude with a whole-class discussion in which we point out that many ancient cultures (including the Babylonians, Greeks, Indians, and Chinese) discovered that this ratio is constant and that it is so important that mathematicians have given it a special name—the Greek letter π. Alternatively, students could research the history of π as a homework assignment or a larger project. They could find out how π was estimated and applied in different ancient cultures. This activity can later be extended into a real-world context by showing how π is used to compare areas of pizzas with diameters, as shown in Figure 16.2: Area of a Circle.

Here are some other examples of open-ended mathematical problems that can be incorporated into lessons:

- **Find the least common multiple.** On a piece of paper, list the first 10 multiples of 6 and the first 10 multiples of 9. What is the first number that appears on both lists? If you list all multiples of 6 and 9 that are less than 100, how many numbers appear on both lists? What other patterns do you see?

- **Reason with linear equations.** List several pairs of numbers whose sum is 10. Graph each pair on the coordinate plane (make the first number in each pair the x-coordinate and the other number the y-coordinate). What do you notice about the set of graphed points? (Gurl, Artzt, & Sultan, 2013, pp. 13-14).

- **Explore transformation of trigonometric functions.** Graph the equations $y = \sin(x)$, $y = 2\sin(x)$, $y = 3\sin(x)$, and $y = 0.5\sin(x)$. How does changing the value of the coefficient a affect the graph of the equation $y = a\sin(x)$?

Many additional examples of open-ended mathematical questions can be found in the figures in Chapter 5: Making Mathematical Connections. We discuss other benefits and uses of open-ended questions in Chapter 7: How to Plan Lessons.

Of course, sometimes a real-world question may be easier to understand. Students may not have the prior knowledge to answer a mathematical question, or their life experiences may make them more familiar with certain situations. We find that learning about students' cultures and backgrounds (which we talk about in Chapter 2: Culturally Responsive Teaching) helps us determine what contexts are most appropriate for our students.

As students get more experience answering nonroutine questions in classwork, we may give longer assignments that span 1 or 2 days. Depending on the topic and our students' prior knowledge, we may or may not use a problem with a real-world context. In Figure 16.3: Point Lattice Assignment (appropriate for Algebra I), students work in pairs to review solving systems of linear equations and inequalities graphically. Students begin by recalling important terms discussed throughout the unit, such as *coordinate axes* and *intersection point*. This review is particularly important for English Language Learners (ELLs), who may also need to write definitions in their own language. Students then draw lines that pass through lattice points (points that have integer coordinates) and write their equations. In coordinate geometry, students are typically given a set of coordinate axes before solving a problem. This assignment is nonroutine because students must choose where to place the axes before they can write the equation. This question can encourage students to think about why we use coordinate axes and which location for the axes would make the most sense for a given problem.

OPEN-ENDED HOMEWORK PROBLEMS

As we said in Chapter 8: How to Plan Homework, we occasionally incorporate open-ended questions into our daily homework. These questions may help students find a pattern that can't be completed in class or transfer learning from one situation to another. Since we can't provide as much assistance for homework as we can for classwork, we don't assign questions that require a great deal of prior knowledge. Instead, we limit open-ended homework assignments to questions that we think students can complete on their own in about 15–30 minutes.

Figure 16.4: Paint a Room contains a homework assignment in which students measure the walls of a room in their home or school and calculate the number of gallons needed to paint it. Other examples of open-ended problems in homework include the figures in Chapter 8: How to Plan Homework as well as the following:

- **Ratios:** Use information from a newspaper, magazine, or the internet to calculate and compare the unit prices of items sold at different supermarkets. Which supermarket offers the best deal?

- **Coordinate geometry:** Take a picture or find an image of a parabola used in architecture in your neighborhood, town, or state. What is the equation for the parabola? Use an online graphing tool to help you. (Students can see that this equation can be used by architects in their design.)

- **Statistics:** If a fair coin is flipped 100 times, how many times would you expect it to come up tails four or five times in a row? Find evidence for your claim by flipping a fair coin 100 times, recording the outcome each time. (Students can also be asked to find a real-world application of this problem, such as a situation that requires the use of a coin flip.)

For many open-ended homework assignments, each student's response is different. To review homework in class, we often ask students to work in groups to compare answers and check each other's work for clarity and originality. Students can share the best or most interesting answer in each group or describe noteworthy solution methods or common errors. Groups can write their responses on chart paper or whiteboards, use computers to type up responses and post them online, or work together to present their results to the class. We can then have a whole-group discussion comparing each group's responses. The homework review should take no more than 15 minutes—about the same amount of time as a review of a more conventional homework consisting of practice problems.

PROJECTS

Projects can take many forms—ranging from highly structured assignments in which the elements of the task are explicitly stated to more open-ended assignments with a clear goal but minimal guidelines (Gregory & Chapman, 2007, p. 146). No matter what form they take, projects should do the following:

- Focus on a particular question or challenge that is meaningful to students.
- Allow students to apply real-world skills, such as planning a timeline or using presentation software.
- Enable students to engage in a prolonged inquiry with frequent reflection and revision.
- Give students some choice in what they create or how they work.
- Create a public product (Larmer & Mergendoller, 2010).

Our projects have components that are similar to the parts of our lesson plans:

- **Scope:** Define the goals, relevant standards, and available resources.
- **Question:** Identify the problem statement that students will address in the project.
- **Instruction:** Provide direct instruction, guided questions, or other work that gives students appropriate background knowledge to answer the question.

- **Feedback:** Give students feedback to help them reflect on and revise their work before the project's final submission.
- **Differentiation:** Allow modifications for students with unique needs (ELLs, students with learning differences, and advanced students).
- **Summary:** Grade the paper, presentation, or some other product that shows what students have learned.

We discuss each of these components below.

Defining the Project's Scope

To design a project, we start by thinking about how it aligns with our course's goals. Typically, we assign no more than one project per unit, so we look at the goals or standards for a unit to identify any that may be appropriate for a long-term assignment. Procedural goals like solving equations or simplifying fractions tend not to work well for projects. However, goals that ask students to *compare* different procedures (such as solving quadratic equations by completing the square, using the quadratic formula, factoring, or graphing), *apply* procedures in real-world situations, or *understand* a concept more accurately embody project goals.

We set aside time in class for us to explain the project and for students to work on it. Factors like exam dates, parent-teacher conferences, the availability of school computers, holidays, and the topic's level of difficulty can affect when we give a project and its size. Many times, some factors (such as district standardized tests or schoolwide goals that must be met by a certain time) restrict our ability to assign projects. We then either reduce the project's scope or don't assign them.

We like to give students as much autonomy as possible when working on projects. Students could choose their final product from a list of options (write a paper, make a video, create a trifold board, or give a presentation). We may give them a goal (find the best mobile phone plan) and give them freedom to find appropriate evidence to support their conclusion, as long as they meet the project's learning objectives (include appropriately labeled graphs). Sometimes, we may provide the structure (independent research) but give students freedom to choose a topic. The amount of autonomy we give depends on many factors, such as students' level of readiness for independent work and the amount of available time.

Selecting a Question

After identifying appropriate goals, we think about the driving question that students will answer while working on the project. We also create a rough plan of how students can answer the question.

We try to design a question that is relevant to students and relates to the world around them. Teachers or students can create a question by extending a problem discussed in class. Asking "what-if" questions or considering other cases can lead to a pattern (we say more about the importance of finding patterns in Chapter 1: Motivating Students). For example, in the activity in Figure 16.2: Area of a Circle, students could use the table of prices for a small and large pizza to make up their own questions and answer them. Students can also use a protocol like What Do You Notice? What Do You Wonder? (we talk about protocols more in Chapter 17: Cooperative Learning) to generate questions.

Designing a driving question from scratch can be challenging. Fortunately, we find that we can use many sources to help us. Many textbooks have challenging open-ended questions that can be modified for projects. News articles that describe a new study or highlight a pressing need in the community may also provide culturally responsive project ideas that are relevant to students. Websites like the ones we describe in the Technology Connections sections of this chapter often have projects that we can adapt for our students.

We also consider whether the project will be an individual or group project. Each has its advantages and disadvantages:

- **Individual projects give students the most flexibility.** We find that individual projects work best if we want to give students more freedom and we feel confident that students can complete the project on their own. In our experience, individual projects allow students to show their creativity. Many times, students pleasantly surprise us with the effort and originality that they invest in their work! Unfortunately, individual projects don't give students opportunities to collaborate and require more time for teachers to give feedback.

- **Group projects allow students to work on more complicated questions and share the workload.** However, students may not distribute the work evenly or fairly, and communication or collaboration problems can impede the group's progress (Pavitt, 1998). Chapter 17: Cooperative Learning describes the benefits and challenges of group work in more detail. We find that having students work in pairs for smaller projects allows them to have meaningful mathematical conversations. Students can form groups of three or four to answer more complicated questions that require a larger workload.

An example of a project appears in Figure 16.5: Project—Bus Redesign Plan. Algebra I students can complete it over several weeks or as an end-of-year assignment. The project is culturally responsive because it is relevant to our students' communities. Students use data on bus routes, population, and traffic to design bus routes that transport people to and from our school more efficiently. For an additional challenge, students can find and analyze local demographic data to design routes that benefit not just students from our school but people throughout the neighborhood. A project like this one can be completed as students work through a unit on statistics. Students will need additional instruction on how to create and interpret tables, histograms, boxplots, and scatterplots.

Table 16.1: Project Ideas shows brief descriptions of other projects.

Table 16.1 Project Ideas

Goal	Project
Fractions: Divide fractions by fractions.	Create a real-world context that models a given set of fraction division problems and illustrate each problem with an appropriate representation.
Ratios: Use proportional relationships to solve ratio and percent problems.	Find the location of local community resources (such as schools, hospitals, or firehouses). Combine it with demographic data to determine the population density of resources. Determine if the community has adequate resources.
Measurement: Compute scale drawings. Graph proportional relationships and interpret the slope of the graph.	Design a plan to retrofit an existing local building to meet Americans with Disabilities Act regulations. Include scale diagrams of ramps with an appropriate slope with pathways with appropriate measurements.
Functions: Write a function that expresses the relationship between two quantities and interpret it in context.	Graph and write the equations of mathematical functions to determine the cost-effectiveness of different health insurance plans.
Constructions: Make and apply geometric constructions.	Create artwork (or re-create teacher-drawn artwork) using geometric constructions with a compass and straightedge.
Polynomials: Graph polynomial functions and interpret characteristics of the graphs in context.	Write a function to represent a can's surface area in terms of its radius. Use technology to graph the function and determine its relative minimum. Determine the dimensions of a cylindrical can with a given volume that saves money in production costs by using the least amount of material.

Providing Appropriate Instruction

Once we determine the essential goals, we think about the instructional activities that students need to complete in order to answer the project's main question.

For larger projects that are done over several weeks, we create an outline that includes *milestones*—important stages or accomplishments for the project. Examples of milestones include sections of a written report (such as an abstract, introduction, data, analysis, limitations, and conclusion) or drafts of a presentation. We post a project description online and give students paper copies if necessary. When we first assign the project, we go over the major parts of the project, including the requirements and due dates for each milestone.

Many times, we provide direct instruction before starting a project or as students work on it to make sure that they have enough mathematical background knowledge. Depending on the project, this instruction can range from a short explanation at the beginning of the period (if the project is relatively self-explanatory) to several days or weeks (if the project is a summary of a unit). Throughout this instruction, we point out how the lesson relates to the project.

We give additional support to ensure that students use their time more effectively. We post due date reminders online and show students how to create reminders on their phones. In class, we frequently mention upcoming deadlines and ask students about their progress as we check homework or monitor student practice during a lesson. To limit the possibility that students get overwhelmed while working on projects, they should get less homework. If necessary, we provide additional instruction on any technology that they may need, such as managing shared documents online.

To promote effective teamwork for group projects, we start by creating a classroom environment in which students feel safe. Building meaningful relationships with students and establishing clear rules and routines (which we discuss in Chapter 11: Building a Productive Classroom Environment) can help them focus on their work. Group projects should be complex enough so that each student has a meaningful amount of challenging work within the group. Each student should have specific tasks (we generally allow groups to assign roles to their members). Other strategies for promoting student collaboration appear in Chapter 17: Cooperative Learning.

Giving Feedback

As students complete a project, we provide *feedback*, which consists of specific suggestions for what students should do next. Feedback that aligns with student goals can significantly improve student success (Cooper, 2016).

While students are working on projects in class, we may circulate around the room to answer questions or address concerns. If necessary, we have individual or group conferences, either inside or outside of class. Verbal feedback works well for informal conversations or comments that would be difficult to explain in writing (for example, if we want to show students how to do something or if we want to discuss a sensitive topic that is more easily discussed in person). Meeting with students also allows us to have a dialogue with them, which is particularly helpful when we need to clarify ambiguous writing or explore next steps. However, it requires a great deal of class time. For longer projects, we often keep a log of our comments, noting the meeting dates and important points of each conversation.

We also provide written feedback. If students are producing a document as a final product, we ask them to create a shared one so that we can insert written comments in the margins. Using a shared document also enables us to keep a record of our suggestions and students' changes. In our experience, written feedback is often more flexible for us since we don't have to take up as much class time. Many ELLs may find written feedback more helpful than verbal feedback since they can take additional time to process and translate our comments. However, we find that written feedback takes more time and often feels less personal than talking to students in person.

We offer positive feedback that praises students' effort ("I can see that you really worked on this draft") instead of their intelligence ("You're a great writer"). Praise can strengthen a growth mindset by showing them that working harder will improve their skills. We talk more about promoting a growth mindset while grading student writing in Chapter 4: Promoting Mathematical Communication.

Another way to provide feedback is to allow students to comment on each other's work. Students can use the project's rubric or scoring guidelines to give verbal or written comments. Peer editing and commenting also can ease our workload by enabling students to receive more feedback. We describe ways for students to edit each other's work in Chapter 17: Cooperative Learning.

Differentiating Projects

English Language Learners can benefit from working on projects. Working on projects can relieve some of the anxiety that some ELLs may feel by allowing them to communicate in a less structured and more relaxed setting. We find that many ELLs have skills that they can demonstrate more effectively by working on a project. For example, they may be skilled in using technology or may know another way to solve a problem. They may know how to conduct research in their first language and transfer that knowledge to English. We discuss appropriate modifications for ELLs more in Chapter 15: Differentiating for Students with Unique Needs.

Pairing students with someone who is fluent in English or a student who speaks their language can also help them feel more connected to their work. Some ELLs may need extra time to process information or translate instructions into another language. ELLs who don't feel comfortable speaking in front of the class during a presentation may want to take a less public role in a group project. Instead of giving a live presentation, ELLs could make their own video so that they can take more time to record and edit it. Some presentation software, such as the free Google Slides, allows users to add captions.

Like ELLs, students with learning differences can also benefit from the added flexibility that arises when they work on projects. We find that many students with learning differences enjoy taking on different roles that enable them to use their strengths. For example, many autistic students have excellent artistic abilities, so they may want to create relevant graphs or illustrations for a group paper or presentation (Armstrong, 2012). However, students with learning differences may need additional support. We may need to pair them with students who can complement their skills and help them complete the project.

Working on projects can also help advanced students. They can work on questions that extend material learned by other students. For example, if the class is learning how to write the equation of a parabola or circle, advanced students could teach themselves how to write the equation of a hyperbola or ellipse. Students could research how ancient cultures wrote the equation of a conic section. They could write an independent study, exploring a topic of interest to them with little direct supervision. We remind ourselves that advanced students can lose focus and motivation if their work is not sufficiently challenging.

We discuss other useful strategies in Chapter 15: Differentiating for Students with Unique Needs.

Grading Projects

When we grade projects, we typically use rubrics. Unlike the scoring guidelines for constructed-response test questions (which we discuss in Chapter 9: How to Plan Tests and Quizzes), rubrics describe performance levels for the project's criteria. Typically, these criteria focus on the mathematical content, the clarity of the language used in the final product, and the amount and quality of the effort that students demonstrate while working.

Table 16.2: Basic Project Rubric shows a simple rubric that we use to grade short written projects. We divide rubrics into several levels, such as "below standard," "meets standard," and "above standard." Performance can also be separated into four levels: "below standard," "approaching standard," "meets standard," and "exceeds standard." In general, we look to see if students used correct mathematics,

Table 16.2 Basic Project Rubric

Category	Below standard	Meets standard	Above standard
Mathematics	Provides only partially clear and correct calculations *with significant errors*	Provides generally clear and correct calculations with proper mathematical language and notation *and only minor errors*	Provides clear, concise, and correct calculations with proper mathematical language and notation
Communication	Provides only partially clear or correct explanations *with significant errors*	Provides clear and correct explanations *with only minor errors*	Provides clear, concise, and correct explanations
Effort	Rarely asked others for help and didn't help others	Asked others for help when necessary	Asked others for help when necessary *and helped others*

clearly explained their thinking, and made significant effort to give or receive help when necessary.

Larger projects often have more complex rubrics, such as the one in Figure 16.5: Project-Bus Redesign Plan. This rubric is divided into several parts, one for each major section of the final product. Each part could be divided further, depending on what criteria we want to emphasize.

For oral presentations, we evaluate additional criteria, as shown in Table 16.3: Oral Presentation Rubric. To encourage students to use class time effectively, we give each student or group a limited amount of time to present. We also judge the quality of their visual aids and their ability to answer questions from students or us effectively.

Teachers can convert student rubric grades to a number by assigning a point value to each level. For example, in Table 16.2: Basic Project Rubric, a "below standard" grade could be a 1, "meets standard" grade could be a 2, and "exceeds standard" grade could be a 3. Since there are three categories, there are 9 possible

Table 16.3 Oral Presentation Rubric

Category	Below standard	Meets standard	Above standard
Time management	Fails to convey important ideas	Effectively conveys *most* of the important ideas in the allotted time	Effectively conveys *all* important ideas in the allotted time
Mathematics	Mathematics contains substantial errors or is incomplete	Correctly uses appropriate mathematics to support claims, *with only minor errors*	Correctly uses appropriate mathematics to support claims, *with no errors*
Language	Speaks and writes with *weak* professionalism, confidence, and passion	Speaks and writes with *adequate* professionalism, confidence, and passion	Speaks and writes with *strong* professionalism, confidence, and passion
Visual aids	Has *limited use* of text and other visuals to convey ideas	Effectively uses *text* to convey ideas	Effectively uses *text and other visuals* to convey ideas
Questions	Demonstrates *weak* knowledge of topic while answering questions	Demonstrates *knowledge* of topic while answering questions (student can adequately explain work but struggles to answer hypothetical questions)	Demonstrates *mastery* of topic while answering questions (effectively answers hypothetical questions)
Effort	Rarely asked others for help or didn't help others	Asked others for help when necessary	Asked others for help when necessary *and helped others*

points in all. If some categories are weighted more than others, then the point total would change accordingly. Each student's grade could then be calculated by dividing each student's number of points earned by the total number of possible points. Teachers who give letter grades can convert each level to a corresponding letter grade.

What Could Go Wrong

Using project-based learning can often seem intimidating and time-consuming. Under pressure to complete the curriculum or prepare students for end-of-year assessments, many teachers may feel that they don't have time for project-based learning assignments. We recommend starting with smaller examples, such as a problem-based introductory activity for a lesson (such as the examples in Chapter 5: Making Mathematical Connections) or a discovery-based homework (we have several examples in Chapter 8: How to Plan Homework). In our experience, incorporating projects even once or twice a semester can provide a welcome break from our regular routine. We also find that working with a colleague to create a project or modify an existing one helps to make the workload more reasonable.

We have often made the mistake of not providing enough support. We are often tempted to sit back and relax while students are working on projects, a result of the mistaken idea that students must learn everything independently without any guidance from us. In our experience, this idea erodes student motivation and focus. Simply working on a project doesn't automatically lead to discovering its solution. Forcing students to discuss a topic without enough background knowledge wastes their time. Research indicates that direct or assisted discovery instruction often leads to better academic outcomes than unassisted or pure discovery learning (Marzano, 2011; Mayer, 2004, p. 17). When students discover new content with proper scaffolding, then their learning can improve dramatically (Alfieri, Brooks, Aldrich, & Tenenbaum, 2011, p. 13; Hmelo-Silver, Duncan, & Chinn, 2007, pp. 104–105; Kirschner, Sweller, & Clark, 2006, p. 84).

Sometimes, we find that our project description omits an important idea or misleads students into doing something different from what we intended. We may not have realized that students need to access information that is not accessible. Students may have found a more efficient method than the one described in the project description. We find that monitoring student progress closely as they work by periodically talking with them is critical. We tell our students that checking in with them to see how they're progressing on the project is as much for *our* benefit as for theirs. If necessary, we adjust the project to reward innovative solutions.

Students may have difficulty working with each other in a group project. We discuss strategies for facilitating group work in Chapter 17: Cooperative Learning.

Student Handouts and Examples

Figure 16.1: Discovering Pi

Figure 16.2: Area of a Circle

Figure 16.3: Point Lattice Assignment

Figure 16.4: Paint a Room

Figure 16.5: Project—Bus Redesign Plan

Technology Connections

The Project Management Institute Educational Foundation (www.pmief.org), the Center for Innovation in Engineering and Science Education (http://www.ciese.org/materials/k12), and PBLWorks (http://my.pblworks.org) have many resources on project-based learning, including sample rubrics, project planners, and successful strategies.

Websites with lessons that incorporate project-based learning include Illustrative Mathematics (http://www.illustrativemathematics.org/curriculum), the Math Assessment Project (http://www.map.mathshell.org), and STatistics Education Web (http://www.amstat.org/ASA/Education/STEW).

Data sources for class projects can be found at the US Census Bureau website (www.census.gov). Census at School (http://ww2.amstat.org/censusatschool) is an international site that allows students to complete an online survey, analyze their class results, and compare the results to data from the United States or other countries.

Kathy Schrock has many rubrics available on her website (http://www.schrockguide.net/assessment-and-rubrics.html).

Figures

Using the string and ruler in your group, measure the distance around the edge of each circle given to your group (this distance is called the *circumference*). Then using the ruler, measure the distance across the widest part of the circle (this distance is called the *diameter*). Record your results in the table below. Round all distances to the nearest millimeter.

CIRCUMFERENCE	DIAMETER	CIRCUMFERENCE ÷ DIAMETER

For each circle, divide the circumference by the diameter. Compare your group's answers with the answers from other groups in the class. What do all of your answers have in common?

Figure 16.1 Discovering Pi

Maria's Pizzeria sells small and large pizza pies. The following table states the sizes and prices of their pies.

Size	Number of Slices	Diameter (in.)	Price
Small	6	12	$15
Large	8	16	$20

1. *To the nearest tenth of a square inch*, calculate the area of a small pie and a large pie.

    Area of small pie: _____

    Area of large pie: _____

2. *To the nearest tenth of a square inch*, calculate the area of one slice from a small pie and one slice from a large pie.

    Area of one slice of small pie: _____

    Area of one slice of large pie: _____

3. Jamal wants to buy a small pizza, but none are available. He can buy slices from a large pizza for $2.75 each.

    a. What is the minimum number of slices that he must buy in order to get at least the equivalent of a small pizza?

    b. Assuming small pizzas are now available, which would cost Jamal more—buying a small pizza or buying the number of slices you calculated in part *a*? Justify with appropriate calculations.

Figure 16.2 Area of a Circle

Algebra I

POINT LATTICES   Names: _____ _____

A *point lattice* is a set of points that are evenly spaced. In this project, you will work with a partner to draw lines on a point lattice and answer questions about them. This will review many of the concepts that we discussed in coordinate geometry.

VOCABULARY REVIEW: Using your notes or other sources, write the definitions of these terms:

Coordinate axes:	
Horizontal lines:	
Vertical lines:	
Diagonal lines:	
Intersection point:	
System of equations:	
System of inequalities:	

1. For the accompanying point lattice, do the following:

   a. Draw as many lines as you can that pass through at least two points.

   b. List as many questions as you can based on your drawing.

Figure 16.3 Point Lattice Assignment

2. For the accompanying point lattice, do the following:

  a. Draw and label a set of coordinate axes. (You may put the axes anywhere on the lattice. They should pass through points in the lattice.)

  b. Draw 2 horizontal, 2 vertical, and 2 diagonal lines. Each line must pass through at least two points.

  c. Write the equation of each line.

3. For the accompanying point lattice, do the following:

  a. Draw and label a set of coordinate axes. (You may put the axes anywhere on the lattice. They should pass through points in the lattice.)

  b. Draw two intersecting lines. Each line must pass through at least two points.

  c. Write the equation of each line.

  d. What is the solution to this system of equations? Explain your reasoning.

Figure 16.3 (Continued)

4. For the accompanying point lattice, do the following:

   a. Copy the coordinate axes and lines from #3 onto the lattice at right.

   b. Modify the equations to make a system of inequalities.

   c. Graph the solution set and label it S.

5. **EXTRA CREDIT:** Create a word problem that is modeled by the system of equations you created in #3 or the system of inequalities that you created in #4.

Figure 16.3 (Continued)

Pick a room with rectangular walls. This could be a room at school, your home, or a friend's or relative's home. In this assignment, you will calculate how much paint you need to paint this room.

1. On a separate sheet of graph paper (attach it to this work), draw all of the walls in the room. Your drawing should do all of the following:
   - Include all windows but not the floor or ceiling.
   - Do not include anything hanging on the walls (such as pictures) or electrical outlets.
   - Include the measurements of all walls and windows.
2. Calculate the areas (in square feet) of all walls in the room that need to be painted. Write the areas on your diagram.

Wall	Length	Width	Area

3. Add up the square footage of all the walls that need to be painted. Write your answer below:

4. If 1 gallon of paint can cover 350 square feet and you paint each wall twice, how many gallons of paint do you need to paint the walls of your room? Round your answer *up* to the nearest whole number. Show your work below:

Figure 16.4 Paint a Room

Project: Redesign the Buses Around Bayside High School

Do you wish taking the bus would be better? Here's your chance to do something about it!

The Metropolitan Transportation Authority (MTA), which runs New York City's public transportation system, wants to redesign bus routes and place bus stops so that people spend less time waiting and traveling. The MTA has already created the Bronx Bus Redesign Plan to improve service in the Bronx. Your job is to design better bus service for people traveling to and from Bayside High School (BHS). Many Bayside students who take the bus have to transfer at least once and spend over an hour traveling to and from school.

In this group project, you will use statistical data to write a proposal with redesigned bus routes that improve service for students traveling to and from Bayside High School. Due dates for each part will be posted online. Your paper must be clearly organized with no grammatical or spelling mistakes. See the rubric for details. You must show substantial effort in writing this proposal, including meeting adequately with your teacher.

Here is a summary of each part of the project. The project rubric appears on the next page.

Introduction: Summarize the Bronx Bus Redesign Plan *in your own words*, explain why it is important, and explain why we need a similar redesign in Queens County. Read about the plan here: http://new.mta.info/sites/default/files/2019-05/410_19_BBNR%20Existing%20Report_Final_2019.pdf

Methods: Describe the data you use (including its source) and the statistical methods that you use to analyze it. Include the MTA's Northeast Queens Bus Study (http://www.mta.info/sites/default/files/northeast_queens_bus_study_-_final_9-28-15.pdf) and an explanation of the MTA's bus redesign projects (https://new.mta.info/bronxbusredesign/about).

Results: Summarize your data in narrative form. Do not analyze or interpret your data in this section. Include relevant charts, graphs, and other visual aids to make your points clearer.

Conclusion: Explain why your results could have occurred, state the importance of your conclusion, and recommend possible new routes.

Figure 16.5 Project—Bus Redesign Plan

References: Indicate the sources of your information. Include both citations (in your text) and a bibliography (at the end of your paper) using APA format, which is explained in these links:
- https://owl.purdue.edu/owl/research_and_citation/apa_style/apa_formatting_and_style_guide/general_format.html
- http://www.citationmachine.net/apa/cite-a-book

Oral presentation: You will have a chance to share your findings to our class in an oral presentation (5 minutes or less) and submit your proposal to the MTA bus redesign project team.

RUBRIC FOR WRITTEN PAPER:

Category	Below standard	Meets standard	Above standard
Introduction	*Fails to clearly or briefly summarize* the Bronx Bus Redesign Plan	Briefly summarizes the Bronx Bus Redesign Plan	Briefly summarizes the Bronx Bus Redesign Plan
		Adequately explains how BHS students will benefit from a similar redesign	Clearly explains how BHS students *and Queens residents* will benefit from a similar redesign
Methods	*Fails to clearly explain* how MTA bus data will be used and analyzed in the study.	Adequately explains how MTA bus data will be used and analyzed in the study.	Clearly explains how MTA bus data *and data from other sources* will be used and analyzed in the study

Figure 16.5 (Continued)

Category	Below standard	Meets standard	Above standard
Results	*Fails to summarize* data in narrative form *Fails to include* at least one relevant table, histogram, boxplot, and scatterplot	Summarizes data in narrative form Includes *one* relevant example of each of the following: table, histogram, boxplot, and scatterplot	Summarizes data in narrative form Includes *at least two* relevant examples of each of the following: tables, histograms, boxplots, and scatterplots
Conclusion	*Fails to clearly describe* locations of redesigned bus routes and their stops *Fails to use statistical analysis* to show how BHS students benefit from your plan	Clearly describes locations of redesigned bus routes and their stops Uses statistical analysis to show how BHS students benefit from your plan	Clearly describes locations of redesigned bus routes and their stops Uses statistical analysis to show how BHS students *and Queens residents* benefit from your plan
References	Includes *only one or two* sources OR sources *not* properly cited in APA format	Includes *three or four* sources, properly cited in APA format	Includes *at least five* sources, properly cited in APA format
Format	*Not* properly formatted in APA format OR has *major mistakes* in spelling, grammar, and style	Properly formatted in APA format with *only a few minor mistakes* in spelling, grammar, and style	Properly formatted in APA format with *no mistakes* in spelling, grammar, and style
Effort	*Fails to* talk with teacher at least once a week to get feedback Submits *fewer than three* drafts OR submits drafts that do *not* have substantial improvements	Talks with teacher *once a week* to get feedback Submits *three* drafts, each with substantial improvements	Talks with teacher *more than once a week* to get feedback Submits *at least four* drafts, each with substantial improvements

Figure 16.5 (Continued)

CHAPTER 17

Cooperative Learning

What Is It?

In Chapter 16: Project-Based Learning, we discussed ways in which students can work for an extended period of time on a real-world question, culminating in a final product. Whether done individually or in groups, projects are typically done over extended periods of time. However, teachers do not have to give a long-term assignment to get many of the benefits of project-based learning. *Cooperative learning*, in which students learn by working together in small groups to complete a task, can be incorporated into many lessons. While many pedagogical techniques seem to come and go, cooperative learning has grown over the last few decades to become popular around the world (Johnson & Johnson, 2009, p. 365).

Simply putting students in groups to solve routine problems does not make an assignment into cooperative learning. According to researchers David and Roger Johnson (2002, pp. 96–97), successful cooperative learning must have five basic elements:

1. **Positive interdependence.** Students see that individual success is linked with that of other group members.
2. **Individual accountability.** Each individual's performance is assessed and the results are given back to the group and the individual.
3. **Face-to-face promotive interaction.** Individuals encourage each other to achieve the group's goals.
4. **Social skills.** Students must learn the interpersonal and group skills (such as trust, communication, decision making, and conflict resolution) that will enable group members to work together.
5. **Group processing.** Group members reflect on how well the team worked together and what could be done to improve future work.

Why We Like It

Like almost every mathematician we know, we find that our mathematical understanding improves tremendously when we work with others. When friends and colleagues critique our work, they can point out problems and help us make connections to other ideas. In Mathematics, like in many other disciplines, discussing and bouncing around ideas is a necessary component of understanding the problem and finding the true or best solution.

Supporting Research

Years of research indicate that cooperative learning can improve students' academic achievement (Marzano, Pickering, & Pollock, 2001, p. 86; Qin, Johnson, & Johnson, 1995, p. 136; Ruffalo, 2018, p. 13; Slavin, Leavey, & Madden, 1984, p. 418; Srougi & Miller, 2018, pp. 326–327). Some of their conclusions about cooperative learning include the following:

- **It supports constructivism.** We describe this in Chapter 16: Project-Based Learning by promoting shared learning and allowing students to construct a deeper understanding of the material. When students work with others, they can often refine their ideas more easily and precisely than they could if working alone (Artzt & Newman, 1997, p. 2; De Lisi & Golbeck, 1999, p. 5).

- **It fosters improved attitudes toward math.** Students gain confidence that they can improve, thus reducing their math anxiety (Artzt & Newman, 1997, pp. 3–4).

- **It gives students a chance to practice social-emotional learning skills.** This includes supporting and listening to each other (VanAusdal, 2019).

- **It helps to prepare students for life after school.** In the workforce, they must know how to interact and collaborate with others (De Lisi & Golbeck, 1999, p. 4).

- **It supports culturally responsive teaching.** We discuss this in Chapter 2: Culturally Responsive Teaching. When students with different abilities and backgrounds work together, they can build a common ground for discourse, which reduces artificial barriers and prejudices (Artzt & Newman, 1997, p. 3).

Cooperative learning is related to a similar technique called *collaborative learning*. Researchers like Kenneth Bruffee (1995, p. 15) argued that the two strategies are different. Cooperative learning clearly defines roles for each student and elicits teacher-generated ideas, while collaborative learning gives students more freedom to define group roles and construct their own knowledge, leading to answers that cannot be predicted by teachers. Others differentiate between cooperative and collaborative learning according to the amount and type of individual work done. They argue that students work *collaboratively* when they make individual progress in conjunction with others (for example, by solving problems individually and then comparing each other's work) but *cooperatively* when they work separately on individual parts of a task (for example, by individually writing one part of a group explanation) ("Collaborative learning," 2017; Ferlazzo, 2016). Research indicates that people who work separately and together on a task perform better than those who only work together (Bernstein, Shore, & Lazer, 2018, p. 8737).

In contrast, others argue that the cooperative and collaborative learning are more similar than different (Jacobs, 2014). They point out that in both settings, students work in groups with little teacher direction to complete specific tasks and share their conclusions with other groups. Both strategies encourage students to learn by discovery and construct knowledge in a social context (Rockwood, 1995, pp. 8–9).

We believe that trying to determine if a task should be labeled cooperative or collaborative is not as useful as thinking about the amount of student autonomy and foundational knowledge in the activity. In the activities that we describe in this chapter, we often let students decide how to divide up the work and encourage them to work independently before coming together. We use the term *cooperative learning* in this book because it is more commonly used in research on math education, but we use it to indicate a wide variety of cooperation and collaboration.

Common Core Connections

Cooperative learning can help students persevere in solving problems (MP.1) and critique the reasoning of others (MP.3) (NGA & CCSSO, 2010, pp. 6–7). It can help students understand or explain many concepts emphasized in the Common Core, such as random sampling (7.SP.A.1), congruence (8.G.A.2), zeros of a polynomial function (A-APR.B.2), and functions (F-IF.A.1) (NGA & CCSSO, 2010, pp. 50, 55, 64, 69).

Application

In this section, we first describe important points to remember while creating and managing cooperative learning tasks. We then discuss some of our favorite cooperative learning strategies.

GENERAL TECHNIQUES

When we design cooperative learning tasks, we think about the ways in which students are assigned to a group, the role played by each student, and the ways in which students are assessed.

Forming Groups

One simple way to form cooperative learning groups is to allow students to work with their immediate neighbors. Since we frequently let students choose their own seats (as we said in Chapter 11: Building a Classroom Environment), we also let students choose their own groups. As we get to know students better, we may adjust their seats. We may wait until a new unit or marking period to change seats so that groups can build cohesiveness over several weeks. Students who know that they will have to work with the same people for some time may be more motivated to communicate effectively to complete tasks (Artzt & Newman, 1997, p. 6). At the same time, changing groups periodically allows us to adjust them based on our observations of student interactions. For variety, we may sometimes tell students to form groups with others that do not sit next to each other. Students can count off according to the number of groups that need to be formed. If we need eight groups, then students count off from one to eight so that all students with the same number are in the same group.

We usually form heterogeneous groups, mixing students with different performance levels together so that all students can work productively. In heterogeneous groups, students can easily help each other. However, advanced students may resent constantly having to explain work to others and may wind up just doing the work themselves, while struggling students may get frustrated at being unable to complete tasks without help. Differentiating roles within groups by quantity and complexity can minimize these problems and allow students to learn according to their level of readiness (Schniedewind & Davidson, 2000).

Sometimes, we may not have enough diversity in performance levels to form heterogeneous groups (for example, we may have a handful of students whose performance levels are very different from the rest of the class). In these situations, we may group students more homogeneously and differentiate between groups instead of within groups. Some groups may then receive more difficult tasks than others, but we try to make tasks sufficiently challenging for all students.

Our experience confirms existing research that groups should typically have between two and four students (Burns, 1990, p. 25; Frey & Fisher, 2010; Slavin, 2014). We find that pairs work well for relatively small tasks. Groups with five or more students can often get unwieldy since large groups increase the likelihood that some students will not have enough work. The optimal group size for a task depends on the learning goal and the number of people required to complete the task (Hess, 2019). More complex tasks require more people in each group.

Assigning Roles for Group Members

To make cooperative learning effective, students also need to feel that their work matters. Each person should have a clearly defined role (such as taking notes or doing part of a problem) that contributes toward the group's final product. We rarely specify which student in each group should take a role. Allowing students to assign jobs themselves gives them some autonomy over their learning and helps them choose the role that works best for them.

Individuals also need to be held accountable for the group's work—in order for the group to succeed, all of its members should meet the task's learning goals (Slavin, 2014). Doing so can minimize the likelihood of situations in which some group members do most of the work while others become bystanders. For example, we may randomly select one student from each group to explain the group's work, or we may ask each student to complete one part of the task so that the entire group can make a conclusion together based on everyone's work.

We often use protocols to facilitate student communication. *Protocols* are structured guidelines that help promote effective conversations among group members. They give people time to listen actively and provide feedback while allowing everyone in the group to participate equally. This structure can reduce the likelihood of extraneous conversation or off-task behavior (Venables, 2015). While protocols can sometimes constrain students and require practice, we find that groups can use protocols to help refocus students who get distracted ("according to our protocol, nobody can talk until she finishes") without personalizing criticism ("you should not talk right now"). We describe some of our favorite protocols later in this section.

Assessing Students

Since cooperative learning tasks are complex, they require multiple and different methods of assessment (Artzt & Newman, 1997, p. 9). When we assess a group, we not only look at its final product but also how its members worked together to create it. As students work, we circulate around the room—not just to answer questions, but also to monitor their behavior and provide assistance so that groups operate smoothly.

We ask students to reflect on their group's work so they can identify strengths that should be continued and weaknesses that need improvement. Students can reflect both on their individual (so that they can be better members of their group) or their group's performance. Students can think about their mathematical understanding and social interaction in the group by answering questions like the following:

- To what extent did I listen respectfully to others and respond to their questions or comments?
- To what extent did I contribute effectively toward the group's goal?
- What math did I learn after completing the task? (Artzt & Newman, 1997, pp. 10–12)

These questions can be adjusted for a reflection on group work. For example, the first question could be changed to this: "To what extent did the group listen respectfully to each other and respond to each other's questions or comments?"

Although we consider good behavior to be an important characteristic of teamwork, we do not like to overemphasize it in rubrics. We feel that doing so may make students think that we are more interested in controlling them than making sure that they stay on-task. To help students to stay focused, we encourage students to respect each other, complete their individual tasks correctly and on time, and work together to reach a common conclusion.

Table 17.1: Self-Assessment Rubric can be used to evaluate cooperative learning.

DIFFERENTIATING FOR STUDENTS WITH UNIQUE NEEDS

When planning cooperative learning, English Language Learners (ELLs), students with learning differences, and advanced students have unique issues that require special consideration.

English Language Learners

Cooperative learning can benefit ELLs in many ways. Research indicates that it can develop their language skills, reduce their anxiety, foster interdependence with their peers, and promote a more positive classroom environment (Alrayah, 2018, p. 30; Gagné & Parks, 2012, pp. 9–10; Suwantarathip & Wichadee, 2010, p. 56). As we said in Chapter 15: Differentiating for Students with Unique Needs, they often have different perspectives (such as alternate ways of doing math) that can be helpful when working cooperatively. We find that ELLs often express themselves with more confidence when they talk with their peers in small groups than when they speak in front of the entire class or to us. We try not to segregate ELLs who speak the same language into one group. We encourage them to speak in English as much as possible. However, if they feel more comfortable speaking in their own language, we allow them to do so and pair them with others who speak both English and their language—if possible. ELLs can also have group "buddies" who agree beforehand to provide extra social, emotional, language, or mathematical support. These buddies do not have to speak the same language.

To help ELLs understand group tasks more clearly, we put directions in writing and explain them verbally. This allows all students, not just ELLs, to follow along more easily as we read the directions and to ask any clarifying questions. We also review important vocabulary terms or ask students to go over them before starting a task. In our experience, getting the entire class' attention once students start working in groups can be difficult. This challenge makes it very important for us to do our best to ensure that all students understand the task before they begin.

Table 17.1 Self-Assessment Rubric

Criteria	Below standard	Meets standard	Exceeds standard
Participation	*Does not complete* required individual tasks on time. *Does not help* the group to discuss ideas and reach consensus on a conclusion	Completes required individual tasks on time. Takes a *minor* role in helping the group to discuss ideas and reach consensus on a conclusion	Completes required individual tasks on time. Takes a *major* role in helping the group to discuss ideas and reach consensus on a conclusion
Respect	*Rarely* speaks and listens politely to others	*Generally* speaks and listens politely to others	*Always* speaks and listens politely to others. Helps to resolve conflicts among group members
Mathematics	Shows *generally incorrect* mathematical work. Demonstrates *weak* understanding of mathematical concepts behind the lesson	Shows *generally correct* mathematical work that helps the group complete its task. Demonstrates *moderate* understanding of mathematical concepts behind the lesson	Shows *completely correct* mathematical work that helps the group complete its task. Demonstrates *strong* understanding of mathematical concepts behind the lesson
Communication	*Does not* articulate ideas clearly using mathematical language	Clearly articulates ideas using a mix of *mathematical and informal* language	Clearly articulates ideas using correct *mathematical* language

Students with Learning Differences

Cooperative learning has many advantages for students with learning differences. Group roles can allow students to use their strengths in ways that can benefit others in a group setting. For example, autistic students who excel at recognizing repeating patterns may see connections in the work of other students (Baron-Cohen, Ashwin, Ashwin, Tavassoli, & Chakrabarti, 2009, p. 1377). Many dyslexic individuals excel at three-dimensional visualization and seeing the "big picture" instead of individual details (Ehardt, 2009, p. 364; von Károlyi, Winner, Gray, & Sherman, 2003, p. 430). Cooperation between students with learning differences and their peers strengthens diversity awareness and helps break down barriers (Palmer, Peters, & Streetman, 2018).

Students with learning differences may need additional accommodations when working in groups. One simple technique that helps all students, and not just those with learning differences, is to set up tasks that allow students to move around. This strategy can be particularly helpful for students with ADHD who may get restless easily. To help make tasks more concrete, we often provide other aids, such as graph paper, manipulatives (discussed in Chapter 15: Differentiating for Students with Unique Needs), and technology (discussed in Chapter 19: Using Technology).

Some students with learning differences may be so far behind the rest of the class that they may get frustrated trying to keep up with others. Other students may spend so much time trying to help them that the group's task may not be completed. We speak to these students privately to see if they would like to work separately. Putting them together allows us to give them more attention than if they were scattered in different groups around the room. To avoid stigmatizing these students, we only use this strategy sparingly and as a last resort.

Advanced Students

Advanced students can benefit from cooperative learning. Answering questions from other group members can help improve their understanding. However, asking advanced students to simply serve as peer teachers, waiting for the rest of the class to catch up to them, fails to address their learning needs. Group tasks need to be differentiated so that advanced students are sufficiently challenged (Tomlinson, 2014, p. 39). For example, we may include questions with different levels of difficulty so that students can choose an appropriately challenging problem.

If we have advanced students that are very far ahead of the rest of the class, we may group them together and give them a different assignment. As with students who are very far behind the rest of the class, we speak to these advanced students privately beforehand to see if they prefer to work separately.

EXAMPLES

Here are some of our favorite cooperative learning techniques.

Think-Pair-Share

Think-Pair-Share (similar to the *turn-and-talk* strategy) is a simple instructional technique in which teachers ask a question, wait several seconds while students think about it by themselves, discuss the question in pairs, and then share their responses with the rest of the class (McTighe & Lyman, 1988, p. 19). A student who happens to be sitting alone can turn to a pair to form a group of three. As we said in Chapter 4: Promoting Mathematical Communication, Think-Pair-Share can engage all students in a mathematical conversation.

We like Think-Pair-Share for many reasons. It fosters wait time (discussed more in Chapter 4: How to Promote Mathematical Communication), which helps give students time to process information and helps improve understanding and long-term memory (Gregory & Chapman, 2007, p. 120). It requires almost no setup. No matter how students are arranged in class, they almost always can turn to a partner. This strategy is also flexible for teachers. If students do not answer a question that we ask during a lesson, we often break the silence by asking students to discuss it with a partner. Usually, we feel a noticeable sense of relief in the room as they turn to someone to discuss their thoughts. While we typically have students sit while they talk, teachers could allow students to stand in order to break up the classroom routine a little bit. Think-Pair-Share holds students accountable by giving all of them a chance to participate in a mathematical conversation in a less stressful environment.

We find that Think-Pair-Share questions work best when we set a short time limit of 30–60 seconds so that we can move the lesson along and students can stay on task. Think-Pair-Share questions can require a short answer (for example, "Who is right and why?") or a longer one (for example, "What's the mistake in this work?"), but the question should be narrow enough so that pairs can reach a conclusion in less than a minute.

While students are discussing the question, we walk around the room to listen to student conversations. Circulating around the room is important to do throughout the year, and is particularly useful in August and September because it gives us a chance to learn how students work with each other, enables us to get to know them better, and helps us identify any students that may need additional assistance.

We ask students to share in different ways. We may ask students to raise their hands and share their thoughts verbally with the rest of the class. If we have mini-whiteboards (which we sometimes use for test review, as we discussed in Chapter 9: How to Plan Tests and Quizzes), we ask pairs to write responses on them and then share with the class.

Notice and Wonder

Notice and Wonder (sometimes called *What Do You Notice? What Do You Wonder?*) is a technique in which students look at a prompt (such as a picture, word problem, pattern, or set of phrases) and state what they notice and what they wonder about it (Rumack & Huinker, 2019, p. 397). If necessary, teachers can provide additional prompts like the following:

- What quantities do you see?
- What can you count? What can you measure?
- What relationships do you see?
- What is the same? What is different?
- How else can you represent the given information? What diagram or table can you make?
- What calculations can you do ("Understand the Problem," n.d.)?

The Notice and Wonder routine is similar to a Think-Pair-Share, as described in the following directions:

1. Think silently about the prompt for a minute.
2. Write down three things that you notice and three things that you wonder.
3. Turn to a partner and briefly discuss and refine your ideas.
4. Modify your writing based on your discussion with your partner.
5. Be ready to share your ideas with the rest of the class.

We may ask students to repeat steps 3 and 4 to give them more of a chance to work independently. After students complete these steps, we record their findings on the board and use it as a launch point for the rest of the lesson.

Notice and Wonder prompts can also help students who are stuck on a challenging question, such as a word problem or a proof. Teachers can show students the given scenario from a word problem but omit the question. After asking students what they notice and wonder, teachers could then reveal what needs to be found ("Beginning to Problem Solve," n.d.).

Table 17.2: Notice and Wonder gives several examples of prompts and possible notice and wonder questions that could be elicited from those prompts:

In our experience, Notice and Wonder helps students understand the context of the problem, explain it in their own words, identify what they need to find, and give a reasonable estimate for the answer. This strategy can also help them generate "what-if" questions for math research or independent study projects.

Table 17.2 Notice and Wonder

Prompt	Notice	Wonder
Scientific notation: The width of a piece of paper is 0.1 mm. 1 m = 1000 mm The average American woman is 1.6 m tall. One World Trade Center is 541 m tall.	The width of a piece of paper is measured in millimeters. The average American woman is much shorter than One World Trade Center. The height of people and buildings are measured in meters.	What is the width of a piece of paper in meters? How many sheets of paper must be stacked in order to reach the height of One World Trade Center? About how many American women would have to be stacked in order to reach the height of One World Trade Center?
Area of a triangle: 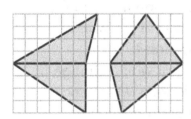	The triangles have the same base. The triangles have the same height. The triangles are not congruent. The triangles have the same area.	Do any triangles with the same base and height have the same area? What would happen if the height changed but the base remained the same? How does the area of a triangle relate to the area of a rectangle?
Sequences: 1, 1, 2, 3, 5, 8, 13, 21, 34, . . .	Each number is the sum of the previous two numbers. There is no common difference between consecutive numbers. The ratio between consecutive numbers approaches 1.618034.	What is the next number in the sequence? What formula can be used to represent this sequence? How would the sequence change if the first two numbers were different?

Table 17.2 (Continued)

Prompt	Notice	Wonder
Circles:	$\angle ACB$ and $\angle ADB$ appear to be congruent. $\angle CAD$ and $\angle DBC$ appear to be congruent. $\triangle AEC$ and $\triangle DBE$ appear to be similar.	Can we prove that $\triangle AEC$ and $\triangle DBE$ are similar? What is the relationship between $m\angle ACE$ and $m\angle ADB$? What would change if point E were moved to point O?
Geometric series:	Each white square is divided into fourths. Each gray square is 1/4 of a larger white square. The diagram has three sets of squares that get smaller. This pattern could continue infinitely.	If the pattern of gray squares continues infinitely, is their total area infinite? What is the relationship between the areas of the gray squares and the areas of the white squares? What would change if triangles were used instead of squares?

Jigsaws

A *jigsaw* is a cooperative learning technique invented by Elliot Aronson in the 1970s (The Jigsaw Classroom, n.d.). Jigsaws typically consist of three parts:

1. Students first form "jigsaw groups," in which each student chooses a different part of the task (such as solving a different problem or choosing a different solution method for the same problem). At this stage, they can work individually, which helps them process the task independently (Butman, 2014, p. 55).

2. Students then form temporary "expert groups" by having one student from each jigsaw group join students from other groups that did the same part of the task. The expert groups work together to agree on a solution and give feedback on individual work.

3. Students then return to their jigsaw groups to share their work. The jigsaw groups compare and write their conclusions on chart paper.

Afterward, the class can do a *gallery walk*, in which students circulate the room to look at each group's work, writing comments or questions on sticky notes and attaching them to the chart papers. Each group can then respond to comments or address any questions or concerns.

Jigsaws can be used to practice problems in class. Each jigsaw group is given a set of problems, one for each person in the group. We typically limit these groups to three or four people to keep the group discussion manageable. Students then form expert groups and solve the problem for that group. We find that using jigsaw as practice works best for lengthy or complicated questions (such as word problems or geometric proofs), where most students might ordinarily complete only one or two problems in class. Since the jigsaw groups discuss several problems when they reconvene, students are exposed to more questions than they would in a whole-class discussion.

Figure 17.1: Jigsaw as Practice can be used at the beginning of a unit or course that uses mathematical writing (such as a Geometry or independent research course). Students are first divided into groups of four and assigned a number from 1 to 4 (this could be done by counting off from 1 to 4 or using numbered cards from a deck) corresponding to a numbered problem. To get a quick understanding of the problems, students first individually match each picture with a mathematical statement. They then form expert groups for that problem and write an explanation of how the picture illustrates the statement. Students can quickly organize themselves into expert groups since all of the students who are working on the same problem can meet in a different corner of the room. Finally, they reconvene into their original jigsaw groups to share their explanations.

We also use jigsaws to introduce a concept through discovery. Asking each student to find a pattern by completing many examples can take up too much class time. Instead, we divide students into expert groups so that each group does one problem simultaneously. (We usually choose problems that are roughly equal in difficulty so that groups can finish in the same amount of time. However, teachers could also vary the difficulty level of the problems to differentiate.) Each group then shares its answer with the rest of the class, typically by writing answers on the board. When students see all answers at once, we then have a whole-class discussion to elicit the pattern.

Figure 17.2: Jigsaw as Discovery shows the beginning of an Algebra II lesson on the Remainder Theorem. In this jigsaw activity, students examine the relationship between the remainder and the divisor of polynomial division. After splitting into expert groups and reconvening into jigsaws, students should see that a polynomial $P(x)$ that is divided by $x - a$ has remainder $P(a)$.

Stations

Stations (sometimes called *learning stations*) are sets of activities on the same theme that are placed in multiple locations around the classroom (Andreasen & Hunt, 2012, p. 242). Since each station relates to a different aspect of the same idea, students need to reach all of the stations in order to achieve the lesson's goals. For example, if the theme for stations is solving word problems involving quadrilaterals, we may create separate stations for problems with parallelograms, rectangles, squares, rhombi, trapezoids, and isosceles trapezoids.

Typically, we have four to six different stations with between one and five problems per station. To prevent too many students from working at the same station, we may duplicate stations. The number of questions per station varies according to their complexity. The problems at each station have roughly the same level of difficulty, although each one may have a slightly harder question as a challenge (Martin, 2019; Tomlinson, 2014, p. 123). We clearly label each station with its topic and, if applicable, its level of difficulty so that students can monitor their understanding.

Although students can theoretically work at stations independently, we prefer that they work in groups of two or three so that they can discuss any issues that they encounter with someone instead of struggling on their own or constantly asking us for help. We typically allow students to choose their own partners, but we may occasionally arrange groups if we want certain people to work together. For example, we may want an ELL to work with someone who is more fluent in English.

One way to organize stations is to write or print questions on large pieces of paper and place them around the room. Next to each station, we attach a model example so students can recall how to solve the problems. Before students start working, we ask them to fold a blank piece of paper into fourths or eighths, depending on the complexity of the problems (teachers could also provide a pre-made solution sheet). They then show all work on this paper. This method saves us the trouble of printing out separate copies for students. However, since students do not have the questions on their paper, this method is better for short questions like simplifying expressions than long questions like word problems. Another way to organize stations is to place labeled folders around the room. Each folder contains copies of the problems for the station as well as a model example. Students can take a copy of the problems and work directly on the paper. Using folders requires more work on our part. However, it keeps student work more organized and allows students to mark up the problem.

Before students start working, we point out which topics are covered in the stations and remind students that they should visit each one. We prefer to assign a starting station for each pair to ensure that students are roughly evenly distributed throughout the room (this can be done quickly by asking students to count off from 1 to 4 or to pick numbered cards—just as we do with jigsaws). We also review the procedures for station work. To help students pace themselves while working, we use a timer and announce when they should move to another station. The amount of time for each station varies depending on the complexity of the problems and the number of stations. If we have four to six stations with between one and five problems per station, then we typically allow 5–10 minutes per station in a 45-minute period.

After completing the stations, students can return to their seats and generalize or reflect on their work. Teachers can give the class a specific question to answer or allow students to choose.

Students could answer questions individually if they are asked to comment on their performance. They could also answer a summary question in groups—each person can alternate writing sentences in a different color (each color should be clearly labeled so we know who is writing). If other colors are not available, then we ask students to write their initials near their work so we can identify the author of each sentence. Here are some examples of questions that students can answer after finishing stations:

- **Which station do you think is most challenging? Which is least challenging? Why?** This question works particularly well when each station has a different method of solving the same type of problem.

- **What do all of the stations have in common?** This question works best if the stations are designed so that students can discover a new idea that is illustrated by all of the stations.

- **What are some common errors or important things to remember while doing these problems?** This question helps students engage in metacognitive activity by thinking about their thinking.

Figure 17.3: Factoring Station shows an example of a station on factoring polynomials completely. This station consists of several problems on factoring completely that are divided into levels of difficulty (we discuss tiered activities more in Chapter 14: Differentiating Instruction). This activity has five stations, each representing different types of factoring—the difference of two squares, factoring by grouping, trinomials, factoring completely, and a mixed practice. We ask each group to complete any five examples from each station (students may choose which problems to complete). When working in pairs, students often divide the task and check each other's work afterward. The problems in all stations are numbered consecutively so that we can easily check answers. When possible, we make copies of our answer key available so that students can check their own work.

We find that stations work particularly well for review and practice. They give students an opportunity to learn from their peers and collaborate on problems. In our experience, many students have been sitting at desks for most of the day, so they enjoy being able to move around the classroom and break up the typical school routine!

Centers

Centers (sometimes called *learning centers*) are sets of activities on *different* themes that are placed in multiple locations around the classroom. Unlike stations, centers are typically not closely related to each other. Instead, they are designed to help students who need additional practice on a particular topic (Martin, 2019; Tomlinson, 2014, p. 123).

One way to use centers is to review several topics at the end of a unit or semester. Each center can consist of problems from several lessons or stations. Each center could then be structured around a major theme of the course, such as ratios, coordinate geometry, statistics, and algebraic expressions. Teachers can also put manipulatives or technology tools in centers to help students complete problems.

Peer Editing

Peer editing, in which students provide feedback on each other's work, ranges from nonevaluative feedback to a summative assessment that results in a grade (Lui & Andrade, 2015). Research indicates that peer feedback that affirms students' work and suggests improvements can enhance their learning (Anker-Hansen & Andrée 2019, p. 16-17; Brookhart, 2017, p. 83; Min, 2006, p. 133).

Before allowing students to edit each other's work, we establish and review rules with students. We find that the following guidelines help students engage in productive conversations:

- Read the work carefully.
- Note what is good about the work as well as anything that should be improved.
- Be fair and respectful.
- Comment on the work, not the person.
- Make specific suggestions to improve the work (Brookhart, 2017, p. 84).

In our version of peer editing, students work in groups of two to four students to solve a problem or create a concept map. Students alternate between working individually and collaboratively. Here are the steps in our version:

1. **Read the problem silently to get a basic understanding.** Write down any thoughts or questions that you want to share with the group.

2. **Share.** In groups, share thoughts and strategies on how to solve the problem.

3. **Get into groups.** Using markers, select one color for each group member. Write down one step or sentence from the solution using the marker of your color. Pass the paper to someone else in your group, who will write the next step and edit any existing ones. This continues until the solution is complete.

4. **To summarize, your group will post your work on the classroom wall.** The class will do a gallery walk and comment on the work of other groups.

Figure 17.4: Peer Editing shows peer editing for a system of equations. In this example, students numbered their steps and wrote their initials to facilitate discussion.

Students can use peer editing to solve several shorter problems at once. Each student in a group writes a solution to one problem on a piece of paper. Students then rotate their papers around their group, making suggestions and corrections until all group members have commented on all papers.

Relays

A variation of a peer editing activity is a *relay*, in which groups of five or six students sit in rows (one student behind the other) to work together on several problems simultaneously. We typically use relays as a fun culminating activity after students have done individual practice.

We start by distributing a set of problems to each group. Each problem should be roughly equal in difficulty and should have roughly the same number of steps so that the work is evenly distributed among all group members. Each problem is printed on a different piece of paper since students will work individually.

For the relay, we give students the following instructions:

1. Work in groups of five or six students. Each group will sit in a row.

2. Your group will get a set of problems, one for each student. You will be given a few minutes to look over them together and discuss how to solve the problems and who will do each problem.

3. At the signal, sit down at your desk and write *only the first step* of the solution to each problem. Write your name next to your work.

4. When you are done, pass your paper back to the person behind you. If you are the last person in the row, bring your paper to the person in the front of your row.

5. At your desk, write *only the next* step to the problem that was just passed back to you.

6. Repeat steps 4 and 5 until all problems are completed.
7. Your group will meet to review the work for all problems before submitting them.

We find that relays can be high-energy activities, which is not surprising since we model our relays on the ones used in math team competitions. The relay can be structured as a race to make it more competitive. Relays can be fun for students who enjoy fast-paced work. Since students generally do a different step for each question and review steps from other problems, they wind up doing parts of several problems cooperatively while contributing individually to each solution.

However, some students may feel overly pressured to work quickly, especially if they feel that the rest of their group is waiting for them. To reduce the stress that students may experience when participating in relays, we encourage students who get stuck on a problem to get help from others in the group. We do not recommend this activity if you think your students may be uncomfortable in hectic classroom settings. We suggest being sensitive to your students' preferences before organizing a relay.

Task Cards

To practice skills like simplifying fractions or solving equations, we sometimes use task cards. A *task card* is a small card that contains an activity or question. Task cards are more flexible than worksheets because they can be easily differentiated simply by reordering or replacing (Cox, n.d.). Students can use task cards to do *card sorting* activities, in which they group cards according to what makes sense to them (Chauvot & Benson, 2008, p. 391). Card sorts enable students to form their own conclusions and construct their own mathematical knowledge (Raymond, 2015, p. 380).

Our task cards consist of sets of 10–20 problems. Students work together in groups of four. Each group gets a set of cards similar to the one in Table 17.3: Task Cards. We use this activity to review three algebraic methods of solving systems of linear equations: substitution, addition, and multiplication. Each of the systems is best solved using one of these methods.

Here are the instructions that we give students before starting the activity in Table 17.3:

1. Work together in groups of four students.
2. Your group will be given a set of task cards. Sort the cards and group them by the most appropriate method for solving the system of equations on each card: substitution, addition, or multiplication.

Table 17.3 Task Cards

A	B	C
$y = 2x$ $y = -6x + 4$	$4x - 3y = 6$ $2x + 3y = 12$	$2x + 5y = 34$ $x + 2y = 14$

D	E	F
$m = -2n - 3$ $m = n$	$5x + y = 16$ $-5x + 3y = 8$	$2x - y = 5$ $5x + 2y = 17$

G	H	I
$3x + 4y = 18$ $y = 2x - 1$	$2x + 3y = 9$ $x - 3y = -3$	$4x - 3y = 11$ $3x - 5y = -11$

J	K	L
$c = 3d - 5$ $2d + 5c = 60$	$2q + r = 7$ $3q - r = 3$	$6x + 3y = 33$ $5x + 2y = 17$

3. Select one problem from each group and solve it on a separate piece of paper.

4. Once you have solved all of your problems, exchange papers with someone else in your group. Make corrections and suggestions to that person's work.

5. Based on these comments, modify your work. Exchange papers again if necessary until you believe that your work is correct.

6. As a group, answer the following summary question: Which solution method is most appropriate for each type of system? Illustrate your answer with examples that you solved. (You can divide your work so that each person in your group rotates writing your explanation.)

7. Your individual solutions and your group summary response will be collected.

What Could Go Wrong

One mistake that we have made is to assume that putting students together in groups will make them work together. Simply giving students a number of problems to solve will not make them work together, as shown in the following example:

1. Graphically determine if the lines represented in each pair of equations are parallel, perpendicular, or neither.
   a. $y = 4x + 3$ and $y = 4x - 5$
   b. $y = \frac{4}{5}x + 1$ and $y = -\frac{5}{4}x + 2$
   c. $y = 5$ and $x = 4$
   d. $y = -2x + 4$ and $y = 3 - 1$

Students can simply split up the work of solving each problem and not collaborate at all. Modifying the task slightly with a summary question that can only be answered by looking at the solutions to all of the problems would force students to work collectively, as shown here:

1. Graphically determine if the lines represented in each pair of equations are parallel, perpendicular, or neither.
   a. $y = 4x + 3$ and $y = 4x - 5$
   b. $y = \frac{4}{5}x + 1$ and $y = -\frac{5}{4}x + 2$
   c. $y = 5$ and $x = 4$
   d. $y = -2x + 4$ and $y = 3x - 1$

2. Based on your work in #1, explain how you can determine if the lines are parallel, perpendicular, or neither without graphing.

Another problem occurs when we feel that *every* lesson has to incorporate cooperative learning. In our experience, some lessons may work better if students work individually. Cooperative learning can be challenging to design and manage. Sometimes, we may not have the energy to do it on a regular basis. We find that effective group work, like many of the other strategies in this book, is more sustainable when we do it occasionally.

Sometimes, students lose focus and stray from their designated task. This usually happens when our tasks are not clearly defined or organized properly. For example, if we do not clearly define roles for everyone, some students may not have anything to do, which can quickly lead to problems. As students are working, we monitor their performance to detect any potential issues or concerns. If several groups are making the same mistake or are asking us the same question, we stop the class (getting their attention by using a pre-established signal like raising our hand) and clarify our instructions. If necessary, we may modify or even stop the activity. We may also ask some students to model what we expect. Whatever we do, we try not to blame students for their confusion. Instead, we publicly acknowledge to them that we may not have designed the activity properly.

Students may sometimes finish early or not finish the task in the allotted time. This often happens when we fail to distribute the work evenly among groups, such as when we give some groups more challenging problems than others. We may, however, differentiate the work by giving harder or easier tasks to certain groups based on their level of readiness. This may be challenging to implement since many groups have a mix of students. Instead, designing tasks so that group members can choose from roles with a mix of difficulty levels (such as tiered activities, which we discuss in Chapter 14: Differentiating Instruction) is more effective.

Many times, we encounter difficulties when asking students to work with each other. Some students do not feel comfortable working with some of their classmates. Some may not feel comfortable working with *anyone*. Some may be so involved with the task that they wind up monopolizing the work, intentionally or unintentionally excluding others. Others may have dominant personalities that intimidate their classmates. If the problem is limited to a group, we speak to those students to determine the problem and try to act as mediators. If necessary, we may ask students to change groups or allow them to complete the task independently. We try to avoid these problems before class by thinking about the method used to form groups. To avoid potential conflicts, we may have to rearrange seats or change the procedures for group formation. As we said earlier, we also try to make sure that *all* students in our class have clearly defined roles in cooperative learning activities. For example, introverts may prefer to act as scribes and ELLs may be more comfortable managing the group's time.

Student Handouts and Examples

Table 17.1: Self-Assessment Rubric

Table 17.2: Notice and Wonder

Figure 17.1: Jigsaw as Practice

Figure 17.2: Jigsaw as Discovery

Figure 17.3: Factoring Station

Figure 17.4: Peer Editing

Table 17.3: Task Cards

Technology Connections

Vanderbilt University's Center for Teaching and Learning has an overview of cooperative learning on its website (http://cft.vanderbilt.edu/guides-sub-pages/setting-up-and-facilitating-group-work-using-cooperative-learning-groups-effectively).

Rubrics for cooperative learning activities can be found on Kathy Schrock's website (www.schrockguide.net).

Resources for Notice and Wonder resources can be found on the website of the National Council of Teachers of Mathematics (http://www.nctm.org/Classroom-Resources/Problems-of-the-Week/I-Notice-I-Wonder) and the Math Forum (http://mathforum.org/pow/support/activityseries/understandtheproblem.html). The 5280 Math website (http://www.5280math.com/noticing-and-wondering) has a number of Notice and Wonder prompts.

The Jigsaw Classroom (www.jigsaw.org) has a collection of resources for jigsaw activities.

The National School Reform Faculty (https://nsrfharmony.org/protocols) has a large collection of protocols that can be used to organize cooperative learning activities.

CHAPTER 17: COOPERATIVE LEARNING 401

Figures

Name: _____

Aim #5: How can we explain mathematical concepts in words?

PART 1: Form jigsaw groups. In your groups, match each picture below with the mathematical statement illustrated by it and write that statement in the box to the right to the picture. Each statement matches with exactly one picture.

	Picture	Statement
1.		
2.		
3.		
4.		

Statements:

In a right triangle, $leg_1^2 + leg_2^2 = hypotenuse^2$.	$a^2 - b^2 = (a-b)(a+b)$
The sum of the interior angle measures of a triangle is 180.	Area of a trapezoid $= \frac{1}{2}h(b_1 + b_2)$

PART 2: Form expert groups. Pick one of the problems above. Make sure that everyone in your group picks a different problem. Meet with all the other students in the class who picked the same problem as you. In your new "expert group," write an explanation using correct mathematical language of how the picture illustrates the statement.

PART 3: Share with your jigsaw group. Go back to your original jigsaw groups. Share your results with the others in your group.

Figure 17.1 Jigsaw as Practice

Algebra II Name: _____

Aim #1.25: What is the Remainder Theorem?

DO NOW

1. Let $f(x) = x^2 + 5x + 1$.

   a. Evaluate $f(1)$.　　　　　　　　b. Divide $f(x)$ by $x - 1$.

2. Let $f(x) = x^2 + 4x - 18$.

   a. Evaluate $f(3)$.　　　　　　　　b. Divide $f(x)$ by $x - 3$.

3. Let $f(x) = x^2 + 2x + 2$.

   a. Evaluate $f(2)$.　　　　　　　　b. Divide $f(x)$ by $x - 2$.

4. Let $f(x) = x^2 - 8x + 15$.

   a. Evaluate $f(5)$.　　　　　　　　b. Divide $f(x)$ by $x - 5$.

5. Use your answers above to fill in the blanks below:

   Remainder theorem: If a polynomial $P(x)$ is divided by $x -$ ___, where a is real, then the remainder equals $P($___$)$.

Figure 17.2 Jigsaw as Discovery

Algebra I   Name: _____

Station #4: Factoring Completely

QUICK REVIEW

NOTE: To **factor completely** means to write an *expression* (not an equation!) of *prime factors* (each quantity is multiplied together, and there are no common factors in the terms of the quantity).

Factor $y^3 + 12y^2 + 35y$ completely.
  $= y(y^2 + 12y + 35)$ Factor the greatest common factor (GCF) from all terms.
  $= y(y + 5)(y + 7)$ Factor all other remaining polynomials into primes.

Level 1: *Factor completely.*

49. $2x^2 + 20x + 48$ 50. $3x^2 - 30x + 63$ 51. $3x^2 - 9x - 120$ 52. $7h^2 - 35h - 42$

Level 2: *Factor completely.*

53. $8p^3 + 12p^2 + 4p$ 54. $18x^3 - 6x^2 - 24x$ 55. $2x^3 - 3x^2 - 5x$ 56. $12k^3 + 6k^2 - 18k$

Level 3: *Factor completely.*

57. $4y^3 - 4y$ 58. $8x^3 - 50x$ 59. $50y^2 - 8$ 60. $36y^2 - 4p^2$

Level 4: *Factor completely.*

61. $256x^4 - 81$ 62. $w^4 - 625n^8$ 63. $3x^4 - 3$ 64. $81x - 16x^5$

Figure 17.3 Factoring Station

$$4x - 6y = 12$$
$$2x + 2y = 6$$

Step #	Work	Initials
1	$4x - 6y = 12$ $2(2x + 2y = 6)$	A.L
2	$4x - 6y = 12$ $4x + 4y = 12$	S.F
3	$-10y = 24$	
4.)	$\frac{-10y}{-10} = \frac{24}{-10} = -12$ ~~-2.4~~ $y = -12$	H.S.
5	$2x + 2y = 6$	A.L.
6	$2x + 2(-12) = 6$	S.A / S.F
7)	$2x - 24 = 6$ $+24 \quad +24$	
8.)	$\frac{2x = 30}{2} = 15 \quad x = 15 \quad$ Solution: $(15, -12)$ ← Write parentheses!	H.S.

Figure 17.4 Peer Editing

CHAPTER 18

Formative Assessment

What Is It?

Formative assessment refers to the process of gathering information about students' understanding in order to improve teaching and learning (Black & Wiliam, 1998, pp. 7–8). Unlike *diagnostic assessment* (which gathers information about student knowledge before instruction) and *summative assessment* (which gathers information about student learning after instruction is complete), formative assessment takes place *during* instruction.

Formative assessment has the following important characteristics:

- It is a process, not simply a pretest.
- It helps teachers and students adjust what they do in order to improve student achievement.
- It uses information taken from the classroom to provide feedback on student learning.
- Since it happens during instruction, it is an assessment *for* learning, not an assessment *of* learning (Marzano, 2015, pp. 10, 22).

Why We Like It

We find that monitoring students' reaction to the lesson and adjusting accordingly improves our teaching. When we pay attention to students' emotional *and* mathematical responses, our students are generally more motivated and focused. Formative assessment not only helps us but also helps students by enabling them to take more control over their learning.

In addition, many formative assessment strategies are simple and flexible. They often don't require elaborate protocols or group work structure. They can easily be incorporated into almost any lesson, including ones with direct instruction.

Supporting Research

For decades, researchers have concluded that formative assessment can build student confidence and improve academic performance (Baron, 2016, p. 52; Black & Wiliam, 1998, p. 61; Beesley, Clark, Dempsey, & Tweed, 2018, p. 13; Bloom, 1968, p. 6). When integrated into mathematical instruction, formative assessment strategies can provide useful information (Accardo & Kuder, 2017, p. 358; Cisterna & Gotwals, 2018, p. 216; Coomes & Lee, 2017, p. 367). Several recent meta-analyses have called for more high-quality studies to determine the precise effect of formative assessment on achievement, attitudes, and motivation (Briggs, Ruiz-Primo, Furtak, Shepard, & Yin, 2012, p. 16; Kingston & Nash, 2011, p. 35; McMillan, Venable, & Varier, 2013, p. 6). Nevertheless, most of the existing research concludes that formative assessment can benefit both students and teachers.

Students can use formative assessment strategies to develop important social and emotional skills, such as recognizing and managing one's emotions, understanding the perspectives of diverse individuals, building constructive relationships with others, and making responsible choices about personal behavior (CASEL, 2013, p. 9). Formative assessment can help students monitor their own learning by tracking their progress and identifying what they need to do to improve (Marzano, 2015, p. 338).

Common Core Connections

Formative assessment can give students effective feedback and help them make sense of problems (MP.1) and strengthen their reasoning skills. It can also improve students' ability to monitor their own learning so they can use appropriate tools strategically and critique the reasoning of others (MP.5, MP.3) (NGA & CCSSO, 2010, pp. 6, 7).

Application

In a familiar classroom routine repeated in many classrooms, teachers ask a question, call on students who raise their hands until someone gives a correct answer, and then repeat the process with another question. Many teachers use this *initiate-response-evaluate* model (Medina, 2001) to assess student learning, even though it has several serious flaws:

- **Teachers may call on some students more than others.** They may have a hidden bias in favoring boys over girls (Sadker & Sadker, 1995, p. 3) or white students over students of color (Barshay, 2018).
- **By making participation voluntary, it engages primarily confident students.** Their success further boosts their confidence, while those who don't raise their hands often become less engaged and confident (Wiliam, 2014).
- **Only one student at a time can answer a question.** If teachers ask two questions per minute for a 20-minute lesson, then students would answer an average of only one question per lesson—clearly not enough to gauge understanding.
- **Class discussions are challenging for some students.** Some children with learning differences, such as autistic students, may have difficulty participating in class discussions (Charania et al., 2010, p. 493).

We sometimes use the initiate-response-evaluate model in class. We find it to be a convenient way to get some information on student learning, especially when we provide direct instruction during lessons. Asking *some* questions in class is better than simply lecturing to students. However, we recognize that the model has serious limitations. When possible, we use other ways to get feedback from students.

In this section, we focus on ways to ask the right kinds of questions, elicit student answers, and address those responses. We conclude this section by listing other methods of formative assessment and references to other chapters, where we describe them in more detail.

ASKING THE RIGHT QUESTIONS

We ask different types of formative assessment questions during a lesson. Table 18.1: Formative Assessment Questions summarizes the most common types.

Although we can't include every formative assessment question type in every lesson, we try to include as many as we can to get a better sense of student understanding.

Table 18.1 Formative Assessment Questions

Question types	Examples
Factual questions ask students to recall definitions, formulas, theorems, and other basic ideas from prior knowledge (Anderson & Krathwohl, 2001, p. 45).	What is the definition of slope? How do we multiply powers with the same base? What is an exponential equation?
Procedural questions ask students to implement a procedure or algorithm from the lesson.	Solve for x: $3(2 - x) + 5 = 12$. Factor $x^3 - 100x$ completely. Graph the equation $2x + y = 8$.
Conceptual questions go beyond factual and procedural questions to assess student understanding of ideas. They help teachers clarify concepts and address common misconceptions (Bruff, n.d.).	Is the sequence defined by the formula $a_n = n^3$ arithmetic, geometric, or neither? Explain. In quadrilateral $ABCD$, $AB = CD$, $BC = AD$, and $m\angle B = 90$. What type of quadrilateral is $ABCD$? Explain your answer. Which statement about observational studies is most accurate? (1) They determine cause and effect relationships. (2) They are used when imposing a treatment on subjects would be unethical. (3) In observational studies, researchers attempt to influence outcomes. (4) They minimize the effect of bias.
Application questions ask students to implement and apply mathematical processes to real-world situations (Bruff, n.d.).	A round pizza has a diameter of 16 inches. Determine its area to the nearest tenth of a square inch. A taxi service charges \$3 per ride, \$2.50 per mile traveled, and \$0.50 per minute spent waiting. What is the total cost of a 20-mile ride to the airport with 3 minutes of waiting? The function $f(x) = 1.3(1.023)^x$ can be used to model the population, in millions, of a large city x years after 2000. Use the model to determine the year in which the city's population reaches 2,000,000.

Table 18.1 (continued)

Question types	Examples
Summary questions ask students to summarize the main ideas of the lesson. If a lesson's aim is stated as a question, students can simply answer the aim as a summary.	Explain the steps involved in solving quadratic equations by completing the square. What is the relationship between inscribed angles and central angles in a circle? How can you use radians to measure angles?
Metacognitive questions ask students to evaluate the appropriateness of a solution method, connect individual ideas to larger conceptual frameworks, or assess their own understanding (Anderson & Krathwohl, 2001, pp. 55–56).	How confident are you in your ability to divide fractions by fractions correctly? How do you determine whether a quadratic-linear system has 0, 1, or 2 solutions just by looking at its graph? How is adding polynomials similar to adding whole numbers?

ELICITING STUDENT RESPONSES

Students can respond to formative assessment questions in many ways. We describe several of them in this section.

Self-assessment Indicators

To help students communicate how well they understand our lesson, we often use self-assessment indicators, such as the following:

- **Colored cups or cards:** Each student gets a set of three cards or cups—one red, one yellow, and one green. Red means "I'm lost," yellow means "I'm not sure what you're saying," and green means "I understand you." Students have them on their desk and display the color that indicates their level of understanding during the lesson.
- **Thumbs up or down:** A thumbs-up sign means "yes" or "I get it," a thumbs-down sign means "no" or "I don't get it," and a thumb pointing

sideways means "I'm not sure." Since students can't hold their hands up during the entire lesson, we typically use this strategy when we ask a specific question.

- **Mini-whiteboards:** Students can write their self-assessment on mini-whiteboards. This allows students to say more than a one-word or one-phrase response.

If we use cups or cards, we distribute them at the beginning of the period (or ask students to take them out and put them on their desk) as part of our opening routine so that they can start using them immediately. We tend to pass out whiteboards right before using them since we find that some students can't resist the temptation to draw pictures if we're not specifically asking that they write math responses!

Self-assessment indicators are especially useful for direct instruction because they provide a simple and unobtrusive way to monitor student thinking. Students can indicate that they are having trouble without calling attention to themselves, which can be particularly helpful for shy individuals. As we teach, we periodically pause and allow everyone to reflect on what was just said so that they can change their indicator if necessary.

Random Selection

One simple way to assess learning is to call on students randomly instead of asking for volunteers. When students can clearly see that a random selection method is used, they are more likely to pay attention (Wiliam, 2014). After students answer a question, we typically allow them to be chosen for future questions so that they continue to pay attention during the lesson.

Many interactive whiteboards and mobile apps offer the ability to choose names randomly from a list. One low-tech option is to write each student's name on a card or wooden ice-cream stick and put the cards or sticks in a jar or box. To make the selection process more personal, students can decorate their own cards or sticks. We find that allowing students to draw names adds an element of fun to the process! Students who don't know the answer can either select another name randomly or ask a classmate for help. Calling on a friend can be particularly helpful for English Language Learners (ELLs) who may not feel comfortable speaking to the entire class.

To put less pressure on individuals, we may group students into pairs. Random selection could also be used to review group work. Teachers could use a different set of cards or sticks in these situations, but we find that using our set of individual names is usually easiest.

We like to use random selection because we can quickly and easily incorporate it into almost any type of lesson, including one that uses direct instruction. We may modify our selection to make it more purposeful, as shown by the following examples:

- We may avoid calling on students for various reasons (they may not be feeling well or may be going through a difficult personal issue) simply by removing their name.
- To increase the probability of calling on certain students (such as those who may be a distraction), we may add some extra copies of their names. We emphasize to students throughout this process that being asked to answer a question is *not* punishment but simply an opportunity to determine where they need additional help.
- We may call on struggling students who know the answer to a particular question even if their names aren't selected because we don't want to pass up an opportunity to support their effort.

Classroom Response Systems

Many times, we use a *classroom response system* (sometimes called a *student response system*, *electronic polling system*, or *clickers*) to monitor student responses. A classroom response system works as follows:

1. A teacher poses a question, typically shown on a computer projector or the board.
2. Students record their responses electronically. Usually, students use an electronic device, which could be a stand-alone device called a clicker or a student's mobile phone. Some systems have the ability to read cards with special barcodes that students hold up to indicate their selection.
3. The system records and summarizes student responses.
4. The teacher adjusts instruction as necessary, depending on student responses (Bruff, n.d.).

We typically use a classroom response system for multiple-choice questions since it can easily record and summarize the percentages of people who selected each choice. Many systems can display the percentages of each student who selected each choice or didn't answer the question. Since each multiple-choice distractor represents a different error, we can then have a conversation about why students chose that answer ("We can see that 20% of you selected choice (A), which is incorrect. What do you think they did wrong?"). Although many classroom response systems allow students to type answers, we find that the systems often can't summarize student responses. We only ask constructed-response questions that have short answers ("What method can be used to prove these two triangles congruent?" or "Solve for x.").

Teachers don't need to rely on an expensive electronic classroom response system to reap its benefits. One simple low-tech substitute is to ask students to hold up the number of fingers that correspond to their answer for a multiple-choice question. Another option is to use mini-whiteboards (which we also use for test review, as we said in Chapter 9: How to Plan Tests and Quizzes). We like mini-whiteboards because they give us immediate feedback on what students are thinking. As we teach a lesson, we periodically ask questions and allow students to write answers on whiteboards (as with a Think-Pair-Share, we encourage students to work in pairs so that they can share ideas with a partner). These questions can range from simple yes-no questions ("Is this equation in slope-intercept form?") to more open-ended questions ("How can we rewrite this equation to put it in slope-intercept form?"). After writing their responses, students hold up their boards so that everyone can see each other's work. We then have a brief discussion about the answers and either move on with the lesson or clarify any remaining confusion.

RESPONDING TO STUDENT ANSWERS

What we do after asking a question depends on student responses. If we see that most students answered our question correctly, then we can move on with the lesson. If students give a variety of correct or incorrect responses, or if they show that they don't understand our lesson, we try one or more of the following ideas:

- Show various student answers and ask students to determine which work is correct and why.
- Re-explain or clarify a point that we made earlier.
- Ask several students to explain their work or explain someone else's work.

When we discuss student mistakes, we try to depersonalize them by focusing on the error instead of the person. We refer to "this step" or "this proof," not "Jen's step" or "Kwame's proof."

OTHER METHODS OF FORMATIVE ASSESSMENT

We discuss additional methods of formative assessment in other parts of this book:

- An exit slip is a response to a question related to the lesson, which students typically complete at the end of class and return to the teacher as they leave the room. Exit slips can give us feedback on how well they understood the lesson. If a lesson's aim is stated as a question, students can simply answer the aim as a summary. We discuss exit slips more in Chapter 7: How to Plan Lessons.

- To assess students' progress while they work, we often circulate around the room. We discuss situations in which we circulate in Chapter 14: Differentiating Instruction, Chapter 16: Project-Based Learning, and Chapter 17: Cooperative Learning.

- We use Think-Pair-Share, quick-writes, and cooperative learning activities (like jigsaws and stations) to help students practice mathematical communication and monitor student understanding. We discuss these strategies more in Chapter 4: Promoting Mathematical Communication and Chapter 17: Cooperative Learning.

- Students can often assess themselves using scoring guidelines (which we discuss in Chapter 4: Promoting Mathematical Communication and Chapter 9: How to Plan Tests and Quizzes) or rubrics (which we discuss in Chapter 14: Differentiating Instruction and Chapter 16: Project-Based Learning).

DIFFERENTIATING FORMATIVE ASSESSMENT

When we teach ELLs, we often monitor their progress in both mathematical content and language. Here are some questions that we think about as we work with them:

- **How do they interact with other students?** As we said in Chapter 17: Cooperative Learning, giving ELLs opportunities to work with other students can improve their language skills. We find that observing ELLs as they work with others gives us valuable insights into their thinking that we don't see as often when they work alone.

- **What strengths do they demonstrate in class?** We find that many ELLs may have hidden assets (such as mathematical competence in an area) that they have difficulty showing because they lack fluency in English.

- **What additional support could benefit them?** Sentence starters, glossaries, classmate "buddies," and manipulatives are just some of the supports that can benefit ELLs.

- **How clear is the language that we use in formative assessment?** Using unclear or overly complex language can result in an assessment of language instead of content (Hill, 2016).

When we work with students with learning differences, we make sure that our formative assessment is structured to benefit all students, using techniques like the following:

- **Clearly state our learning goals throughout the lesson.** This can help students understand its purpose and assess their own learning more effectively.
- **Clearly state what students must do to improve their work.** We provide explicit steps for future improvement.
- **Empower students to assess themselves.** Giving students appropriate scoring guidelines or rubrics for formative assessment tasks enables them to monitor their learning more closely and adjust their strategies appropriately (Brookhart & Lazarus, 2017, pp. 16, 19).

What Could Go Wrong

We sometimes confuse formative assessments with diagnostic or summative assessments. This usually happens when we decide to grade them (for example, when we count exit tickets as a quiz). Formative assessment during a lesson helps us modify our instruction and helps students monitor their progress. In our experience, grading formative assessments (which essentially turns them into summative assessments) may sometimes be useful for encouraging students to stay on-task. However, we find that frequently grading formative assessments generally backfires since it erodes the sense of trust that we build with our students by reducing the freedom that students have to make mistakes. We prefer to keep these assessments formative by not grading them.

One common mistake that we've made is to go through the motions of formative assessments but fail to adjust our teaching appropriately. If most of the class doesn't understand what we just taught, moving on to the next topic is obviously inappropriate. We find that moving on with the lesson when a significant portion of the class is confused can quickly make *all* students feel frustrated. Students can tell when their peers are being ignored, and they may fear that they may soon be next. One way to clear up confusion during a lesson is to re-explain the idea and use a Think-Pair-Share discussion. If time permits, we may also try to group together students who need more help and answer their questions.

Technology can often fail. We try to keep our formative assessment strategies as simple as possible, often relying on techniques that don't require elaborate technology (such as asking students to indicate answer choices by holding up their fingers instead of using an electronic classroom system).

Technology Connections

Many mobile phone apps and interactive whiteboard software give teachers the ability to select names randomly from a predetermined list. Searching for "random name selector" in your mobile phone app store should yield a wide variety of free options. We prefer software that allows us to import names so that we don't have to type all of them into the software.

Some interactive whiteboards have software and hardware that can be used as classroom response systems to quickly quiz students. Other options include the free Google Classroom (http://classroom.google.com), Google Forms (http://forms.google.com), Kahoot! (www.kahoot.com), Quizlet (www.quizlet.com), Geogebra (http://www.geogebra.org/materials), and Desmos Classroom Activities (http://teacher.desmos.com), all of which allow teachers to create their own questions (or upload questions from other sources) and collate student responses. ZipGrade (www.zipgrade.com) allows teachers to grade multiple-choice questions quickly. Nearpod (http://nearpod.com) is an interactive lesson builder that allows teachers to embed quizzes directly into presentations.

The Plickers website (www.plickers.com) has special cards with barcodes that students can hold up in class to indicate their answer. Teachers can use a special app on their phone to scan and analyze student responses from Plickers cards, so students don't need special devices. Although Plickers requires a monthly subscription, many of its features are available for free.

Edpuzzle (www.edpuzzle.org) allows teachers to create and upload videos, insert quizzes, and monitor student responses. Flipgrid (www.flipgrid.com) is a video discussion platform that allows students to record and share videos.

Other online lists of formative assessment tools can be found on the Edutopia (http://www.edutopia.org/blog/5-fast-formative-assessment-tools-vicki-davis) and Shake Up Learning (http://shakeuplearning.com/blog/20-formative-assessment-tools-for-your-classroom) websites.

CHAPTER 19

Using Technology

What Is It?

Most math teachers use some form of technology in their classroom. Technology includes the following:

- Stand-alone devices (such as graphing calculators) used exclusively for math
- Computer software (such as Minitab and Mathematica) that performs mathematical functions
- Web-based mathematical tools (such as Desmos, Geogebra, and DeltaMath)
- Hardware and software (such as interactive whiteboards, online videos, document cameras, computer projectors, and Google Classroom) designed to support all types of classrooms

Throughout this book, we discuss the many ways in which we use technology to support various aspects of our teaching. Each chapter also has a Technology Connections section with links to relevant online resources. In this chapter, we talk about some of the broader issues related to technology.

Why We Like It

We find that technology enables us to find patterns and perform routine procedures more quickly in our instruction. This ability saves time in class, which we can then use for more meaningful tasks like additional practice or more challenging questions.

In addition, technology often makes math more comfortable for students, since many of them have grown up with electronic devices. In fact, they usually figure out how to operate our technology before we do! Having control over the tech can provide increased autonomy and feelings of competence—and both can improve motivation.

Supporting Research

Research indicates that using technology in math instruction can improve students' academic outcomes (Cheung & Slavin, 2013, p. 100; Young, 2017, p. 26) and motivation (Higgins, Huscroft-D'Angelo, & Crawford, 2017, p. 26). Technology can especially benefit English Language Learners (ELLs) (Freeman, 2012, p. 60) and students with learning differences (Li & Ma, 2010, p. 232; Mulcahy, Maccini, Wright, & Miller, 2014, p. 159).

Common Core Connections

Technology is an integral part of the Common Core standards. Using appropriate tools strategically (MP.5) is one of the Standards for Mathematical Practice (NGA & CCSSO, 2010, p. 7). Other standards that call for the use of technology include drawing geometric shapes with given conditions (7.G.A.2), interpreting scientific notation (8.EE.A.4), finding approximate solutions to the equation $f(x) = g(x)$ (A-REI.D.11), identifying key features of functions (F-IF.C.7), and computing linear correlation (S-ID.C.8) (NGA & CCSSO, 2010, pp. 50, 54, 66, 69, 81).

Application

In this section, we discuss some of the ways that technology has affected the way that we organize our class and present mathematical content.

CLASSROOM ORGANIZATION

In other chapters of this book, we discuss ways in which we use technology to organize our classroom, including typing lesson plans, posting homework, building relationships with parents, and implementing cooperative learning. Here are some general principles and additional strategies that relate to technology and classroom organization.

Providing Reliable Information Online

One of the most important advantages of having an online classroom presence is that students can rely on it as a resource. A class website can be much more than simply a place to post homework. We try to provide a one-stop resource of information related to our course, such as:

- Course description
- Grading policy
- Our contact information
- Dates and topics of upcoming assessments
- Notes from previous lessons
- Homework assignments
- Links to tutorials and other relevant resources

In our experience, putting useful content in one place can help build trust with students and parents, many of whom expect to find information online. We try to post only information that we feel we can update regularly. This not only helps to keep website maintenance more manageable, but also makes our information more reliable to students and parents.

Online Discussions

Many online classroom management systems have discussion tools that mimic popular social media platforms. This allows students to have online discussions about the math that we talk about in class. For example, we encourage students to post questions about homework or classwork online. Many of our students prefer posting a message online to other means of communication. Allowing students to respond to each other also encourages broader participation.

To maintain order and keep discussions respectful and focused, we review appropriate online behavior both at the beginning of the school year and periodically after that. Like our classroom rules (which we discuss in Chapter 11: Building a Productive Classroom Environment), we try to keep our guidelines short and simple. Here are some of our favorite guidelines, along with brief explanations of each:

1. **Treat others as you want to be treated.** Since online discussions aren't done face-to-face, they can give students a false sense of anonymity that leads them to make inappropriate or rude comments. Remembering that a human with feelings is behind every online comment can help (Shea, 1994, p. 35).

2. **Think and read before you send.** Would you want your family to see your comment? Spoken comments can be quickly forgotten, but online comments can last much longer. Online comments also can't capture the additional meaning that we convey with our tone, facial expressions, or gestures, so

reading before sending to see if anything could be misinterpreted is also critical (Brooks, 2019).

3. **Write as if you were speaking in class.** Many students adopt a more casual tone in online communication that may obscure their meaning or even offend someone. If you shouldn't say it in class, you shouldn't say it online ("Beyond Emily," 2008).

4. **Respect other people's time.** People often set their mobile phones to receive a notification every time a message is sent. This can quickly get annoying if users stray into a rapid-fire barrage of short responses ("Yes!" "True that." "Me too." "Totally with that!") that can clog people's phones and patience.

5. **Forgive other people's mistakes.** Think twice before responding to a mistake. Responding angrily can easily lead to flame wars and hurt feelings. If a response is warranted, give them the benefit of the doubt and try not to be arrogant or self-righteous (Shea, 1994, p. 45).

Teachers can use a variety of methods to discuss these guidelines with students. One way is to conduct an online discussion in class. Some students could be assigned to model *inappropriate* behavior so that others can determine the best way to respond. Another possibility is to share examples of online comments made by previous students (with names removed, of course) and ask students to talk about their appropriateness. We sometimes share some of our worst online mistakes to remind students that we're human.

We also remind students that online communication has certain limits. We encourage them not to send messages very late at night and to wait a reasonable amount of time before receiving a response. Unfortunately, some students expect an immediate response when they post an online comment and may even get upset if nobody responds quickly enough. We model this by telling students that we won't respond to messages sent late at night until the next morning and may take extra time to respond to messages sent on weekends.

In addition, we adhere to our guidelines in our interactions with students by not being overly friendly with them online and thinking before we send a message. We never publicize our personal email addresses and social media profiles, and we use separate addresses for our professional work.

Flipped Classrooms

Another instructional technique that often relies heavily on technology is a *flipped classroom* (sometimes called *flipped learning*). In a flipped classroom, in-class lessons and take-home assignments are "flipped"—students receive direct instruction individually and then meet in class to apply the newly learned ideas ("Definition of Flipped Learning," 2014). Students typically view online videos and complete online practice outside of class and then do what would have been their homework in class the next day with teacher guidance (Petty, 2018).

We suggest creating a webpage or online folder with a video, text webpages, and practice problems for each lesson. Linking to webpages in addition to videos helps students who may find reading a text to be faster or more efficient than scrolling through a video, especially if they are looking for something specific. Practice problems can come from websites that provide answers and explanations so students can check their progress (we suggest some resources in the Technology Connections section of this chapter).

Flipped classrooms have several benefits:

- Many students feel more in control of their studying since they can review material at their own pace and practice basic skills, which can improve student motivation (Graziano & Hall, 2017, p. 12).

- Teachers can spend more time interacting with students in class, which improves their ability to understand their social and emotional needs (Goodwin & Miller, 2013).

- Putting material online enables students who are absent or had difficulty with particular lessons to catch up without holding the rest of the class back.

- Students become more motivated to take good notes and get help from their peers, which are important life skills that we try to teach students.

- Teachers can differentiate instruction by linking to videos that use other languages, showing various solution methods, or even providing separate videos for students with different levels of readiness.

However, flipped classrooms also have important limitations:

- **Students who lack the time or reliable internet access won't be prepared for class.** If students can't view online tutorials at home, teachers could make DVDs of lessons, allow students to watch videos in class, or use resources like textbooks that don't require internet access (Gough, DeJong, Grundmeyer, & Baron, 2017, p. 400).

- **Monitoring student progress on lessons can be challenging since students watch videos outside of class.** We discuss some tools to help track student progress in the Technology Connections section of this chapter.

- **Teachers must find relevant videos or make their own for each lesson.** Some teachers may lack the time or confidence to make their own videos, and videos made by others may not present material clearly or accurately enough.

- **Students can't ask questions about the lesson until they return to class.** One possible remedy is to encourage students to post questions online or contact teachers directly (Graziano & Hall, 2017, p. 12). However, such responses may not be sufficiently timely or complete and could leave students frustrated.
- **Flipped classrooms may widen student achievement gaps.** One study found that white, male, high-performing students did better in flipped classrooms, perhaps because they feel more comfortable asking teachers for help (Setren, Greenberg, Moore, & Yankovich, 2019, p. 19; Sparks, 2019).
- **Teachers can no longer create suspense about the goal of the lesson.** We find that this sense of discovery can often keep students focused since they want to know where the lesson is headed. However, other students may appreciate the certainty of knowing the lesson's objectives in advance.

We believe that flipped classrooms offer great potential for students to take more control over their learning. They work well for classes that have large blocks of time (such as a double-period) since students can start the class by watching videos and then collaborate with others to work on problems. Teachers can also incorporate elements of a flipped classroom by implementing an *in-class flip,* in which they make videos available as a station that students can visit during classwork (Gonzalez, 2014). Flipped classrooms also work best when most students are willing and able to access online lessons. We find that if too many students don't watch the lesson online, we have to spend time in class going over the lesson *and* doing practice, which reduces the ability of a flipped classroom to save time.

In short, we think that flipped lessons can work best for teachers as an occasional change of pace from a regular lesson. Flipped learning can be good for review lessons or for classes of repeater students, where students may already be familiar with concepts and would benefit more from practice than discovery. Teachers with more time who have students with consistent internet access can consider implementing flipped lessons more regularly.

MATHEMATICAL CONTENT

In this section, we discuss some of the ways that technology has changed both the mathematical content that we teach and the way that we teach it.

Supporting Mathematical Procedures

One obvious advantage of technology is that it helps students and us perform routine procedures more quickly and accurately. Instead of factoring a polynomial expression $p(x)$ algebraically, students can graph the function $y = p(x)$ on a calculator. Instead of solving a system of equations algebraically, students can graph each equation and find the intersection points. As a result, students no longer need the ability to calculate quickly and accurately in order to do higher mathematics (Devlin, 2019, p. 10).

However, just because we include technology in our instruction doesn't mean that we avoid teaching procedures altogether. We frequently remind students that digital tools can't always find accurate solutions to all problems. For example, students can't solve an equation with imaginary roots by graphing on a calculator. In these cases, students need to know other solution methods.

We find that technology gives students another option for checking their work on mathematical procedures. It can also empower struggling students by allowing them to solve more complicated problems without having to master cumbersome procedures. In short, technology can be powerful, but it has its limitations.

Supporting Pattern Recognition

As we said in Chapter 1: Motivational Strategies, finding and understanding patterns is central to mathematical thinking and is a powerful motivational strategy. Technology enables students to see patterns from multiple examples quickly and efficiently. Here are some examples:

> Figure 19.1: Simulation of 1,000 Coin Flips shows a spreadsheet graph of the percentage of tails in 1,000 coin flips. The graph shows that after a few coin flips, the percentage of tails can vary considerably but approaches 50% as the number of coin flips increases. This graph illustrates the Law of Large Numbers—the principle that as the number of trials of an experiment increases, the actual ratio of outcomes approaches the theoretical ratio of outcomes. Teachers can use a spreadsheet to record and graph a class simulation (in which each student flips a coin about 30 times and all students record their results on a shared spreadsheet). To save time, teachers can use a spreadsheet formula to simulate 1,000 coin flips and create a graph from the results.
>
> Figure 19.2: Transformations of Functions shows a jigsaw activity in which students use technology to graph horizontal and vertical transformations of the absolute value function. Graphing with technology allows students to see several examples in seconds, allowing them to quickly determine how adding a constant affects the graph of a function.

Figure 19.3: Centroid of a Triangle shows an activity in which students use digital tools (such as Desmos or Geogebra) to construct a triangle and find the point P where its three medians intersect. Since technology enables students to explore the relationship between the intersection point and the coordinates of the vertices, they can see that the coordinates of point P are the average of the coordinates of the vertices.

In short, technology can be a powerful tool for finding patterns without getting bogged down in excessive calculations (Nirode, 2018, p. 184).

Supporting Deliberate Practice

Technology can be a powerful tool for supporting *deliberate practice*, an activity in which motivated students repeatedly perform challenging tasks and get feedback on their performance so they can improve (Ericsson, Krampe, & Tesch-Römer, 1993, p. 367).

Using websites for deliberate practice has several advantages. Technology allows students to work at their own pace, which helps build their confidence and motivation (we discuss the idea of individualized work more in Chapter 14: Differentiating Instruction). Digital tools also enable us to collate student responses and analyze student work, which helps us to customize future instruction. A website can give immediate feedback by showing students the correct answer, so it can help them learn procedural skills.

When creating online practice assignments, we try to strike a balance between assigning too *many* problems (the repetition may bore or discourage them) and too *few* problems (the lack of practice may not build skills sufficiently). Requiring students to solve 50 equations to demonstrate mastery is too much, but asking them to solve only two is not enough. For many topics, having students solve between three and five problems on each topic correctly may be enough. Asking for and listening to student feedback can help teachers determine the optimal number and type of problems that we give in student assignments.

However, we believe that technology is no substitute for a teacher. A website does a poor job of explaining what students did wrong, grading complex problems, or evaluating the efficiency of their solution. A website can't build relationships with students, acknowledge their individual needs, or offer meaningful encouragement. As we said in Chapter 11: Building a Productive Classroom Environment, positive teacher-student relationships can improve students' self-esteem and motivation. In order to promote deliberate practice, students need meaningful and consistent interaction with teachers who can build on their strengths while improving their weaknesses.

USING TECHNOLOGY FOR CULTURALLY RESPONSIVE TEACHING

As we said in Chapter 2: Culturally Responsive Teaching, we try to customize our instruction to align with students' interests and prior knowledge. Here are some ways that technology can make our teaching more culturally responsive:

- Several classroom management systems and mobile apps allow teachers to build relationships with families (which we discuss in Chapter 12: Building Relationships with Parents).

- Online resources allow us to provide more personalized assignments (which we discuss in Chapter 14: Differentiating Instruction).

- Collaborative software and online resources allow students to work together on projects that affect their community (which we discuss in Chapter 16: Project-Based Learning).

- Students can use technology while working at stations or centers (which we discuss more in Chapter 17: Cooperative Learning).

USING TECHNOLOGY TO DIFFERENTIATE INSTRUCTION

In Chapter 15: Differentiating for Students with Unique Needs, we discussed various ways that we modify our instruction to meet the needs of English Language Learners, students with learning differences, and advanced students. Technology can help facilitate many of these accommodations.

For ELLs, we like to use tools that can deepen their mathematical understanding without requiring fluency in English (such as virtual manipulatives and math websites in their native language) as well as tools that can improve their English language skills (such as online dictionaries). As ELLs use these tools, we typically allow them to speak or write in any language that is comfortable for them. We rely on classmates or online translation tools to help us translate as necessary.

For students with learning differences, we often rely on virtual manipulatives, graphing software, and online drawing tools to help students visualize abstract mathematical concepts. Text-to-speech tools can help students who need text to be read aloud to them, although we find that they often have difficulty with mathematical symbols, which may make them less useful in certain situations.

What Could Go Wrong

One of the greatest challenges that we face when using technology in the classroom is that not all students have equal access to it. Many schools lack the resources to give a laptop or tablet to every child. Many students, particularly those in black,

Hispanic, or rural households, don't have advanced mobile phones or high-speed internet access at home (Anderson, Perrin, Jiang, & Kumar, 2019). When possible, we try to accommodate students by providing offline alternatives (such as printouts or DVDs of online material) or encouraging them to work on a school device (if available) during a free period or before or after school.

Sometimes, we get so caught up in the allure of technology that we forget to consider whether it's truly necessary for our lesson. Fancy animations or online games may be entertaining, but students could also find them distracting or confusing. In our experience, simpler is usually better.

Technology often behaves in unexpected ways. School filters block websites, laptop batteries die, cords get unplugged, USB drives get lost, students forget passwords, teachers forget keystrokes—the list of possible ways that technology can go wrong is practically endless. Whenever we use technology in the classroom, we practice it repeatedly until we feel confident that we know how it should behave. We write down step-by-step procedures and keystrokes, which we also make available to students. When possible, we try to have some type of nontechnological backup available, even if it's just another explanation using chalk and a blackboard. To prevent files from getting lost, we save them on cloud-based file storage systems. In addition, we readily admit to students that we're also learning the technology. Admitting that we're not perfect frees them up to correct our mistakes and help us find shortcuts that we often didn't even realize existed.

We try to be flexible in our adaptation of technology. We may get so fond of a particular app, website, or graphing calculator model that we become reluctant to consider more efficient alternatives simply because we don't want to learn *another* set of procedures. Both of us have been teaching long enough to have found from experience that most digital tools get updated or become obsolete within a few years. Instead of fighting this trend, we have come to accept it as a necessary "evil" of technology. When we learn a new or updated piece of technology, we focus on learning how it can perform common tasks (such as graphing a line, accessing a table of values, or finding a point of intersection) that we were able to complete with other tools. Doing an online search for a particular function ("How do I restrict a domain of a function in Desmos?") or simply asking our students usually yields an answer more quickly than looking through an official manual. Maintaining a sense of flexibility and humor also helps!

A common source of frustration that students encounter with technology is that it usually requires a great deal of precision and accuracy. They may mistype an expression or fail to write an ordered pair with parentheses, which may result in an error. Students who don't see their mistake may then get annoyed when they don't get their desired result. We tell students that they should carefully review what they typed before proceeding to the next step. We ask students to look for and use visual aids built into the technology. For example, a fraction tool enables students to avoid typing a fraction like $\frac{1}{3+2}$ as 1/(3 + 2), which is both hard to read and easy to mistype as 1/3 + 2. We encourage students to be patient and careful, pointing out that attention to detail is an important part of communication in any language but especially so in math. (We discuss other issues with symbols and mathematical precision in Chapter 3: Teaching Math as a Language.) Having students work in pairs when using technology so that one student can check the other student's entries can also reduce frustration.

In addition, students often blindly trust their calculator's output and fail to consider whether it's reasonable. Figure 19.4: Two Views of a Graph Using Technology shows two views of the function $f(x) = x(x + 1)(x - 11)$. The graph on the left shows a function on a default 20 x 20 grid. Students who don't see that $f(x)$ is a cubic function may mistakenly think that the graph is a parabola. The graph on the right shows a more accurate graph of $f(x)$. To obtain this view using technology, students must recognize that the function's zeros are 0, −1, and 11, so the x-axis needs to include these x-intercepts. In addition, they need to know that the graph has two turning points, so adjusting the maximum and minimum y-values in the view results in a graph that more clearly shows that $f(x)$ is a cubic function.

To prevent these mistakes from occurring, we emphasize the importance of understanding the general behavior of the functions being used. This includes knowing the general characteristics of functions that we graph. In addition, we frequently remind students to ask themselves if the answers that they got using technology are reasonable. If not, they should check their work to see if they made a mistake.

When students use technology, they can easily get distracted by going onto another website or opening up another app. We design tasks that keep students as busy as possible. We also frequently circulate around the room, immediately addressing any instances of off-task behavior with a gentle reminder to get back to work.

Student Handouts and Examples

Figure 19.2: Transformations of Functions
Figure 19.3: Centroid of a Triangle

Technology Connections

Many school districts have classroom management or website software, such as Google Classroom (http://classroom.google.com), available for educators. We recommend that you check with your school or district to see what is available.

Virginia Shea's Rules of Netiquette, first published in 1994 (http://www.albion.com/netiquette/book/index.html), may seem outdated, but many of its basic rules about online behavior still hold up today.

The Flipped Learning Network (http://flippedlearning.org) provides resources related to flipped learning. Edpuzzle (http://edpuzzle.com) allows teachers to embed formative assessments into videos and monitor student progress. Padlet (http://padlet.com) allows teachers to create a wall for each unit or lesson and allows students to ask questions or respond to others' questions.

Teachers can use virtual manipulatives as motivations for lessons. Didax (http://www.didax.com/virtual-manipulatives-for-math) has a wide variety of virtual manipulatives, including two-color counters, pattern blocks, spinners, and algebra tiles. The National Library of Virtual Manipulatives (http://nlvm.usu.edu) has a wide range of online tools for pre-K to grade 12 math.

DeltaMath (www.deltamath.com) provides hundreds of individualized questions by topic for middle and high school math. We use this site frequently for classwork, homework, and assessments.

Both the Desmos (www.desmos.com) and Geogebra (www.geogebra.org) online math software offer the ability to change the language, which can be useful for ELLs. NASA CONNECT (http://www.knowitall.org/series/nasa-connect-math-simulations) has several short videos that visualize ratios and algebra for middle school students. Google Translate (http://translate.google.com) provides translation in several languages.

Google (http://cloud.google.com/text-to-speech) provides limited text-to-speech capability for free online.

Many school districts have social media guidelines for educators. Our district's guidelines are located on our union's website (http://www.uft.org/files/attachments/doe-social-media-guidelines.pdf and http://www.uft.org/teaching/doe-social-media-guidelines). We strongly recommend that you check with your district to see what policies are in place.

We save our school-related files on cloud-based file storage systems like Google Drive (http://drive.google.com), SugarSync (www.sugarsync.com), Dropbox (www.dropbox.com), or Apple's iCloud (www.icloud.com).

Figures

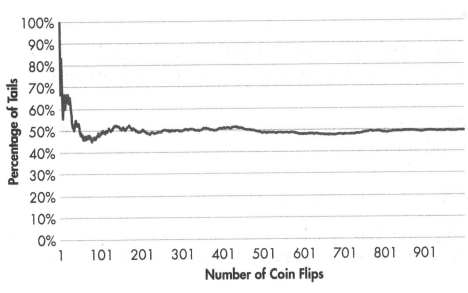

Figure 19.1 Simulation of 1,000 Coin Flips

Algebra I

AIM #43: HOW DO TRANSFORMATIONS OF THE ABSOLUTE VALUE FUNCTION AFFECT ITS GRAPH?

You will work in groups to complete a jigsaw activity.

1. Form "jigsaw groups" of four students. Each of you will work independently on the problems for Group 1, 2, 3, or 4.

2. You will then form temporary "expert groups"—all Group 1 members will meet together, all Group 2 members, etc. The expert groups work together to agree on a solution and provide feedback on individual work.

3. Return to your jigsaw groups and share your work. Answer the summary questions together.

GROUP 1

1. Using technology, graph $y = |x| + 1$, $y = |x| + 2$, $y = |x| + 3$, and $y = |x| + 4$. Sketch the graphs on the same coordinate plane.

2. Without using technology, describe how the graphs of $y = |x| + 5$ and $y = |x|$ compare.

Figure 19.2 Transformations of Functions

GROUP 2

3. Using technology, graph $y = |x| - 1$, $y = |x| - 2$, $y = |x| - 3$, and $y = |x| - 4$. Sketch the graphs on the same coordinate plane.

4. Without using technology, describe how the graphs of $y = |x| - 5$ and $y = |x|$ compare.

GROUP 3

5. Using technology, graph $y = |x + 1|$, $y = |x + 2|$, $y = |x + 3|$, and $y = |x + 4|$. Sketch the graphs on the same coordinate plane.

6. Without using technology, describe how the graphs of $y = |x + 5|$ and $y = |x|$ compare.

GROUP 4

7. Using technology, graph $y = |x - 1|$, $y = |x - 2|$, $y = |x - 3|$, and $y = |x - 4|$. Sketch the graphs on the same coordinate plane.

8. Without using technology, describe how the graphs of $y = |x - 5|$ and $y = |x|$ compare.

SUMMARY

9. Based on your group's work, fill in the blanks:

$y = |x| + c$ **is the graph of** $y = |x|$ **shifted** _____ **units** _____.
$(c > 0)$

$y = |x| - c$ **is the graph of** $y = |x|$ **shifted** _____ **units** _____.
$(c > 0)$

$y = |x + b|$ **is the graph of** $y = |x|$ **shifted** _____ **units to the** _____. $(b > 0)$

$y = |x - b|$ **is the graph of** $y = |x|$ **shifted** _____ **units to the** _____. $(b > 0)$

Figure 19.2 (Continued)

CHAPTER 19: USING TECHNOLOGY   431

1. Use technology to construct a triangle whose vertices have integer coordinates on a coordinate plane. Plot your triangle below.

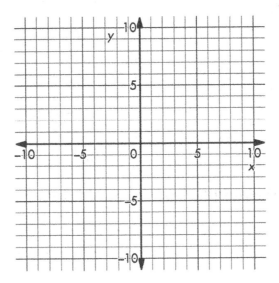

2. Find the midpoints of each side and draw all three medians of the triangle.
3. State the coordinates of the point P where all three medians meet.
4. Change the coordinates of the triangle. How are the coordinates of point P related to the coordinates of the triangle's vertices?

Figure 19.3 Centroid of a Triangle

432 THE MATH TEACHER'S TOOLBOX

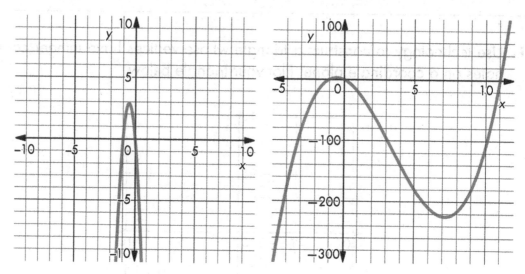

Figure 19.4 Two Views of a Graph Using Technology

CHAPTER 20

Ending the School Year

What Is It?

The end of the school year can be a challenge for both students and teachers. Students may be thinking more about the summer or the upcoming school year than what we're discussing in class, and we may have trouble maintaining our energy. The end of the school year can be a time for us to look back on what we accomplished and look ahead to the future.

Why We Like It

At the end of the year, having the right activities can keep us focused. Once school starts, we often get so busy that we have few opportunities to think about what we or our students have learned. Knowing that we have a plan in place at the end gives us a meaningful goal toward which we can work throughout the year. Devising a good plan is also helpful for students who may have several weeks of the school year remaining after taking a state test or an Advanced Placement exam.

Supporting Research

Human brains naturally forget information that appears trivial in order to make room for meaningful experiences. Research indicates that if students recall and rehearse important information periodically, they are more likely to save it into long-term memory (Sousa, 2017, p. 131). Review activities can also reinforce a sense of *closure*, the process by which students' short-term memories summarize what has been learned. Closure increases the probability that they will be permanently stored in long-term memories (Sousa, 2017, pp. 79, 335).

End-of-year activities may also reinforce what psychologist and economist Daniel Kahneman called the *peak-end effect* (sometimes called the *peak-end rule*), the theory that people form memories of an experience based on how they felt at its most intense moments (the peak) and the last moments (the end) rather than the sum or average of all moments (Hoogerheide, Vink, Finn, Raes, & Paas, 2017, p. 1120; Kahneman, Fredrickson, Schreiber, & Redelmeier, 1993, p. 403). Since people make future decisions based on those feelings, how students feel at the end of the year can affect how they feel about school in the future (Finn, 2015, p. 378). Although some studies suggest that the peak-end effect may lose its power over time and may not be as strong for complex experiences (Strijbosch et al., 2019), the research generally indicates that what happens at the end of an experience has at least *some* effect on how people remember it.

Common Core Connections

Year-end review and reflection can support several of the Common Core's Standards for Mathematical Practice, including persevering in solving problems (MP.1), using appropriate tools strategically (MP.5), making use of structure (MP.7), and finding regularity in repeated reasoning (MP.8) (NGA & CCSSO, 2010, pp. 6–8).

Application

We like to use the end of the school year to review topics that students discussed throughout the year, reflect on their learning, and recognize their achievements. In this section, we describe some of our favorite strategies for each of those goals. We then talk about the importance of continuing relationships that we build with students after they leave our class.

REVIEW

Since summarizing and reviewing what we've taught is critical for reinforcing content, we set aside time at the end of the school year for this process in every course we teach, even those that don't end in a final exam.

Planning for Review

We incorporate year-end review when we start planning over the summer. We begin by looking at our calendars and setting aside 4–6 weeks, which usually gives us enough time to go over all of the units in our course. Looking at our pacing calendars, which we take from such sources as the teacher's edition of our textbook or our district's curriculum, we map out approximately how much time we can spend on each unit. Doing this mapping at the beginning of the year—and modifying our plan as necessary while we teach—helps us stay on track and increases the likelihood that we will have enough time to review at the end.

We don't wait until we finish teaching all of our lessons to review. (We discuss our unit and lesson plans in Chapter 6: How to Plan Units and Chapter 7: How to Plan Lessons.) Throughout the year, we always include at least one question from previous topics in our homework (discussed more in Chapter 8: How to Plan Homework) to facilitate retrieval practice, which increases the likelihood that students will remember the information later and apply it in new situations (Lemov, 2017; Smith & Weinstein, 2016). We discuss retrieval practice more in Chapter 4: Promoting Mathematical Communication. As the end of the year approaches, we incorporate additional review questions into classwork and homework, especially before teaching a lesson that requires prior knowledge.

Connecting to Other Topics

As we cover more material and students' skills and confidence improve, we include more challenging problems that require higher-order thinking and connect to other mathematical topics. For example, Algebra II students can solve the equations $3(x+1)+8=16$, $3(x+1)^2+8=16$, and $3(4^{x+1})+8=16$ side-by-side. After solving, we ask students to compare similarities (all involve isolating the expression with the variable and undoing the operations being performed on it) and differences (quadratic equations require taking the square root of both sides, exponential equations require taking the logarithm of both sides, while linear equations require neither).

As we review at the end of the year, we show students how different units are connected—something we may not do during the year because students haven't learned enough material or we don't have the time. Chapter 3: Teaching Math as a Language and Chapter 5: Making Mathematical Connections contain other relevant examples and strategies.

Test-taking Strategies

If our students are taking a year-end exam, we discuss test-taking strategies in our review. We're not fond of it—after all, we try not to teach to a particular test, but we find that it is part of the reality of teaching. As with other strategies in this chapter, we don't wait until May or June to start sharing them with students. We introduce these strategies in our teaching early in the year and emphasize them more as the year continues.

Table 20.1 Using Technology for Multiple-Choice Questions

Question	Strategy
Which expression is equivalent to $x^2 - 4$? (1) $(x + 2)(x - 2)$ (2) $(x - 2)(x - 2)$ (3) $(x + 2)(x + 2)$ (4) $(x + 4)(x - 1)$	Instead of factoring, students can solve this question graphically by setting each of the answer choices equal to y to see which equation has the same graph as $y = x^2 - 4$.
What is the value of x in the equation $2(x + 8) = 30$? (1) 7 (2) 11 (3) 23 (4) 52	Instead of solving the equation algebraically, students can substitute each of the answer choices into the equation to see which one makes a true statement.

One of our favorite strategies is to use a graphing calculator to compare choices for multiple-choice questions, as shown in Table 20.1: Using Technology for Multiple-Choice Questions:

One limitation of using technology is that many of the test-taking techniques that we describe here don't work for constructed-response questions, which do not give answer choices. We frequently remind students about this limitation and encourage them to use other methods when necessary.

For constructed-response questions, we spend time talking about the rubrics or scoring guidelines that are used to grade them. We share scoring guidelines and anonymous student responses so that students understand how different levels of work are graded and learn how to evaluate their own work.

We also share general test-taking strategies for students, such as the following:

- **Immediately after getting the test paper, write down any formulas and other facts that are not given.** This can serve as a handy reference and free up mental energy during the exam.
- **Devise a plan of action for the test before taking it.** What topics will be most challenging? Answering the easiest questions first can build confidence, while answering the hardest questions first allows students to have enough time.

- **Show work, even for multiple-choice questions.** Writing down work often can earn partial credit, even if the answer is wrong. While multiple-choice questions usually don't require students to show work, we suggest that they write down something that explains what they were thinking. These notes can enable them to more easily check their answer later.

- **Show your reasoning clearly.** The more disorganized and illegible students' written work is, the less likely they are to receive credit, even if some of it is correct. We point out that graders often lose patience reading through muddied writing and may not grade as fairly as we should. We also share some of our own experiences of getting fatigued while grading many papers.

When we review, we also encourage students to share strategies with the rest of the class. Working in pairs or small groups, they can talk about their methods for solving problems or flush out possible misconceptions.

We describe other strategies relevant for taking and reviewing for tests in Chapter 9: How to Plan Tests and Quizzes and Chapter 19: Using Technology.

Maintaining Student Engagement

Maintaining student engagement at the end of the year can be a challenge. However, year-end review doesn't have to be limited to distributing worksheets of test questions, especially for courses that don't lead to a standardized test or have a culminating exam several weeks before school ends! Using a variety of activities can help keep students focused and break up routine.

Here are some other engaging techniques for reinforcing material:

- **Cooperative learning activities:** While we frequently use cooperative learning, we find them especially helpful at the end of the year. Relays, stations, centers, task cards, and other activities can help students recall prior knowledge while maintaining a higher level of energy and focus (Chapter 17: Cooperative Learning has more details). Students can work together to prepare review lessons for the rest of the class.

- **Student-generated material:** Students can make review sheets, vocabulary charts, or class presentations that summarize key ideas from the year (Chapter 14: Differentiating Instruction has more details).

- **Projects:** Projects allow students to showcase talents that often aren't used in class, learn important life skills (like time management and collaboration), and synthesize information learned at different times of the year (Chapter 16: Project-Based Learning has more details).

- **Field trips:** Out-of-classroom activities can allow students to see mathematical ideas in the real world and give them an incentive at the end of the year. However, they require a great deal of coordination with students, parents, and school officials (Heyck-Merlin, 2015). Teachers who can't arrange a field trip could also ask students to travel to places on their own or plan a "virtual" field trip (Borovoy, 2015). See the Technology Connections section for resources on "virtual" field trips.

The days or weeks of school after testing ends can be an excellent time to experiment with new techniques. We don't have new required material to teach, so we often have more flexibility. We can try a new technique or technology tool with little penalty.

REFLECTION

By reflecting on their work at the end of the school year, students can improve their ability to use metacognition, the process of thinking about thinking (which we discuss in Chapter 4: Promoting Mathematical Communication). Metacognition helps students develop important life skills (Cavilla, 2017, p. 4). Since reflection can also reduce stress, a major impediment in math instruction, we find that doing reflective activities at the end of the year can help to set a positive tone for next year. Here are some ways that we encourage student year-end reflection:

- **Complete an anonymous class survey to evaluate the class.** We try to keep these surveys short, simple, open-ended, and anonymous so that students can provide honest feedback without fear of repercussion. Students can write about a time that they experienced success or a challenge as well as what we could do to improve our teaching. We discuss ways to solicit student opinion in Chapter 11: Building a Productive Classroom Environment and Chapter 18: Formative Assessment.

- **Assemble a portfolio.** A portfolio is a collection of student work that shows student growth over time. Students can include examples of work that they are proud to show others, struggled to understand well, and show progress in their learning. Portfolios can be especially helpful for students with learning differences to document their progress over the year. This is especially important for school personnel who need to modify yearly Individualized Education Plans (IEPs) for students with learning differences. We discuss portfolios more in Chapter 14: Differentiating Instruction.

- **Hold teacher-student conferences.** Teachers can schedule private conferences with students to discuss their performance. These meetings can take place outside of class if time permits or in class if students are working on another activity, such as a final project. Teachers and students can discuss final grades at these conferences. Students could even use them as opportunities to propose final grades with appropriate justifications.

- **Make a video reflection.** Students can work in small groups to create a video that describes their favorite topics or memories from the class. These videos can not only be shared with the rest of the class at the end of the year but can be shown to future classes (Cook, 2015). Video reflections can be helpful for English Language Learners (ELLs) who may feel more comfortable speaking than writing.

- **Write a letter to future students.** Students can write a short letter to future students with tips on how to succeed and why success is important. Teachers can then collect these letters and give them to students next year (Cook, 2015).

We reflect not just on our students' performance but also our own. Doing so at the end of the year instead of the summer allows us to write down thoughts while they're still fresh in our minds. If we collaborate with another teacher, we also reflect on how we worked together (we talk about collaboration in Chapter 13: Collaborating with Other Teachers). In addition, periodically reassessing beliefs, relationships with students, and other aspects of our teaching is a critical element of being a culturally responsive teacher (discussed more in Chapter 2: Culturally Responsive Teaching).

RECOGNITION

Recognizing students' accomplishments can improve their achievement and motivation (Elias, 2018). Unfortunately, recognition is often limited to a public award given only to the highest performing students. This leaves the majority of students feeling incompetent and questioning whether their effort can make a difference (Elias, 2018). As we said in Chapter 1: Motivating Students, building students' self-efficacy can help alleviate their math anxiety and can improve their motivation and their academic performance. We prefer ways to recognize all students' progress in ways that can be public or private.

One way that we recognize student achievement is by giving feedback on their work (such as classwork, tests, and projects) throughout the year. Providing feedback on students' year-end reflections helps us summarize their accomplishments. To facilitate this feedback, teachers can keep a log of student progress during the year, briefly recording any noteworthy accomplishments ("Had good explanation for solving exponential equations," "Asked why sum of deviations is always zero—led to good discussion!"). In our feedback, we praise their effort and make specific suggestions so that they can see that their work can lead to improvement and show that we have high expectations for them. We discuss ways to give meaningful feedback in more detail in Chapter 2: Culturally Responsive Teaching and Chapter 4: Promoting Mathematical Communication.

When we have time, we contact parents or guardians to acknowledge student accomplishments. We rarely have time to call or write to the parent of every student, so we focus on reaching the families of students that have made the most improvement or those that we think could benefit from receiving extra encouragement. For example, one of Bobson's students started out the school year quietly, rarely speaking up and barely passing his first quiz. After the student received a poor grade on his first test, he started working much harder in class (Bobson later found out that his parents had gotten a tutor for him and encouraged him to work harder), asking Bobson and other students for help. As he gained more confidence, he began answering his classmates' questions and explaining homework problems. His grades also improved, culminating in a high grade on our state exam. In addition to praising his work throughout the year, Bobson called his parents at the end of the year to thank the student's parents for their support of him and "validate" their encouragement. Chapter 12: Building Relationships with Parents has other techniques for working with parents and guardians.

MAINTAINING RELATIONSHIPS WITH STUDENTS

Spending months teaching and building a relationship with students only to walk away from it at the end of the school year seems strange! We try to stay in touch with as many of our students as possible after they leave our class. This can be done with simple things like greeting them by name in the hallway or occasionally emailing or sending an online message after they've graduated. Maintaining relationships with students also helps us find monitors and teaching assistants, many of whom are former students of ours.

Compiling a list of students' accomplishments and keeping in touch with them also helps us write recommendations and give them guidance. With our encouragement, many of our former students have gone on to succeed in college and beyond, working in education, business, STEM, medicine, and other fields. Some have even gone on to become our colleagues and friends.

DIFFERENTIATING YEAR-END ACTIVITIES

When we plan year-end activities, we think about how we can differentiate them to meet the needs of ELLs and students with learning differences.

ELLs may benefit from doing projects, portfolios and videos, where they can showcase talents that are not often used in typical math instruction. They may need additional time and support to process information. For example, they can use dictionaries or work with classmates who can speak to them in their native language. When possible, we assess ELLs' knowledge and language skills separately, using goals established by us, our school, or our district.

Students with learning differences often benefit from cooperative learning. Many have skills that work best in group activities, which also give them a chance to socialize with—and learn from—their peers in a setting that is frequently more relaxed than a whole-group lesson. Differentiating the final product of year-end activities (for example, by giving students the option to reflect on their work in an essay or a video) can make them more accessible to students with learning differences.

Chapter 15: Differentiating for Students with Unique Needs contains strategies and tips that we have for customizing our instruction.

What Could Go Wrong

In this section, we discuss three common problems that occur at the end of the school year.

YEAR-END FATIGUE

Maintaining focus can be challenging as the calendar approaches May and June. We find that almost any child can experience year-end fatigue—advanced students may have done enough work to earn their desired grade, struggling students may think that they have no chance of passing, and seniors may have already decided what they'll be doing next year. In addition, we often get tired ourselves.

To combat stress and fatigue, we try to stay focused on our class's year-end goals (such as an end-of-year exam or final project) with constant reminders. For example, telling students how many days remain until they take a year-end exam or submit a final project can keep them on-task. We try to avoid giving students the message that we want the school year to end. Countdowns of the number of school days remaining may make students feel like we don't want to be around them anymore, which can decrease their willingness to work and lead to further problems. We also try to vary our routine as much as possible with the various activities that we describe throughout this book, such as games, relays, stations, or centers. In addition, admitting to students that we're also tired or stressed can make us appear more human and strengthen our rapport with them.

"WHAT CAN I DO TO PASS?"

Other students have the opposite problem—they skip class or do little work until May, when they suddenly become concerned about their grades. "What can I do to pass?" is a question we hear all too often in the spring! No matter how frustrated we may feel, we try not to respond with comments that may seem sarcastic ("Stay awake in class!") or angry ("Where have you been the last eight months?"). We find that such responses may convey our frustration but are usually counterproductive.

Instead, we try to prevent them from occurring in the first place. Developing meaningful relationships with students and their parents usually enables us to identify issues *before* they become irreparable. Seeing all students as individuals who are products of their cultures and surroundings (which is not just a critical element of culturally responsive teaching but good teaching in general) can put students' actions in context and help us resolve issues. We may offer help more quickly to students who appear to be struggling or reach out to parents more regularly.

When students come to us to discuss their grade, we first acknowledge their concerns ("I'm glad we're talking—it took courage to admit you need help") while politely pointing out the dangers of procrastination ("I wish we had had this conversation sooner because there isn't much room for error now"). We then discuss with them the work that they have done and the work that needs to be completed. We also review the grading policy with them and walk them through the necessary calculations to determine what work they can make up and when it's due.

In these situations, we often face the dilemma of what to do when students have missed a great deal of work. In our experience, many factors affect our decision, such as our school's or district's grading policy, our school's culture, the amount of time left before the end of the year, the personal circumstances of each student, and our own sense of fairness. Evaluating our beliefs and challenging them so that we can identify potential sources of bias can help us re-evaluate our assumptions and determine the right course of action. We discuss issues related to grading in Chapter 10: How to Develop an Effective Grading Policy.

Some students, despite compelling personal reasons or a sincere change of heart, simply won't pass. We find this to be one of the hardest issues that teachers have to face. In these situations, we talk to students and their parents about possible options. This can include getting as much knowledge as possible before summer school or the following school year. If necessary, we may also suggest practice work to strengthen basic skills.

RUNNING OUT OF TIME

Despite our best efforts to plan ahead, we frequently wind up behind where we hoped we would be at the end of the year. School cancelations, illness, personal issues, or lessons that simply took more time than we had expected can cause even the most detailed plans to go awry.

If we need to prepare students for a year-end test or project, then we look at our curriculum to cut or combine lessons as necessary to fit. In these situations, we try to leave at least 2 weeks just to review. We also try to incorporate more review into each lesson and homework. We explain our plan to our students so that they understand the urgency of the situation and support it.

Technology Connections

Larry Ferlazzo's blog (http://larryferlazzo.edublogs.org/2015/04/17/the-best-ways-to-finish-the-school-year-strong) has links to several resources on ending the school year.

To help students make year-end projects, we suggest resources like Google Classroom (http://classroom.google.com). Teachers can use Google Forms or SurveyMonkey (www.surveymonkey.com) for year-end surveys.

Larry Ferlazzo's blog (http://larryferlazzo.edublogs.org/2009/09/08/the-best-sites-where-students-can-plan-virtual-trips) and Edutopia (http://www.edutopia.org/blog/film-festival-virtual-field-trips) have resources on creating virtual field trips.

For year-end review, we use a variety of websites, including DeltaMath (www.deltamath.com) and Khan Academy (http://khanacademy.com).

Appendix A: *The Math Teacher's Toolbox* Technology Links

Chapter 1: Motivating Students

Description	Link
Center for Self-Determination Theory	http://selfdeterminationtheory.org
ClassDojo	http://www.classdojo.com
"Classes of Donkey" by David Truss	http://pairadimes.davidtruss.com/classes-of-donkeys
Desmos	http://www.desmos.com
Didax	http://www.didax.com/virtual-manipulatives-for-math
Geogebra	http://www.geogebra.org
LiveSchool	http://www.whyliveschool.com
Living Proof: Stories of Resilience Along the Mathematical Journey (Heinrich, Lawrence, Pons, & Taylor, 2019)	http://www.maa.org/press/ebooks/living-proof-stories-of-resilience-along-the-mathematical-journey-2
Math in Daily Life	http://www.learner.org
Math in Movies (Oliver Knill)	http://www.math.harvard.edu/~knill/mathmovies/index.html
Mathalicious	http://www.mathalicious.com

Description	Link
National Library of Virtual Manipulatives	http://nlvm.usu.edu
"On Using and Not Using ClassDojo: Ideological Differences?" by Larry Cuban	http://larrycuban.wordpress.com/2014/03/15/on-using-and-not-using-classdojo-ideological-differences
Proof Blocks	http://www.proofblocks.com
Real World Math	http://realworldmath.org
The Math Lab	http://themathlab.com/toolbox/algebra%20stuff/algebra%20tiles.htm

Chapter 2: Culturally Responsive Teaching

Description	Link
"We the People" National Alliance	http://iris.siue.edu/math-literacy
Cult of Pedagogy	http://www.cultofpedagogy.com/culturally-responsive-misconceptions
	http://www.cultofpedagogy.com/culturally-responsive-teaching-strategies
Culturally Responsive Leadership	http://culturallyresponsiveleadership.com
Edutopia	http://www.edutopia.org/topic/culturally-responsive-teaching
Journal of Urban Math Education	http://ed-osprey.gsu.edu/ojs/index.php/JUME/index
New York State Education Department: Culturally Responsive Sustaining Education Framework	http://www.nysed.gov/common/nysed/files/programs/crs/culturally-responsive-sustaining-education-framework.pdf
Radical Math	http://www.radicalmath.org
Region X Equity Assistance Center	http://educationnorthwest.org/sites/default/files/resources/culturally-responsive-teaching.pdf
Story of Mathematics	http://www.storyofmathematics.com

Description	Link
Teaching Diverse Learners	http://www.brown.edu/academics/education-alliance/teaching-diverse-learners/strategies-0/culturally-responsive-teaching-0
Zaretta Hammond	http://crtandthebrain.com/resources

Chapter 3: Teaching Math as a Language

Description	Link
California Department of Education math glossary	http://www.cde.ca.gov/ci/ma/cf/documents/mathfwglossary.pdf
Classmint	http://www.classmint.com
Common Core Standards mathematics glossary	http://www.corestandards.org/Math/Content/mathematics-glossary
Cram	http://www.cram.com
Google Classroom	http://classroom.google.com
Holt, Rinehart, and Winston glossaries	http://my.hrw.com/math06_07/nsmedia/tools/glossary/msm/glossary.html http://my.hrw.com/math06_07/nsmedia/tools/glossary/aga/glossary.html
Kahoot!	http://www.kahoot.com
Mathwords	http://mathwords.com
Merriam-Webster Dictionary	http://m-w.com
New York State Bilingual Math Glossaries	http://steinhardt.nyu.edu/metrocenter/resources/glossaries
New York State Education Department mathematical glossary	http://www.p12.nysed.gov/assessment/nysaa/2013-14/glossaries/glossarymathrev.pdf
Notecards	http://www.easynotecards.com
Oxford Dictionary	http://www.oxforddictionaries.com
Quizlet	http://www.quizlet.com

Description	Link
Texas Education Agency math glossary	http://www.texasgateway.org/resource/interactive-math-glossary
Using English for Academic Purposes	http://www.uefap.net/speaking/speaking-symbols/speaking-symbols-mathematical
Virginia Department of Education word wall cards	http://www.doe.virginia.gov/instruction/mathematics/resources/vocab_cards

Chapter 4: Promoting Mathematical Communication

Description	Link
EquatIO	http://equat.io
Google Classroom	http://classroom.google.com
Google Docs	http://docs.google.com
Google Slides	http://slides.google.com
Holt, Rinehart, and Winston glossaries	http://my.hrw.com/math06_07/nsmedia/tools/glossary/msm/glossary.html
	http://my.hrw.com/math06_07/nsmedia/tools/glossary/aga/glossary.html
Khan Academy	http://khanacademy.com
MathType	http://store.wiris.com
Microsoft PowerPoint	http://products.office.com/en-us/powerpoint
New York State Bilingual Math Glossaries	http://steinhardt.nyu.edu/metrocenter/resources/glossaries
Random student generator for Google Chrome	http://chrome.google.com/webstore/detail/random-student-generator/kieflbdkopabcodmbpibhafnjalkpkod
Which One Doesn't Belong?	http://wodb.ca
	http://www.stenhouse.com/content/which-one-doesnt-belong

Chapter 5: Making Mathematical Connections

Description	Link
Achieve the Core Common Core coherence map	http://achievethecore.org/coherence-map
Common Core Progressions	http://ime.math.arizona.edu/progressions
Conditional probability (MathBootCamps)	http://www.mathbootcamps.com/conditional-probability-notation-calculation
Connecting Representations protocol (New Visions for Public Schools)	http://curriculum.newvisions.org/math/course/getting-started/instructional-routine-connecting-representations
Exploding Dots (James Tanton)	http://www.explodingdots.org
James Tanton	http://www.jamestanton.com
Mathematical Connections (YouCubed)	http://www.youcubed.org/resources/tour-mathematical-connections
Mathies (online manipulatives)	http://mathies.ca/apps.php#Aa1
Multiple representations of a function (Wikipedia)	https://en.wikipedia.org/wiki/Multiple_representations_(mathematics_education)
New Visions for Public Schools online curriculum	http://curriculum.newvisions.org/math/course/algebra-i/modeling-with-functions
Nix the Tricks (Tina Cardone and the MTBoS)	http://nixthetricks.com
Similarity and proportional reasoning (MathCounts)	http://www.mathcounts.org/resources/video-library/mathcounts-minis/mini-66-similarity-and-proportional-reasoning
Two-way tables (Math and Stats)	http://www.mathandstatistics.com/learn-stats/probability-and-percentage/using-contingency-tables-for-probability-and-dependence
Two-way tables (MathBitsNotebook)	http://mathbitsnotebook.com/Algebra1/StatisticsReg/ST2TwoWayTable.html
Underground Mathematics	http://undergroundmathematics.org
Using the Area Model (Bobson Wong)	http://bobsonwong.com/blog/20-think-inside-the-box-1
	http://bobsonwong.com/blog/21-think-inside-the-box-2

Chapter 6: How to Plan Units

Description	Link
eMathInstruction	http://www.emathinstruction.com
EngageNY	http://www.engageny.org
EquatIO	http://www.texthelp.com/en-us/products/equatio
Google Calendar	http://calendar.google.com
Google Docs	http://docs.google.com
Illustrative Mathematics	http://www.illustrativemathematics.org
MathType	http://store.wiris.com/en

Chapter 7: How to Plan Lessons

Description	Link
Desmos Classroom Activities	http://teacher.desmos.com
EquatIO	http://www.texthelp.com/en-us/products/equatio
Geogebra	http://geogebra.org
Google Docs	http://docs.google.com
Google Slides	http://slides.google.com
Kahoot!	http://kahoot.com
Khan Academy	http://www.khanacademy.com
Math Bits Notebook	http://www.mathbitsnotebook.com
MathType	http://store.wiris.com/en
National Council of Teachers of Mathematics	http://nctm.org
Plickers	http://plickers.com

Chapter 8: How to Plan Homework

Description	Link
California Department of Education math glossary	http://www.cde.ca.gov/ci/ma/cf/documents/mathfwglossary.pdf
Dad's Worksheets	http://www.dadsworksheets.com
DeltaMath	http://www.deltamath.com
Desmos	http://www.desmos.com
Geogebra	http://www.geogebra.org
Google Classroom	http://classroom.google.com
Homeschoolmath.net	http://www.homeschoolmath.net
Kahoot!	http://www.kahoot.com
Khan Academy	http://www.khanacademy.com
Kuta Software	http://www.kutasoftware.com
New York State Education Department mathematical glossary	http://www.p12.nysed.gov/assessment/nysaa/2013-14/glossaries/glossarymathrev.pdf
Quizlet	http://www.quizlet.com
Texas Education Agency math glossary	http://www.texasgateway.org/resource/interactive-math-glossary
ZooWhiz	http://www.zoowhiz.com

Chapter 9: How to Plan Tests and Quizzes

Description	Link
College Board AP Exam questions	http://apcentral.collegeboard.org/courses
College Board PSAT and SAT questions	http://collegereadiness.collegeboard.org/sample-questions
DeltaMath	http://www.deltamath.com
Google Classroom	http://classroom.google.com
JMAP	http://www.jmap.org
Kahoot!	http://www.kahoot.com
New York State Grade 3-8 Exams	http://www.engageny.org
New York State Regents High School Exams	http://nysedregents.org

Description	Link
PARCC sample questions	http://parcc.pearson.com/practice-tests
Plickers	http://plickers.com
Problem-Attic	http://www.problem-attic.com
Quizlet	http://www.quizlet.com
Smarter Balanced sample questions	http://www.smarterbalanced.org/assessments/samples
Social Science Statistics	https://www.socscistatistics.com/descriptive/histograms
ZipGrade	http://www.zipgrade.com

Chapter 10: How to Develop an Effective Grading Policy

Description	Link
Association for Supervision and Curriculum Development	http://www.ascd.org
Credits for Teachers	http://creditsforteachers.com/K12-Standard-Based-Grading-Resources
Dale Ehlert	http://whenmathhappens.com/standards-based-grading
DeltaMath	http://www.deltamath.com
Google Sheets	http://sheets.google.com
Kate Owens	http://blogs.ams.org/matheducation/2015/11/20/a-beginners-guide-to-standards-based-grading
Microsoft Office Templates	http://templates.office.com
Yelena Weinstein	http://questformasteryblog.wordpress.com

Chapter 11: Building a Productive Classroom Environment

Description	Link
Cornell note-taking system (WikiHow)	http://www.wikihow.com/Take-Cornell-Notes
Edutopia	http://www.edutopia.com
Evernote	http://www.evernote.com
Google Forms	http://forms.google.com
Harry K. Wong	http://harrywong.com
Larry Ferlazzo's EdWeek blog	http://blogs.edweek.org/teachers/classroom_qa_with_larry_ferlazzo/2014/08/q_a_collections_best_ways_to_begin_end_the_school_year.html
Learning Strategies Center (Cornell University)	http://lsc.cornell.edu/notes.html
OneNote	http://www.onenote.com
Poorvu Center for Teaching and Learning (Yale University)	http://poorvucenter.yale.edu/FacultyResources/Managing-the-Classroom
Scholastic	http://www.scholastic.com
SurveyMonkey	http://surveymonkey.com
The Mathematicians Project	http://awesome-table.com/-Kq4eNy0oVl-JUj1pK7I/view
ThoughtCo	http://thoughtco.com
Women You Should Know	http://womenyoushouldknow.net/downloadable-stem-role-models-posters
Zapier	http://zapier.com/blog/best-note-taking-apps
Zoho Notebook	http://www.zoho.com/notebook

Chapter 12: Building Relationships with Parents

Description	Link
¡Colorín Colorado!	http://www.colorincolorado.org/article/making-your-first-ell-home-visit-guide-classroom-teachers
Apple iTunes Store	http://itunes.apple.com
Class Dojo	http://www.classdojo.com
Edmodo	http://www.edmodo.com
Edublogs	http://www.edublogs.org
Edutopia	http://www.edutopia.org
Google Classroom	http://classroom.google.com
Google Play Store	http://play.google.com/store
Parent-Teacher Home Visit Project	http://www.pthvp.org
Remind	http://www.remind.com
TalkingPoints	http://talkingpts.org
TeacherVision	http://www.teachervision.com
ThinkWave	http://www.thinkwave.com
ThoughtCo	http://www.thoughtco.com
Weebly for Educators	http://education.weebly.com
WordPress	http://www.wordpress.com

Chapter 13: Collaborating with Other Teachers

Description	Link
Association for Supervision and Curriculum Development	http://www.ascd.org
Chicago Lesson Study Group	http://www.lessonstudygroup.net
College Board	http://secure-media.collegeboard.org/apc/AP-Interdisciplinary-Teaching-and-Learning-Toolkit.pdf
Dropbox	http://dropbox.com
Dummies.com	http://www.dummies.com/business/marketing/social-media-marketing/how-to-track-twitter-hashtags/

APPENDIX A: THE MATH TEACHER'S TOOLBOX TECHNOLOGY LINKS

Description	Link
Edmodo	http://www.edmodo.com
Edublogs	http://www.edublogs.org
Educational Leadership July 2011 focus issue	http://www.ascd.org/publications/newsletters/education-update/jul11/vol53/num07/toc.aspx
Edutopia	http://www.edutopia.org
Edutopia	http://www.edutopia.org/blog/new-teachers-becoming-connected-educators-lisa-dabbs
Google Docs	http://docs.google.com
Google Drive	http://drive.google.com
I Teach Math (#IteachMath)	http://twitter.com/hashtag/ITeachMath
Lesson Study Group (Mills College)	http://lessonresearch.net
Math for America	http://www.mathforamerica.org
Math Twitter Blogosphere (#MTBoS)	http://twitter.com/hashtag/MTBoS
National Council of Teachers of Mathematics	http://nctm.org http://www.nctm.org/Affiliates/Directory
New York State Education Department	http://www.highered.nysed.gov/tcert/resteachers/models.html
SugarSync	http://sugarsync.com
TeachThought	http://www.teachthought.com/education/8-ideas-10-guides-and-17-tools-for-a-better-professional-learning-network
Twitter	http://www.twitter.com
US State Department teacher peer observation template	http://americanenglish.state.gov/files/ae/resource_files/peer_observation_handout.pdf
University of Wisconsin-La Crosse	http://www.uwlax.edu/sotl/lsp/
Weebly for Educators	http://education.weebly.com
WNET Education	http://www.thirteen.org/edonline/concept2class/interdisciplinary/index.html
WordPress	http://www.wordpress.com

Chapter 14: Differentiating Instruction

Description	Link
¡Colorín Colorado!	http://www.colorincolorado.org
DeltaMath	http://deltamath.com
Evernote	http://evernote.com
Google Sites	http://sites.google.com
Howard Gardner	http://howardgardner.com
International Association for the Study of Cooperation in Education	http://www.iasce.net/home/resources
IRIS Center (Vanderbilt University)	http://iris.peabody.vanderbilt.edu
Jigsaw Classroom	http://www.jigsaw.org
Kidblog	http://kidblog.org
Learning Disabilities Online	http://www.ldonline.org/educators
National Association of Special Education Teachers	http://www.naset.org
National Center for Research on Gifted Education	http://ncrge.uconn.edu
Quick Rubric	http://www.quickrubric.com
Rcampus	http://www.rcampus.com/rubricshellc.cfm?mode=gallery&sms=publicrub&sid=21&
TeAchnology	http://www.teach-nology.com/web_tools/rubrics/

Chapter 15: Differentiating for Students with Unique Needs

Description	Link
¡Colorín Colorado!	http://www.colorincolorado.org
Achieve the Core coherence map	http://achievethecore.org/coherence-map
American Mathematics Competition	http://www.maa.org/math-competitions
Center for Teaching (Vanderbilt University)	http://cft.vanderbilt.edu/guides-sub-pages/creating-accessible-learning-environments
Center on Online Learning and Students with Disabilities	http://www.centerononlinelearning.res.ku.edu

APPENDIX A: THE MATH TEACHER'S TOOLBOX TECHNOLOGY LINKS

Description	Link
Do2Learn	http://do2learn.com
Edutopia	http://www.edutopia.org/topic/english-language-learners
Google Translate	http://translate.google.com
Kathy Schrock	http://www.schrockguide.net/assessment-and-rubrics.html
Learning Disabilities Online	http://www.ldonline.org
LINKS Learning	http://www.linkslearning.k12.wa.us/Teachers/1_Math/2_Curriculum_Planning/2_Math_Concept_Maps/index.html
Many Things	http://manythings.org
Math League	http://www.mathleague.org
MathCounts	http://www.mathcounts.org
Mathematics for English Language Learners	http://www.mell.org
National Association for Gifted Children	http://www.nagc.org
National Association of Special Education Teachers	http://www.naset.org
National Center for Research on Gifted Education	http://ncrge.uconn.edu
National Society for the Gifted and Talented	http://www.nsgt.org
New York City Interscholastic Mathematics League	http://www.nyciml.org
New York State Bilingual Math Glossaries	http://steinhardt.nyu.edu/metrocenter/resources/glossaries
Teacher Toolkit	http://www.theteachertoolkit.com/index.php/tool/frayer-model
TeacherVision	http://www.teachervision.com/teaching-strategies/special-needs
ThoughtCo	http://www.thoughtco.com/the-frayer-model-for-math-2312085
Understanding Language (Stanford University)	http://ell.stanford.edu/teaching_resources/math

Chapter 16: Project-Based Learning

Description	Link
Census at School	http://ww2.amstat.org/censusatschool
Center for Innovation in Engineering and Science Education	http://www.ciese.org/materials/k12
Illustrative Mathematics	http://www.illustrativemathematics.org/curriculum/
Kathy Schrock	http://www.schrockguide.net/assessment-and-rubrics.html
Math Assessment Project	http://www.map.mathshell.org
PBLWorks	http://my.pblworks.org
Project Management Institute Educational Foundation	http://www.pmief.org
STatistics Education Web	http://www.amstat.org/ASA/Education/STEW
US Census Bureau	http://www.census.gov

Chapter 17: Cooperative Learning

Description	Link
5280 Math	http://www.5280math.com/noticing-and-wondering
Center for Teaching and Learning (Vanderbilt University)	http://cft.vanderbilt.edu/guides-sub-pages/setting-up-and-facilitating-group-work-using-cooperative-learning-groups-effectively
Jigsaw Classroom	http://www.jigsaw.org
Kathy Schrock	http://www.schrockguide.net
Math Forum	http://mathforum.org/pow/support/activityseries/understandtheproblem.html
National Council of Teachers of Mathematics	http://www.nctm.org/Classroom-Resources/Problems-of-the-Week/I-Notice-I-Wonder
National School Reform Faculty	https://nsrfharmony.org/protocols

Chapter 18: Formative Assessment

Description	Link
Desmos Classroom Activities	http://teacher.desmos.com
Edpuzzle	http://www.edpuzzle.org
Edutopia	http://www.edutopia.org/blog/5-fast-formative-assessment-tools-vicki-davis
Flipgrid	http://www.flipgrid.com
Geogebra	http://www.geogebra.org/materials
Google Classroom	http://classroom.google.com
Google Forms	http://forms.google.com
Kahoot!	http://www.kahoot.com
Nearpod	http://nearpod.com
Plickers	http://www.plickers.com
Quizlet	http://www.quizlet.com
Shake Up Learning	http://shakeuplearning.com/blog/20-formative-assessment-tools-for-your-classroom
ZipGrade	http://www.zipgrade.com

Chapter 19: Using Technology

Description	Link
Google Classroom	http://classroom.google.com
Rules of Netiquette (Virginia Shea)	http://www.albion.com/netiquette/book/index.html
Flipped Learning Network	http://flippedlearning.org
Edpuzzle	http://edpuzzle.com
Padlet	http://padlet.com
Didax	http://www.didax.com/virtual-manipulatives-for-math
National Library of Virtual Manipulatives	http://nlvm.usu.edu
DeltaMath	http://www.deltamath.com
Desmos	http://www.desmos.com
Geogebra	http://www.geogebra.org
NASA CONNECT	http://www.knowitall.org/series/nasa-connect-math-simulations

Description	Link
Google Translate	http://translate.google.com
Google text-to-speech	http://cloud.google.com/text-to-speech
NYC Dept. of Education social media policy	http://www.uft.org/files/attachments/doe-social-media-guidelines.pdf
	http://www.uft.org/teaching/doe-social-media-guidelines
Google Drive	http://drive.google.com
SugarSync	http://www.sugarsync.com
Dropbox	http://www.dropbox.com
Apple iCloud	http://www.icloud.com

Chapter 20: Ending the School Year

Description	Link
DeltaMath	http://www.deltamath.com
Edutopia	http://www.edutopia.org/blog/film-festival-virtual-field-trips
Google Classroom	http://classroom.google.com
Google Forms	http://forms.google.com
Khan Academy	http://khanacademy.com
Larry Ferlazzo	http://larryferlazzo.edublogs.org/2015/04/17/the-best-ways-to-finish-the-school-year-strong
	http://larryferlazzo.edublogs.org/2009/09/08/the-best-sites-where-students-can-plan-virtual-trips
SurveyMonkey	http://www.surveymonkey.com

References

3 ways to memorize the quadratic formula. (n.d.). *wikiHow*. Retrieved from http://www.wikihow.com/Memorize-the-Quadratic-Formula

Abdulrahim, N. A., & Orosco, M. J. (2019). Culturally responsive mathematics teaching: A research synthesis. *The urban review*. Retrieved from http://link.springer.com/article/10.1007/s11256-019-00509-2

Accardo, A. L., & Kuder, S. J. (2017). Monitoring student learning in algebra. *Mathematics Teaching in the Middle School, 22*(6), 352–359. Retrieved from http://www.nctm.org/Publications/Mathematics-Teaching-in-Middle-School/2017/Vol22/Issue6/Monitoring-Student-Learning-in-Algebra

Achieve. (2018a). A framework to evaluate cognitive complexity in mathematics assessments. Retrieved from http://www.achieve.org/mathematics-assessment-cognitive-complexity-framework

Achieve. (2018b). Independent analysis of the alignment of the ACT to the Common Core State Standards. Retrieved from http://www.achieve.org/achieve-act-review

Adams, A. E., Pegg, J., & Case, M. (2015). Anticipation guides: Reading for mathematics understanding. *Mathematics Teacher, 108*(7), 498–504. Retrieved from http://www.nctm.org/Publications/mathematics-teacher/2015/Vol108/Issue7/Anticipation-Guides_-Reading-for-Mathematics-Understanding

Adams, T. L. (2003). Reading mathematics: More than words can say. *The Reading Teacher, 56*(8), 786. Retrieved from http://connection.ebscohost.com/c/articles/9639209/reading-mathematics-more-than-words-can-say

Adkins-Sharif, J. (2017). Beginning again with marginalized parents. *Educational Leadership, 75*(1), 34–38. Retrieved from http://www.ascd.org/publications/educational-leadership/sept17/vol75/num01/Beginning-Again-With-Marginalized-Parents.aspx

Aguirre, J. M., & del Rosario Zavala, M. (2013). Making culturally responsive mathematics teaching explicit: A lesson analysis tool. *Pedagogies: An International Journal, 8*(2), 163–190. Retrieved from http://www.tandfonline.com/doi/full/10.1080/1554480X.2013.768518

Alfieri, L., Brooks, P. J., Aldrich, N. J., & Tenenbaum, H. R. (2011). Does discovery-based instruction enhance learning? *Journal of Educational Psychology, 103*(1), 1–18. Retrieved from http://psycnet.apa.org/doiLanding?doi=10.1037%2Fa0021017

Alkire, S. (2002). Dictation as a language learning device. *The Internet TESL Journal, VIII*(3). Retrieved from http://iteslj.org/Techniques/Alkire-Dictation.html

Allen, K., & Schnell, K. (2016). Developing mathematics identity. *Mathematics Teaching in the Middle School, 21*(7), 398–405. Retrieved from http://www.nctm.org/Publications/Mathematics-Teaching-in-Middle-School/2016/Vol21/Issue7/Developing-Mathematics-Identity

Alrayah, H. (2018). The effectiveness of cooperative learning activities in enhancing EFL learners fluency. *English Language Teaching, 11*(4), 21–31. Retrieved from http://www.ccsenet.org/journal/index.php/elt/article/view/74072

Ames, D. L., & Fiske, S. T. (2013). Intentional harms are worse, even when they're not. *Psychological Science, 24*(9), 1755–1762. Retrieved from http://journals.sagepub.com/doi/abs/10.1177/0956797613480507

Anderson, L. W. (2018). A critique of grading: Policies, practices, and technical matters. *Education Policy Analysis Archives, 26*(49). Retrieved from http://epaa.asu.edu/ojs/article/view/3814

Anderson, L. W., & Krathwohl, D. R. (Eds.). (2001). *A taxonomy for learning, teaching, and assessing: A revision of Bloom's taxonomy of educational objectives: Complete edition.* New York, NY: Longman.

Anderson, M., Perrin, A., Jiang, J., & Kumar, M. (2019, April 22). 10% of Americans don't use the internet. Who are they? Retrieved from http://www.pewresearch.org/fact-tank/2019/04/22/some-americans-dont-use-the-internet-who-are-they

Andreasen, J. B., & Hunt, J. H. (2012). Using math stations for commonsense inclusiveness. *Teaching Children Mathematics, 19*(4), 238–246. Retrieved from http://www.nctm.org/Publications/teaching-children-mathematics/2012/Vol19/Issue4/Using-Math-Stations-for-Commonsense-Inclusiveness

Anker-Hansen, J., & Andrée, M. (2019). Using and rejecting peer feedback in the science classroom: A study of students' negotiations on how to use peer feedback when designing experiments. *Research in Science & Technological Education, 37*(3),

1–20. Retrieved from http://www.tandfonline.com/doi/full/10.1080/02635143.2018.1557628

Annamma, S. A., Anyon, Y., Joseph, N. M., Farrar, J., Greer, E., Downing, B., & Simmons, J. (2019). Black girls and school discipline: The complexities of being overrepresented and understudied. *Urban Education, 54*(2), 211–242. Retrieved from http://journals.sagepub.com/doi/abs/10.1177/0042085916646610

Annenberg Foundation. (n.d.). How does disciplinary literacy differ from content-area literacy? Retrieved from http://www.learner.org/courses/readwrite/disciplinary-literacy/what-is-disciplinary-literacy/4.html

Armstrong, T. (2012). First, discover their strengths. *Educational Leadership, 70*(2), 10–16. Retrieved from http://www.ascd.org/publications/educational-leadership/oct12/vol70/num02/First,-Discover-Their-Strengths.aspx

Aronson, B., & Laughter, J. (2016). The theory and practice of culturally relevant education. *Review of Educational Research, 86*(1), 163–206. Retrieved from http://journals.sagepub.com/doi/abs/10.3102/0034654315582066

Artzt, A., & Newman, C. M. (1997). *How to use cooperative learning in the mathematics class* (2nd ed.). Reston, VA: National Council of Teachers of Mathematics.

Artzt, A. F., Armour-Thomas, E., & Curcio, F. R. (2008). *Becoming a reflective mathematics teacher: A guide for observations and self-assessment*. Studies in Mathematical Thinking and Learning Series. (2nd ed.). New York, NY: Routledge.

August, D., & Shanahan, T. (2006). *Developing literacy in second-language learners: Report of the National Literacy Panel on language-minority children and youth* [Executive Summary] [PDF file]. Mahwah, NJ: Lawrence Erlbaum Associates. Retrieved from http://www.cal.org/content/download/2243/29073/version/3/file/developing-literacy-in-second-language-learners-executive-summary.pdf

Ausubel, D. P. (1963). A teaching strategy for culturally deprived pupils: Cognitive and motivational considerations. *The School Review, 71*(4), 454–463. Retrieved from http://www.journals.uchicago.edu/doi/10.1086/442680

Awan, R.-u.-N., Azher, M., Anwar, M. N., & Naz, A. (2010). An investigation of foreign language classroom anxiety and its relationship with students' achievement. *Journal of College Teaching & Learning (TLC), 7*(11). Retrieved from http://clutejournals.com/index.php/TLC/article/view/249/239

Bacher-Hicks, A., Chin, M. J., Kane, T. J., & Staiger, D. O. (May 2017). *An evaluation of bias in three measures of teacher quality: Value-added, classroom observations, and student surveys* [PDF file]. Retrieved from http://scholar.harvard.edu/files/andrewbacherhicks/files/an_evaluation_of_bias_in_three_measures_of_teacher_quality.pdf

Bae, S. (2017). *It's about time: Organizing schools for teacher collaboration and learning. Teachers' time: Collaborating for learning, teaching, and leading* [PDF file]. Stanford, CA: Stanford Center for Opportunity Policy in Education. Retrieved from http://edpolicy.stanford.edu/sites/default/files/Hillsdale%20Teacher%20Time%20Final.pdf

Baldinger, E., Selling, S. K., & Virmani, R. (2016). Supporting novice teachers in leading discussions that reach a mathematical point: Defining and clarifying mathematical ideas. *Mathematics Teacher Educator, 5*(1). Retrieved from http://www.nctm.org/Publications/Mathematics-Teacher-Educator/2016/Vol5/Issue1/Supporting-Novice-Teachers-in-Leading-Discussions-That-Reach-a-Mathematical-Point_-Defining-and-Clarifying-Mathematical-Ideas

Baldwin, L., Omdal, S. N., & Pereles, D. (2015). Beyond stereotypes. *Teaching Exceptional Children, 47*(4), 216–225. Retrieved from http://journals.sagepub.com/doi/full/10.1177/0040059915569361

Baran-Lucarz, M. (2011). The relationship between language anxiety and the actual and perceived levels of foreign language pronunciation. *Studies in Second Language Learning and Teaching, 1*(4), 491–514. Retrieved from http://files.eric.ed.gov/fulltext/EJ1136575.pdf

Barkman, R. C. (2018, January 18). See the world through patterns. *Psychology Today*. Retrieved from http://www.psychologytoday.com/us/blog/singular-perspective/201801/see-the-world-through-patterns

Baron, L. (2016). Formative assessment at work in the classroom. *Mathematics Teacher, 110*(1), 46–52. Retrieved from http://www.nctm.org/Publications/Mathematics-Teacher/2016/Vol110/Issue1/Formative-Assessment-at-Work-in-the-Classroom

Baron-Cohen, S., Ashwin, E., Ashwin, C., Tavassoli, T., & Chakrabarti, B. (2009). Talent in autism: Hyper-systemizing, hyper-attention to detail and sensory hypersensitivity. *Philosophical Transactions of the Royal Society, B: Biological Sciences, 364*(1522), 1377–1383. Retrieved from http://royalsocietypublishing.org/doi/full/10.1098/rstb.2008.0337

Barrett, M. E., Swan, A. B., Mamikonian, A., Ghajoyan, I., Kramarova, O., & Youmans, R. J. (2014). Technology in note taking and assessment: The effects of congruence on student performance. *International Journal of Instruction, 7*(1), 49–58. Retrieved from http://files.eric.ed.gov/fulltext/EJ1085258.pdf

Barrows, H., & Tamblyn, R. (1980). *Problem-based learning: An approach to medical education*. New York, NY: Springer.

Barrows, H. S. (1996). Problem-based learning in medicine and beyond: A brief overview. *New Directions for Teaching and Learning, 68*, 3–12. Retrieved from http://onlinelibrary.wiley.com/doi/abs/10.1002/tl.37219966804

Barshay, J. (2018). 20 judgments a teacher makes in 1 minute and 28 seconds. The Hechinger Report. May 7, Retrieved from http://hechingerreport.org/20-judgments-a-teacher-makes-in-1-minute-and-28-seconds

Barton, C. (2018). On formative assessment in math: How diagnostic questions can help. *American Educator, 42*(2), 33–38. Retrieved from http://files.eric.ed.gov/fulltext/EJ1182085.pdf

Batchelor, S., Torbeyns, J., & Verschaffel, L. (2019). Affect and mathematics in young children: An introduction. *Educational Studies in Mathematics, 100*(3), 201–209. Retrieved from http://link.springer.com/article/10.1007%2Fs10649-018-9864-x

Bay-Williams, J., Duffett, A., & Griffith, D. (2016). *Common Core math in the K-8 classroom: Results from a national teacher survey* [PDF file]. Fordham Institute: Thomas B. Retrieved from http://fordhaminstitute.org/sites/default/files/publication/pdfs/20160623-common-core-math-k-8-classroom-results-national-teacher-survey0.pdf

Beckmann, S., & Izsák, A. (2015). Two perspectives on proportional relationships: Extending complementary origins of multiplication in terms of quantities. *Journal for Research in Mathematics Education, 46*(1), 17–38. Retrieved from http://www.nctm.org/Publications/journal-for-research-in-mathematics-education/2015/Vol46/Issue1/Two-Perspectives-on-Proportional-Relationships_-Extending-Complementary-Origins-of-Multiplication-in-Terms-of-Quantities

Beesley, A. D., Clark, T. F., Dempsey, K., & Tweed, A. (2018). Enhancing formative assessment practice and encouraging middle school mathematics engagement and persistence. *School Science and Mathematics, 118*(1–2), 4–16. Retrieved from http://onlinelibrary.wiley.com/doi/abs/10.1111/ssm.12255

Beginning to problem solve with "I Notice, I Wonder" (n.d.). Retrieved from http://www.nctm.org/Classroom-Resources/Problems-of-the-Week/I-Notice-I-Wonder

Beilock, S. L., Gunderson, E. A., Ramirez, G., & Levine, S. C. (2010). Female teachers' math anxiety affects girls' math achievement. *Proceedings of the National Academy of Sciences, 107*(5), 1860–1863. Retrieved from http://www.pnas.org/content/pnas/107/5/1860.full.pdf

Ben-Hur, M. (2006). *Concept-rich mathematics instruction: Building a strong foundation for reasoning and problem solving*. Alexandria, VA: ASCD. Available from http://www.ascd.org/publications/books/106008/chapters/Conceptual-Understanding.aspx

Bennett, S., & Kalish, N. (2006). *The case against homework: How homework is hurting our children and what we can do about it*. New York, NY: Three Rivers Press.

Bernstein, E., Shore, J., & Lazer, D. (2018). How intermittent breaks in interaction improve collective intelligence. *Proceedings of the National Academy of Sciences, 115*(35), 8734–8739. Retrieved from http://www.pnas.org/content/115/35/8734

Berry, R. Q., III, & Larson, M. R. (2019). The need to catalyze change in high school mathematics. *Phi Delta Kappan, 100*(6), 39–44. Retrieved from http://journals.sagepub.com/doi/full/10.1177/0031721719834027

Beyond Emily: Post-ing etiquette. (2008, August 13). *Edutopia*. Retrieved from http://www.edutopia.org/netiquette-guidelines

Bill and Melinda Gates Foundation. (2012). *Asking students about teaching: Student perception surveys and their implementation* [PDF file]. Retrieved from http://k12education.gatesfoundation.org/download/?Num=2504&filename=Asking_Students_Practitioner_Brief.pdf

Bixby, M. M. (2018). Effective and efficient use of math writing tasks. *Mathematics Teacher, 112*(2), 143–146. Retrieved from http://www.nctm.org/Publications/Mathematics-Teacher/2018/Vol112/Issue2/Effective-and-Efficient-Use-of-Math-Writing-Tasks

Black, P., & Wiliam, D. (1998). Assessment and classroom learning. *Assessment in Education: Principles, Policy & Practice, 5*(1), 7–74. Retrieved from http://www.tandfonline.com/doi/abs/10.1080/0969595980050102

Blanton, M., Brizuela, B. M., Gardiner, A. M., Sawrey, K., & Newman-Owens, A. (2015). A learning trajectory in 6-year-olds' thinking about generalizing functional relationships. *Journal for Research in Mathematics Education, 46*(5), 511–558. Retrieved from http://www.nctm.org/Publications/Journal-for-Research-in-Mathematics-Education/2015/Vol46/Issue5/A-Learning-Trajectory-in-6-Year-Olds_-Thinking-About-Generalizing-Functional-Relationships

Blitz, C. L., & Schulman, R. (2016). *Measurement instruments for assessing the performance of professional learning communities* (REL 2016-144) [PDF file]. Washington, DC: U.S. Department of Education, Institute of Education Sciences, National Center for Education Evaluation and Regional Assistance, Regional Educational Laboratory Mid-Atlantic. Retrieved from http://files.eric.ed.gov/fulltext/ED568594.pdf

Bloom, B. S. (1968). Learning for mastery. *Evaluation Comment, 1*(2), 1–12. Retrieved from http://files.eric.ed.gov/fulltext/ED053419.pdf

Bonner, E. P. (2014). Investigating practices of highly successful mathematics teachers of traditionally underserved students. *Educational Studies in Mathematics, 86*(3), 377–399. Retrieved from http://link.springer.com/article/10.1007/s10649-014-9533-7

Borgioli, G. M. (2008). Equity for English language learners in mathematics classrooms. *Teaching Children Mathematics, 15*(3), 185–191. Retrieved from http://www.nctm.org/Publications/teaching-children-mathematics/2008/Vol15/Issue3/Equity-for-English-Language-Learners-in-the-Mathematics-Classroom

Borman, G. D., Grigg, J., Rozek, C. S., Hanselman, P., & Dewey, N. A. (2018). Self-affirmation effects are produced by school context, student engagement with the intervention, and time: Lessons from a district-wide implementation. *Psychological Science, 29*(18). Retrieved from http://journals.sagepub.com/doi/abs/10.1177/0956797618784016

Borovoy, A. E. (2015). 5-minute film festival: Virtual field trips [web log comment] *Edutopia.* July 31, Retrieved from https://www.edutopia.org/blog/film-festival-virtual-field-trips

Bottia, M. C., Valentino, L., Moller, S., Mickelson, R. A., & Stearns, E. (2016). Teacher collaboration and Latinos/as' mathematics achievement trajectories. *American Journal of Education, 122*(4), 505–535. Retrieved from http://www.journals.uchicago.edu/doi/abs/10.1086/687274

Bottiani, J. H., Larson, K. E., Debnam, K. J., Bischoff, C. M., & Bradshaw, C. P. (2017). Promoting educators' use of culturally responsive practices: A systematic review of inservice interventions. *Journal of Teacher Education, 69*(4), 367–385. Retrieved from http://journals.sagepub.com/doi/abs/10.1177/0022487117722553

Bouck, E. C., & Park, J. (2018). A systematic review of the literature on mathematics manipulatives to support students with disabilities. *Education and Treatment of Children, 41*(1), 65–106. Retrieved from http://muse.jhu.edu/article/689028

Bovbjerg, K. M. (2006). Teams and collegiality in educational culture. *European Educational Research Journal, 5*(3–4), 244–253. Retrieved from http://journals.sagepub.com/doi/10.2304/eerj.2006.5.3.244

Boyle, J. D., & Kaiser, S. B. (2017). Collaborative planning as a process. *Mathematics Teaching in the Middle School, 22*(7), 406–411. Retrieved from http://www.nctm.org/Publications/Mathematics-Teaching-in-Middle-School/2017/Vol22/Issue7/Collaborative-Planning-as-a-Process

Brain Parade. (2015, June 16). 11 classroom management strategies for children with special needs. Retrieved from http://www.brainparade.com/2015/06/16/11-classroom-management-strategies-for-children-with-special-needs.

Brame, C. J. (2013). *Writing good multiple choice test questions.* Vanderbilt University Center for Teaching. Retrieved from http://cft.vanderbilt.edu/guides-sub-pages/writing-good-multiple-choice-test-questions

Brendefur, J. L., Strother, S., Rich, K., & Appleton, S. (2016). Assessing student understanding: A framework for testing and teaching. *Teaching Children Mathematics, 23*(3), 174–181. Retrieved from http://www.nctm.org/Publications/Teaching-Children-Mathematics/2016/Vol23/Issue3/Assessing-Student-Understanding_-A-Framework-for-Testing-and-Teaching

Briggs, D. C., Ruiz-Primo, M. A., Furtak, E., Shepard, L., & Yin, Y. (2012). Meta-analytic methodology and inferences about the efficacy of formative assessment. *Educational Measurement: Issues and Practice, 31*(4), 13–17. Retrieved from http://onlinelibrary.wiley.com/doi/full/10.1111/j.1745-3992.2012.00251.x

Brimi, H. M. (2011). Reliability of grading high school work in English. *Practical Assessment, Research & Evaluation, 16*(17). Retrieved from http://pareonline.net/getvn.asp?v=16&n=17

Brook, L., Sawyer, E., & Rimm-Kaufman, S. E. (2007). Teacher collaboration in the context of the responsive classroom approach. *Teachers and Teaching, 13*(3), 211–245. Retrieved from http://www.tandfonline.com/doi/abs/10.1080/13540600701299767

Brookhart, S., & Lazarus, S. (2017). *Formative assessment for students with disabilities* [PDF file]. Washington, DC: The Council of Chief State School Officers. Retrieved from http://www.ccsso.org/sites/default/files/2017-12/Formative_Assessment_for_Students_with_Disabilities.pdf

Brookhart, S. M. (2013). *How to create and use rubrics for formative assessment and grading*. Alexandria, VA: ASCD. Retrieved from http://www.ascd.org/publications/books/112001/chapters/What-Are-Rubrics-and-Why-Are-They-Important%C2%A2.aspx

Brookhart, S. M. (2017). *How to give effective feedback to your students* (2nd ed.). Alexandria, VA: ASCD. Retrieved from http://www.ascd.org/ASCD/pdf/siteASCD/publications/books/How-to-Give-Effective-Feedback-to-Your-Students-2nd-Edition-sample-chapters.pdf

Brooks, A. (2019, January 28). 10 netiquette guidelines online students need to know. Retrieved from http://www.rasmussen.edu/student-experience/college-life/netiquette-guidelines-every-online-student-needs-to-know

Brown, B. D., Horn, R. S., & King, G. (2018). The effective implementation of professional learning communities. *Alabama Journal of Educational Leadership, 5*, 53–59. Retrieved from http://files.eric.ed.gov/fulltext/EJ1194725.pdf

Brown, S. (2005). You made it through the test; what about the aftermath? *Mathematics Teaching in the Middle School, 11*(2), 69–74. Retrieved from http://www.nctm.org/Publications/mathematics-teaching-in-middle-school/2005/Vol11/Issue2/You-Made-It-through-the-Test;-What-about-the-Aftermath_

Bruff. D. (n.d.). *Classroom response systems ("clickers")*. Retrieved from http://cft.vanderbilt.edu/guides-sub-pages/clickers

Bruffee, K. A. (1995). Sharing our toys: Cooperative learning versus collaborative learning. *Change: The Magazine of Higher Learning, 27*(1), 12–18. Retrieved from http://www.tandfonline.com/doi/abs/10.1080/00091383.1995.9937722

Bruner, J. S., Goodnow, J. J., & Austin, G. A. (1956). *A study of thinking*. London, UK: Chapman & Hall.

Bruun, F., Diaz, J. M., & Dykes, V. J. (2015). The language of mathematics. *Teaching Children Mathematics*, 21(9), 531–536. Retrieved from http://www.nctm.org/Publications/Teaching-Children-Mathematics/2015/Vol21/Issue9/The-Language-of-Mathematics

Buchheister, K., Jackson, C., & Taylor, C. E. (2019). What, how, who: Developing mathematical discourse. *Mathematics Teaching in the Middle School*, 24(4), 202–208. Retrieved from http://www.nctm.org/Publications/Mathematics-Teaching-in-Middle-School/2019/Vol24/Issue4/What,-How,-Who_-Developing-Mathematical-Discourse

Burns, M. (1990). The math solution: Using groups of four. In N. Davidson (Ed.). *Collaborative learning in mathematics: A handbook for teachers* (pp. 21–46). Menlo Park, CA: Addison-Wesley.

Burriss, K. G., & Snead, D. (2017). Middle school students' perceptions regarding the motivation and effectiveness of homework. *School Community Journal*, 27(2), 193–210. Retrieved from http://www.adi.org/journal/2017fw/BurrissSneadFall2017.pdf

Burstein, J., Flor, M., Tetreault, J., Madnani, N., & Holtzman, S. (2012). *Examining linguistic characteristics of paraphrase in test-taker summaries* [PDF file]. ETS Research Report Series. Retrieved from http://files.eric.ed.gov/fulltext/EJ1109968.pdf

Butler, A. (2017, October 10). Multiple-choice testing: Are the best practices for assessment also good for learning? [web log comment]. Retrieved from http://www.learningscientists.org/blog/2017/10/10-1

Butman, S. M. (2014). A new twist on collaborative learning. *Mathematics Teaching in the Middle School*, 20(1), 52–57. Retrieved from http://www.nctm.org/Publications/mathematics-teaching-in-middle-school/2014/Vol20/Issue1/Mathematical-Explorations_-A-New-Twist-on-Collaborative-Learning

Buttaro, A., Jr., & Catsambis, S. (2019). Ability grouping in the early grades: Long-term consequences for educational equity in the United States. *Teachers College Record*, 121(2), 1–50. Retrieved from http://www.tcrecord.org/Content.asp?ContentId=22574

Byrd, C. M. (2016). Does culturally relevant teaching work? An examination from student perspectives. *SAGE Open*, 6(3). Retrieved from http://journals.sagepub.com/doi/full/10.1177/2158244016660744

Cameron, J., Banko, K. M., & Pierce, W. D. (2001). Pervasive negative effects of rewards on intrinsic motivation: The myth continues. *The Behavior Analyst*, 24(1), 1–44. Retrieved from http://www.ncbi.nlm.nih.gov/pmc/articles/PMC2731358

Carbonneau, K. J., Marley, S. C., & Selig, J. P. (2013). A meta-analysis of the efficacy of teaching mathematics with concrete manipulatives. *Journal of Educational Psychology, 105*(2), 380–400. Retrieved from http://psycnet.apa.org/doiLanding?doi=10.1037%2Fa0031084

Carifio, J., & Carey, T. (2010). Do minimum grading practices lower academic standards and produce social promotions? *Educational Horizons, 884*, 219–230. Retrieved from http://files.eric.ed.gov/fulltext/EJ895689.pdf

Carrabba, C. C., & Farmer, A. (2018). The impact of project based learning and direct instruction on the motivation and engagement of middle school students. *Language Teaching and Educational Research (LATER), 1*(2), 163–174. Retrieved from http://files.eric.ed.gov/fulltext/ED591136.pdf

Cavilla, D. (2017). The effects of student reflection on academic performance and motivation. SAGE Open. Retrieved from http://journals.sagepub.com/doi/10.1177/2158244017733790.

Cazden, C. B., & Leggett, E. L. (1976). Culturally responsive education: A response to LAU Remedies II. National Conference on Research Implications of the Task Force Report of the U.S. Office of Civil Rights, U.S. Department of Health, Education and Welfare. Retrieved from http://files.eric.ed.gov/fulltext/ED135241.pdf

Celeste, L., Baysu, G., Phalet, K., Meeussen, L., & Kende, J. (2019). Can school diversity policies reduce belonging and achievement gaps between minority and majority youth? Multiculturalism, colorblindness, and assimilationism assessed. *Personality and Social Psychology Bulletin*. Retrieved from http://journals.sagepub.com/doi/full/10.1177/0146167219838577

Center for Education Policy Analysis (n.d.). *Racial and ethnic achievement gaps*. Stanford, CA: Stanford University, Stanford Center for Opportunity Policy in Education. Retrieved from http://cepa.stanford.edu/educational-opportunity-monitoring-project/achievement-gaps/race

Charania, S. M., LeBlanc, L. A., Sabanathan, N., Ktaech, I. A., Carr, J. E., & Gunby, K. (2010). Teaching effective hand raising to children with autism during group instruction. *Journal of Applied Behavior Analysis, 43*(3), 493–497. Retrieved from http://www.ncbi.nlm.nih.gov/pmc/articles/PMC2938940

Chauvot, J., & Benson, S. (2008). Card sorts, state tests, and meaningful mathematics. *Mathematics Teaching in the Middle School, 13*(7), 390–397. Retrieved from http://www.nctm.org/Publications/mathematics-teaching-in-middle-school/2008/Vol13/Issue7/Card-Sorts,-State-Tests,-and-Meaningful-Mathematics

Cheng, T. (2016, February 4). *5 ways to use rewards in the classroom* [web log comment]. Retrieved from http://www.learningandthebrain.com/blog/rewards-in-the-classroom

Cherng, H.-Y. S., & Ho, P. (2018). In thoughts, words, and deeds: Are social class differences in parental support similar across immigrant and native families? *Sociological Quarterly, 59*(1), 85–110. Retrieved from http://www.tandfonline.com/doi/abs/10.1080/00380253.2017.1383142

Cheung, A. C. K., & Slavin, R. E. (2013). The effectiveness of educational technology applications for enhancing mathematics achievement in K-12 classrooms: A meta-analysis. *Educational Research Review, 9*, 88–113. Retrieved from http://www.sciencedirect.com/science/article/pii/S1747938X13000031

Cheung, W. M., & Wong, W. Y. (2014). Does lesson study work? *International Journal for Lesson and Learning Studies, 3*(2), 137–149. Retrieved from http://www.emeraldinsight.com/doi/pdfplus/10.1108/IJLLS-05-2013-0024

Chiu, M. M., Chow, B. W.-Y., McBride, C., & Mol, S. T. (2015). Students' sense of belonging at school in 41 countries. *Journal of Cross-Cultural Psychology, 47*(2), 175–196. Retrieved from http://journals.sagepub.com/doi/10.1177/0022022115617031

Cisterna, D., & Gotwals, A. W. (2018). Enactment of ongoing formative assessment: Challenges and opportunities for professional development and practice. *Journal of Science Teacher Education, 29*(3), 200–222. Retrieved from http://www.tandfonline.com/doi/full/10.1080/1046560X.2018.1432227

Civil, M., & Menéndez, J. M. (2010). *Involving Latino and Latina parents in their children's mathematics education* [PDF file]. National Council of Teachers of Mathematics. Retrieved from http://www.nctm.org/uploadedFiles/Research_and_Advocacy/research_brief_and_clips/Research_brief_17-civil.pdf

Clark, E. (2012, April). Sibling in the spotlight. Retrieved from http://www.parents.com/parenting/dynamics/sibling-rivalry/sibling-in-the-spotlight

Clarke, D. (1997). *Constructive assessment in mathematics*. Emeryville, CA: Key Curriculum Press.

Cleaver, S. (2018, June 27). What is a word wall? We Are Teachers. Retrieved from http://www.weareteachers.com/what-is-a-word-wall.

Cleveland, J. (2018, November 16). She said that if a question has... [tweet]. Retrieved from http://twitter.com/jacehan/status/1063521463318265856

Cole, J., & Feng, J. (2015, April 15–16). *Effectiveness strategies for improving writing skills in English language learners*. Chicago, IL: Paper presented at the Chinese American Educational Research and Development Association Annual Conference. Retrieved from http://files.eric.ed.gov/fulltext/ED556123.pdf

Collaborative for Academic, Social, and Emotional Learning (2013). *2013 CASEL guide: Effective social, and emotional learning programs—preschool and elementary school edition*. Chicago: Author. Retrieved from http://casel.org/wp-content/uploads/2016/01/2013-casel-guide-1.pdf

Collaborative for Academic, Social, and Emotional Learning (n.d.). What Is SEL? Retrieved from http://casel.org/what-is-sel

Collaborative learning vs. cooperative learning: What's the difference? (2017, March 5). Retrieved from http://resourced.prometheanworld.com/collaborative-cooperative-learning

Common Core Standards Writing Team (2011, December 26). 6–7, Ratios and proportional relationships. In *Progressions for the Common Core state standards in mathematics (draft)*. Tucson, AZ: Institute for Mathematics and Education, University of Arizona. Retrieved from http://commoncoretools.files.wordpress.com/2012/02/ccss_progression_rp_67_2011_11_12_corrected.pdf

Common Core Standards Writing Team (2013a, March 1). Front matter, preface, introduction. In *Progressions for the Common Core state standards in mathematics (draft)*. Tucson, AZ: Institute for Mathematics and Education, University of Arizona. Retrieved from http://commoncoretools.me/wp-content/uploads/2013/07/ccss_progression_frontmatter_2013_07_30.pdf

Common Core Standards Writing Team (2013b, March 1). Grade 8, High School, Functions. In *Progressions for the Common Core state standards in mathematics (draft)*. Tucson, AZ: Institute for Mathematics and Education, University of Arizona. Retrieved from http://commoncoretools.me/wp-content/uploads/2013/07/ccss_progression_functions_2013_07_02.pdf

Condliffe, B., Visher, M. G., Bangser, M. R., Drohojowska, S., & Saco, L. (2017). *Project-based learning: A literature review*. New York, NY: MDRC. Retrieved from http://www.mdrc.org/publication/project-based-learning/file-full

Cook, C. M., & Faulkner, S. A. (2010). The use of common planning time: A case study of two Kentucky schools to watch. *RMLE Online: Research in Middle Level Education, 34*(2), 1–12. Retrieved from http://files.eric.ed.gov/fulltext/EJ914054.pdf

Cook, C. R., Fiat, A., Larson, M., Daikos, C., Slemrod, T., Holland, E. A., … Renshaw, T. (2018). Positive greetings at the door: Evaluation of a low-cost, high-yield proactive classroom management strategy. *Journal of Positive Behavior Interventions, 20*(3), 149–159. Retrieved from http://journals.sagepub.com/doi/pdf/10.1177/1098300717753831

Cook, G. (2015). The final push before summer. *Education Update, 57*(5). Retrieved from http://www.ascd.org/publications/newsletters/education-update/may15/vol57/num05/The-Final-Push-Before-Summer.aspx

Cook, L., & Friend, M. (1995). Co-teaching: Guidelines for creating effective practices. *Focus on Exceptional Children, 28*(3), 1–16. Retrieved from http://www.researchgate.net/publication/234620116_Co-Teaching_Guidelines_for_Creating_Effective_Practices

Coomes, J., & Lee, H. S. (2017). Empowering mathematical practices. *Mathematics Teaching in the Middle School, 22*(6). Retrieved from http://www.nctm.org/Publications/Mathematics-Teaching-in-Middle-School/2017/Vol22/Issue6/Empowering-Mathematical-Practices

Cooper, H., Robinson, J. C., & Patall, E. A. (2006). Does homework improve academic achievement? A synthesis of research, 1987–2003. *Review of Educational Research, 76*(1), 1–62. Retrieved from http://journals.sagepub.com/doi/abs/10.3102/00346543076001001

Cooper, R. (2016, December 11). 5 simple ways to give students feedback during project based learning #HackingPB [web log comment]. Retrieved from http://rosscoops31.com/2016/12/11/5-simple-ways-give-students-feedback-project-based-learning-hackingpbl

Cornelius, S. (2015, July 8). 5 classroom tips to support ELL students [web log comment]. Retrieved from http://blog.edmentum.com/5-classroom-tips-support-ell-students

Cox, J. (n.d.). Teaching strategies using task cards. Retrieved from http://www.teachhub.com/teaching-strategies-using-task-cards

Cox, T., & Singer, S. (2011). Taking the work out of homework. *Mathematics Teacher, 104*(7), 514–519. Retrieved from http://www.nctm.org/Publications/mathematics-teacher/2011/Vol104/Issue7/Taking-the-Work-out-of-Homework

Cummins, J. (2005, September 23). Teaching for cross-language transfer in dual language education: Possibilities and pitfalls. Paper presented at the TESOL Symposium on Dual Language Education: Teaching and Learning Two Languages in the EFL Setting. Retrieved from http://www.tesol.org/docs/default-source/new-resource-library/symposium-on-dual-language-education-3.pdf

Cunningham, G. (2009). *The new teacher's companion: Practical wisdom for succeeding in the classroom.* Alexandria, VA: ASCD. Retrieved from http://www.ascd.org/Publications/Books/Overview/The-New-Teachers-Companion.aspx

Curcio, F. R., & Artzt, A. F. (2007). Reading, writing, and mathematics: A problem-solving connection. In D. Lapp, J. Flood, & N. Farnan (Eds.). *Content area reading and learning: Instructional strategies* (3rd ed., pp. 257–270). New York, NY: Taylor & Francis.

Danielson, C. (2016). *Which one doesn't belong?* Portsmouth, NH: Stenhouse.

Darling-Hammond, L., & Adamson, F. (2010). *Beyond basic skills: The role of performance assessment in achieving 21st century standards of learning.* Stanford, CA: Stanford University, Stanford Center for Opportunity Policy in Education. Retrieved from http://scale.stanford.edu/system/files/beyond-basic-skills-role-performance-assessment-achieving-21st-century-standards-learning.pdf

Darling-Hammond, L., Austin, K., Shulman, L., & Schwartz, D. (n.d.). Lessons for life—learning and transfer. In *The learning classroom: Theory into practice online course.* Stanford, CA: Annenberg Foundation. Retrieved from http://www.learner.org/courses/learningclassroom/support/11_learning_transfer.pdf

Davern, L. (1996). Listening to parents of children with disabilities. *Educational Leadership, 53*(7). Retrieved from http://www.ascd.org/publications/educational-leadership/apr96/vol53/num07/Listening-to-Parents-of-Children-with-Disabilities.aspx

de Freitas, E., & Zolkower, B. (2011). Developing teacher capacity to explore non-routine problems through a focus on the social semiotics of mathematics classroom discourse. *Research in Mathematics Education, 13*(3), 229–247. Retrieved from http://www.tandfonline.com/doi/abs/10.1080/14794802.2011.624705

De Jong, E., & Commins, N. L. (n.d.). *How should ELLs be grouped for instruction?* ¡Colorín Colorado! Retrieved from http://www.colorincolorado.org/article/how-should-ells-be-grouped-instruction

De Lisi, R., & Golbeck, S. L. (1999). Implications for Piagetian theory for peer learning. In A. M. O'Donnell & A. King (Eds.). *Cognitive perspectives on peer learning.* New York, NY: Routledge. Retrieved from. http://www.taylorfrancis.com/books

De Neve, D., Devos, G., & Tuytens, M. (2015). The importance of job resources and self-efficacy for beginning teachers' professional learning in differentiated instruction. *Teaching and Teacher Education, 47,* 30–41. Retrieved from http://www.sciencedirect.com/science/article/pii/S0742051X14001541

Dean, J. (2011). The Zeigarnik effect. *Psyblog February, 8.* www.spring.org.uk/2011/02/the-zeigarnik-effect.php

Deci, E. L., Ryan, R. M., & Koestner, R. (2001). The pervasive negative effects of rewards on intrinsic motivation: Response to Cameron (2001). *Review of Educational Research, 71*(1), 43–51. Retrieved from http://journals.sagepub.com/doi/abs/10.3102/00346543071001043

Definition of flipped learning. (2014, March 12). Retrieved from http://flippedlearning.org/definition-of-flipped-learning

Delisle, R. (1997). *How to use problem-based learning in the classroom.* Alexandria, VA: ASCD. Retrieved from http://www.ascd.org/publications/books/197166/chapters/What_Is_Problem-Based_Learning%C2%A2.aspx

Dettmers, S., Trautwein, U., Lüdtke, O., Kunter, M., & Baumert, J. (2010). Homework works if homework quality is high: Using multilevel modeling to predict the development of achievement in mathematics. *Journal of Educational Psychology, 102*(2), 467–482. Retrieved from http://www.researchgate.net/publication/44951983_Homework_Works_if_Homework_Quality_Is_High_Using_Multilevel_Modeling_to_Predict_the_Development_of_Achievement_in_Mathematics

Deunk, M. I., Smale-Jacobse, A. E., de Boer, H., Doolaard, S., & Bosker, R. J. (2018). Effective differentiation practices: A systematic review and meta-analysis of studies on the cognitive effects of differentiation practices in primary education. *Educational Research Review, 24*, 31–54. Retrieved from http://www.sciencedirect.com/science/article/pii/S1747938X18301039

Devlin, K. (2012). *Introduction to mathematical thinking*. Palo Alto, CA: Keith Devlin. Retrieved from http://www.mat.ufrgs.br/~portosil/curso-Devlin.pdf

Devlin, K. (2019). *How technology has changed what it means to think mathematically*. Palo Alto, CA: Author. Retrieved from http://web.stanford.edu/~kdevlin/Papers/DanesiChapter.pdf

Dewey, J. (1938). *Experience and education*. New York, NY: Macmillan. Retrieved from http://openlibrary.org/books/OL7264857M/Experience_and_Education

Dixon, F. A., Yssel, N., McConnell, J. M., & Hardin, T. (2014). Differentiated instruction, professional development, and teacher efficacy. *Journal for the Education of the Gifted, 37*(2), 111–127. Retrieved from http://journals.sagepub.com/doi/10.1177/0162353214529042

Dong, Y. R. (2016). Create a responsive learning community for ELLs. *Mathematics Teacher, 109*(7), 534–540. Retrieved from http://www.nctm.org/Publications/Mathematics-Teacher/2016/Vol109/Issue7/Create-a-Responsive-Learning-Community-for-ELLs

Drake, K. N., & Long, D. (2009). Rebecca's in the dark: A comparative study of problem-based learning and direct instruction/experiential learning in two fourth-grade classrooms. *Journal of Elementary Science Education, 21*(1), 1–16. Retrieved from http://files.eric.ed.gov/fulltext/EJ849707.pdf

Dray, B. J., & Wisneski, D. B. (2011). Mindful reflection as a process for developing culturally responsive practices. *Teaching Exceptional Children, 44*(1), 28–36. Retrieved from http://journals.sagepub.com/doi/pdf/10.1177/004005991104400104

Du, F., & Wang, Q. (2017). New teachers' perspectives of informal mentoring: Quality of mentoring and contributors. *Mentoring & Tutoring: Partnership in Learning, 25*(3), 309–328. Retrieved from http://www.tandfonline.com/doi/full/10.1080/13611267.2017.1364841

DuFour, R. (2004). What is a professional learning community? *Educational Leadership*, *61*(8), 6–11. Retrieved from http://www.ascd.org/publications/educationalleadership/may04/vol61/num08/What-Is-a-Professional-Learning-Community.aspx

DuFour, R., & Reeves, D. (2016). The futility of PLC Lite. *Phi Delta Kappan*, *97*(6), 69–71. Retrieved from http://journals.sagepub.com/doi/abs/10.1177/0031721716636878

Dunlosky, J., Rawson, K. A., Marsh, E. J., Nathan, M. J., & Willingham, D. T. (2013). Improving students' learning with effective learning techniques: Promising directions from cognitive and educational psychology. *Psychological Science in the Public Interest*, *14*(1), 4–58. Retrieved from http://www.indiana.edu/~pcl/rgoldsto/courses/dunloskyimprovinglearning.pdf

Duursma, E., Augustyn, M., & Zuckerman, B. (2008). Reading aloud to children: The evidence. *Archives of Disease in Children*, *93*(7), 554–557. Retrieved from http://www.reachoutandread.org/FileRepository/ReadingAloudtoChildren_ADC_July2008.pdf

Dweck, C. S. (2007). The perils and promises of praise. *Educational Leadership*, *65*(2), 34–39. Retrieved from http://www.ascd.org/publications/educational-leadership/oct07/vol65/num02/The-Perils-and-Promises-of-Praise.aspx

Earnest, D. (2017). Clock work: How tools for time mediate problem solving and reveal understanding. *Journal for Research in Mathematics Education*, *48*(2), 191–223. Retrieved from http://www.nctm.org/Publications/Journal-for-Research-in-Mathematics-Education/2017/Vol48/Issue2/Clock-Work_-How-Tools-for-Time-Mediate-Problem-Solving-and-Reveal-Understanding

Ehardt, K. (2009). Dyslexia, not disorder. *Dyslexia*, *15*(4), 363–366. Retrieved from http://onlinelibrary.wiley.com/doi/abs/10.1002/dys.379

Ehlert, D. (2015, September 14). Promoting growth mindset through assessment [web log comment]. *Mathematics Teaching in the Middle School*. Retrieved from http://www.nctm.org/Publications/Mathematics-Teaching-in-Middle-School/Blog/Promoting-Growth-Mindset-through-Assessment

Eisenberger, R., & Shanock, L. (2003). Rewards, intrinsic motivation, and creativity: A case study of conceptual and methodological isolation. *Creativity Research Journal*, *15*, 121–130. Retrieved from http://www.tandfonline.com/doi/abs/10.1080/10400419.2003.9651404

Elias, M. J. (2018, January 5). Nurturing intrinsic motivation in students. *Edutopia*. Retrieved from http://www.edutopia.org/article/nurturing-intrinsic-motivation-students

EngageNY. (2013). A story of functions: A curriculum overview for Grades 9–12. Retrieved from http://www.engageny.org/sites/default/files/resource/attachments/a_story_of_functions_currriculum_map_and_overview_9-12.pdf

Ericsson, K., Krampe, R., & Tesch-Römer, C. (1993). The role of deliberate practice in the acquisition of expert performance. *Psychological Review, 100*(3), 363–406. Retrieved from http://projects.ict.usc.edu/itw/gel/EricssonDeliberatePracticePR93.pdf

Fadel, C., Trilling, B., & Bialik, M. (2015). The role of metacognition in learning and achievement. In Center for Curriculum Redesign: *Four-dimensional education: The competencies learners need to succeed.* Retrieved from http://www.kqed.org/mindshift/46038/the-role-of-metacognition-in-learning-and-achievement

Falout, J. (2014). Circular seating arrangements: Approaching the social crux in language classrooms. *Studies in Second Language Learning and Teaching, 4*(2), 275–300. Retrieved from http://files.eric.ed.gov/fulltext/EJ1134769.pdf

Fan, W., & Williams, C. M. (2010). The effects of parental involvement on students' academic self-efficacy, engagement and intrinsic motivation. *Educational Psychology, 30*(1), 53–74. Retrieved from http://www.tandfonline.com/doi/abs/10.1080/01443410903353302

Fan, X., & Chen, M. (2001). Parental involvement and students' academic achievement: A meta-analysis. *Educational Psychology Review, 13*(1), 1–22. Retrieved from http://link.springer.com/article/10.1023%2FA%3A1009048817385

Farmer, A. (2018). The impact of student-teacher relationships, content knowledge, and teaching ability on students with diverse motivation levels. *Language Teaching and Education Research (LATER), 1*(1), 13–24. Retrieved from http://files.eric.ed.gov/fulltext/ED588829.pdf

Fenner, D. S., Kester, J., & Snyder, S. (n.d.). The five pillars of equitably grading ELLs [web log comment]. *¡Colorín Colorado!* Retrieved from http://colorincolorado.org/blog/five-pillars-equitably-grading-ells

Ferlazzo, L. (2009, May 19). Parent involvement or parent engagement? *Learning First Alliance* [web log comment]. Retrieved from http://learningfirst.org/blog/parent-involvement-or-parent-engagement

Ferlazzo, L. (2013, April 16). Positive, not punitive, classroom management tips [web log comment]. *Edutopia.* Retrieved from http://www.edutopia.org/blog/positive-not-punitive-part-1-larry-ferlazzo.

Ferlazzo, L. (2016). Collaborative learning, Common Core, and ELLs [web log comment]. *Edutopia.* Retrieved from http://www.edutopia.org/blog/collaborative-writing-common-core-ells-larry-ferlazzo-katie-hull-sypnieski

Ferlazzo, L., & Sypnieski, K. H. (2018a). *The ELL Teacher's toolbox: Hundreds of practical ideas to support your students.* San Francisco, CA: Jossey-Bass.

Ferlazzo, L., & Sypnieski, K. H. (2018b, Fall) Teaching English language learners. *American Educator.* Retrieved from http://www.aft.org/ae/fall2018/ferlazzo_sypnieski

Fernandes, M. A., Wammes, J. D., & Meade, M. E. (2018). The surprisingly powerful influence of drawing on memory. *Current Directions in Psychological Science, 27*(5), 302–308. Retrieved from http://journals.sagepub.com/doi/abs/10.1177/0963721418755385

Finn, B. (2015). Retrospective utility of educational experiences: Converging research from education and judgment and decision-making. *Journal of Applied Research in Memory and Cognition, 4*(4), 374–380. Retrieved from http://www.sciencedirect.com/science/article/pii/S2211368115000340

Fiorella, L., & Mayer, R. (2017). Spontaneous spatial strategy use in learning from scientific text. *Contemporary Educational Psychology, 49,* 66–79. Retrieved from http://www.sciencedirect.com/science/article/pii/S0361476X1730005X

Fleming, G. (2019). *Why math is more difficult for some students.* ThoughtCo. Retrieved from http://www.thoughtco.com/why-math-seems-more-difficult-for-some-students-1857216

Flores, A., & Kimpton, K. E. (2012). Multicultural and gender equity issues in a history of mathematics course: Not only dead European males. *Journal of Mathematics Education at Teachers College, 3*(2), 37–42. Retrieved from http://journals.tc-library.org/index.php/matheducation/article/view/870/530

Fong, C. J., Patall, E. A., Vasquez, A. C., & Stautberg, S. (2019). A meta-analysis of negative feedback on intrinsic motivation. *Educational Psychology Review, 31*(1), 121–162. Retrieved from http://link.springer.com/article/10.1007%2Fs10648-018-9446-6

Francis, D. V. (2012). Sugar and spice and everything nice? Teacher perceptions of black girls in the classroom. *The Review of Black Political Economy, 39*(3), 311–320. Retrieved from http://link.springer.com/article/10.1007/s12114-011-9098-y

Frayer, D., Frederick, W., & Klausmeier, H. (1969). *A schema for testing the level of concept mastery.* Madison, WI: Wisconsin Center for Education.

Freeman, B. (2012). Using digital technologies to redress inequities for English language learners in the English speaking mathematics classroom. *Computers & Education, 59*(1), 50–62. Retrieved from http://www.sciencedirect.com/science/article/pii/S0360131511002739

Freeman, B., Higgins, K. N., & Horney, M. (2016). How students communicate mathematical ideas: An examination of multimodal writing using digital technologies. *Contemporary Educational Technology*, 7(4), 281–313. Retrieved from http://files.eric.ed.gov/fulltext/EJ1117603.pdf

Freiss, S. (2008). Great education debate: Reforming the grade system. *USA Today*. Retrieved from http://www.usatoday.com/news/education/2008-05-18-zeroes-side_N.htm

Freund, L. S. (1990). Maternal regulation of children's problem-solving behavior and its impact on children's performance. *Child Development*, 61(1), 113–126. Retrieved from http://onlinelibrary.wiley.com/doi/abs/10.1111/j.1467-8624.1990.tb02765.x

Frey, N., & Fisher, D. (2010). Making group work productive. *Educational Leadership*, 68(1). Retrieved from http://www.ascd.org/publications/educational-leadership/sept10/vol68/num01/Making-Group-Work-Productive.aspx

Friess, S. (2008, May 18). Great education debate: Reforming the grading system. *USA Today*. Retrieved from http://www.usatoday.com/news/education/2008-05-18-zeroes-side_N.htm

Frisby, B., & Martin, M. (2010). Instructor-student and student-student rapport in the classroom. *Communication Education*, 59(2), 146–164. Retrieved from http://www.researchgate.net/publication/232904107_Instructor-Student_and_Student-Student_Rapport_in_the_Classroom/download

Gabriel, R., & Wenz, C. (2017). Three directions for disciplinary literacy. *Educational Leadership*, 74(5). Retrieved from http://www.ascd.org/publications/educational-leadership/feb17/vol74/num05/Three-Directions-for-Disciplinary-Literacy.aspx

Gagné, N., & Parks, S. (2012). Cooperative learning tasks in a grade 6 intensive English as a second language class: Turn-taking and degree of participation. *The Language Learning Journal*, 44(2), 169–180. Retrieved from http://www.tandfonline.com/doi/full/10.1080/09571736.2012.751120

Gallagher, E. (2013). The effects of teacher-student relationships: Social and academic outcomes of low-income middle and high school students. *OPUS*, 5(1), 12–15. Retrieved from http://steinhardt.nyu.edu/appsych/opus/issues/2013/fall/gallagher

Gándara, P., & Santibañez, L. (2016). The teachers our English language learners need. *Educational Leadership*, 73(5), 32–37. Retrieved from http://www.ascd.org/publications/educational-leadership/feb16/vol73/num05/The-Teachers-Our-English-Language-Learners-Need.aspx

Gardner, H. (2006). *Multiple intelligences: New horizons in theory and practice.* New York, NY: Basic Books.

Gardner, H. (2013, October 16). "Multiple intelligences" are not "learning styles." *Washington Post.* Retrieved from http://www.washingtonpost.com/news/answer-sheet/wp/2013/10/16/howard-gardner-multiple-intelligences-are-not-learning-styles

Garfunkel, S., & Montgomery, M. (Eds.). (2016). *GAIMME: Guidelines for assessment and instruction in mathematical modeling education.* Philadelphia: COMAP and SIAM. Retrieved from http://www.siam.org/Publications/Reports/Detail/Guidelines-for-Assessment-and-Instruction-in-Mathematical-Modeling-Education

Garrison, C., & Ehringhaus, M. (n.d.). *Formative and Summative Assessments in the Classroom.* Association for Middle Level Education. Retrieved from http://www.amle.org/BrowsebyTopic/WhatsNew/WNDet/TabId/270/ArtMID/888/ArticleID/286/Formative-and-Summative-Assessments-in-the-Classroom.aspx

Gay, A. S. (2008). Helping teachers connect vocabulary and conceptual understanding. *Mathematics Teacher, 102*(3), 218–223. Retrieved from http://www.nctm.org/Publications/mathematics-teacher/2008/Vol102/Issue3/Helping-Teachers-Connect-Vocabulary-and-Conceptual-Understanding

Gay, G. (2018). *Culturally responsive teaching: Theory, research, and practice* (3rd ed.). New York, NY: Teachers College Press.

Ghousseini, H., Lord, S., & Cardon, A. (2017). Supporting math talk in small groups. *Teaching Children Mathematics, 23*(7), 422–428. Retrieved from http://www.nctm.org/Publications/Teaching-Children-Mathematics/2017/Vol23/Issue7/Supporting-Math-Talk-in-Small-Groups

Goddard, R., Goddard, Y., Sook Kim, E., & Miller, R. (2015). A theoretical and empirical analysis of the roles of instructional leadership, teacher collaboration, and collective efficacy beliefs in support of student learning. *American Journal of Education, 121*(4), 501–530. Retrieved from http://www.journals.uchicago.edu/doi/10.1086/681925

Godfrey, D., Seleznyov, S., Anders, J., Wollaston, N., & Barrera-Pedemonte, F. (2018). A developmental evaluation approach to lesson study: Exploring the impact of lesson study in London schools. *Professional Development in Education, 45*(2), 325–340. Retrieved from http://www.tandfonline.com/doi/full/10.1080/19415257.2018.1474488

Gojak, L. M. (2013, October 3). Making mathematical connections. National Council of Teachers of Mathematics. Retrieved from http://www.nctm.org/News-and-Calendar/Messages-from-the-President/Archive/Linda-M_-Gojak/Making-Mathematical-Connections

Gonzalez, J. (2013, October 13). Open your door: Why we need to see each other teach. *Cult of Pedagogy*. Retrieved from http://www.cultofpedagogy.com/open-your-door

Gonzalez, J. (2014). Modifying the flipped classroom: The "in-class" version. *Edutopia*. Retrieved from http://www.edutopia.org/blog/flipped-classroom-in-class-version-jennifer-gonzalez

Gonzalez, J. (2017, September 10). Culturally responsive teaching: 4 misconceptions. *Cult of Pedagogy*. Retrieved from http://www.cultofpedagogy.com/culturally-responsive-misconceptions

Gonzalez, V. (2018, March 12). Interactive word walls enliven vocab learning. *MiddleWeb*. Retrieved from http://www.middleweb.com/37209/interactive-word-walls-enliven-vocab-learning

Goodall, J., & Montgomery, C. (2013). Parental involvement to parental engagement: A continuum. *Educational Review*, 66(4), 399–410. Retrieved from http://www.tandfonline.com/doi/full/10.1080/00131911.2013.781576

Goodwin, B. (2014). Get all students to speak up. *Educational Leadership*, 72(3), 82–83. Retrieved from http://www.ascd.org/publications/educational-leadership/nov14/vol72/num03/Get-All-Students-to-Speak-Up.aspx

Goodwin, B., & Hein, H. (2017). Learning styles: It's complicated. *Educational Leadership*, 74(7), 79–80. Retrieved from http://www.ascd.org/publications/educational-leadership/apr17/vol74/num07/Learning-Styles@-It's-Complicated.aspx

Goodwin, B., & Miller, K. (2013). Evidence on flipped classrooms is still coming in. *Educational Leadership*, 70(6), 78–80. Retrieved from http://www.ascd.org/publications/educational-leadership/mar13/vol70/num06/Evidence-on-Flipped-Classrooms-Is-Still-Coming-In.aspx

Gordon, E. W., & Wilkerson, D. A. (1966). *Compensatory education for the disadvantaged; programs and practices, preschool through college*. New York, NY: College Entrance Examination Board. Retrieved from http://files.eric.ed.gov/fulltext/ED011274.pdf

Gottfried, A. E., Marcoulides, G. A., Gottfried, A. W., & Oliver, P. H. (2013). Longitudinal pathways from math intrinsic motivation and achievement to math course accomplishments and educational attainment. *Journal of Research on Educational Effectiveness*, 6(1), 68–92. Retrieved from http://www.tandfonline.com/doi/abs/10.1080/19345747.2012.698376

Gough, E., DeJong, D., Grundmeyer, T., & Baron, M. (2017). K-12 teacher perceptions regarding the flipped classroom model for teaching and learning. *Journal of Educational Technology Systems*, 45(3), 390–423. Retrieved from http://journals.sagepub.com/doi/10.1177/0047239516658444

Graham, S., & Morales-Chicas, J. (2015). The ethnic context and attitudes toward 9th grade math. *International Journal of Educational Psychology, 4*(1), 1–32. Retrieved from http://files.eric.ed.gov/fulltext/EJ1111710.pdf

Gray, D. L., Hope, E. C., & Matthews, J. S. (2018). Black and belonging at school: A case for interpersonal, instructional, and institutional opportunity structures. *Educational Psychologist, 53*(2), 97–113. Retrieved from http://www.tandfonline.com/doi/full/10.1080/00461520.2017.1421466

Graziano, K. J., & Hall, J. D. (2017). Flipping math in a secondary classroom. *Journal of Computers in Mathematics and Science Teaching, 36*(1), 5–16. Retrieved from http://www.learntechlib.org/primary/p/178270

Gregory, G. H., & Chapman, C. (2007). *Differentiated instructional strategies: One size doesn't fit all* (2nd ed.). Thousand Oaks, CA: Corwin Press.

Groth, R. E. (2011). Improving teaching through lesson study debriefing. *Mathematics Teacher, 104*(6), 446–451. Retrieved from http://www.nctm.org/Publications/mathematics-teacher/2011/Vol104/Issue6/Improving-Teaching-through-Lesson-Study-Debriefing

Gunderson, E. A., Park, D., Maloney, E. A., Beilock, S. L., & Levine, S. C. (2017). Reciprocal relations among motivational frameworks, math anxiety, and math achievement in early elementary school. *Journal of Cognition and Development, 19*(1), 21–46. Retrieved from http://www.tandfonline.com/doi/full/10.1080/15248372.2017.1421538

Gurl, T. J., Artzt, A. F., & Sultan, A. (2012). *Implementing the Common Core state standards through mathematical problem solving: High school.* Reston, VA: National Council of Teachers of Mathematics.

Gurl, T. J., Artzt, A. F., & Sultan, A. (2013). *Implementing the Common Core state standards through mathematical problem solving: Middle school.* Reston, VA: National Council of Teachers of Mathematics.

Guskey, T. R. (2011). Five obstacles to grading reform. *Educational Leadership, 69*(3), 16–21. Retrieved from http://www.ascd.org/publications/educational-leadership/nov11/vol69/num03/Five-Obstacles-to-Grading-Reform.aspx

Guskey, T. R., & Bailey, J. M. (2010). *Developing standards-based report cards.* Thousand Oaks, CA: Corwin Press.

Guskey, T. R., & Jung, L. A. (2012). Four steps in grading reform. *Principal Leadership, 13*(4), 22–28. Retrieved from http://eric.ed.gov/?id=EJ1002406

Guskey, T. R., & Jung, L. A. (2016). Grading: Why you should trust your judgment. *Educational Leadership, 73*(7), 50–54. Retrieved from http://www.ascd.org/publications/educational-leadership/apr16/vol73/num07/Grading@-Why-You-Should-Trust-Your-Judgment.aspx

Hackenberg, A. J., & Lee, M. Y. (2015). Relationships between students' fractional knowledge and equation writing. *Journal for Research in Mathematics Education, 46*(2), 196–243. Retrieved from http://www.nctm.org/Publications/journal-for-research-in-mathematics-education/2015/Vol46/Issue2/Relationships-Between-Students_-Fractional-Knowledge-and-Equation-Writing

Hammond, Z. (2015). *Culturally responsive teaching and the brain*. Thousand Oaks, CA: Corwin Press.

Hannula, M. S. (2019). Young learners' mathematics-related affect: A commentary on concepts, methods, and developmental trends. *Educational Studies in Mathematics, 100*(3), 309–316. Retrieved from http://link.springer.com/article/10.1007/s10649-018-9865-9

Harvey, E. J., & Kenyon, M. C. (2013). Classroom seating considerations for 21st century students and faculty. *Journal of Learning Spaces, 2*(1). Retrieved from http://files.eric.ed.gov/fulltext/EJ1152707.pdf

Heath, C., & Heath, D. (December 2006). The curse of knowledge. *Harvard Business Review*. Retrieved from http://hbr.org/2006/12/the-curse-of-knowledge

Heinrich, A. K., Lawrence, E. D., Pons, M. A., & Taylor, D. G. (Eds.). (2019). *Living proof: Stories of resilience along the mathematical journey*. Providence, RI: American Mathematical Society. Retrieved from https://www.maa.org/press/ebooks/living-proof-stories-of-resilience-along-the-mathematical-journey-2

Heroux, J. R., Peters, S. J., & Randel, M. A. (2014). Fostering self-determination through culturally responsive teaching. In D. Lawrence-Brown & M. Sapon-Shevin (Eds.). *Condition critical: Key principles for equitable and inclusive education* (pp. 187–203). New York, NY: Teachers College Press.

Hess, K. K. (2019). Deepening student understanding with collaborative discourse. *ASCD Express, 14*(22). Retrieved from http://www.ascd.org/ascd-express/vol14/num22/deepening-student-understanding-with-collaborative-discourse.aspx

Heyck-Merlin, M. (2015, September 25). Fret-free field trips: Create a recipe for success. *Edutopia*. Retrieved from http://www.edutopia.org/blog/fret-free-field-trips-timeline-maia-heyck-merlin

Hibbard, K. M., Van Wagenen, L., Lewbel, S., Waterbury-Wyatt, S., Shaw, S., Pelletier, K., ... Wislocki, J. A. (1996). *A teacher's guide to performance-based learning and assessment.* Alexandria, VA: ASCD. Retrieved from http://www.ascd.org/publications/books/196021/chapters/What_is_Performance-Based_Learning_and_Assessment,_and_Why_is_it_Important%C2%A2.aspx

Hidi, S. (2015). Revisiting the role of rewards in motivation and learning: Implications of neuroscientific research. *Educational Psychology Review, 28*(1), 61–93. Retrieved from http://link.springer.com/article/10.1007%2Fs10648-015-9307-5

Hiebert, J., & Grouws, D. (2007). The effects of mathematics teaching on students' learning. In F. K. Lester, Jr. (Ed.). *Second handbook of research on mathematics teaching and learning* (pp. 371–404). Charlotte, NC: Information Age.

Higgins, K., Huscroft-D'Angelo, J., & Crawford, L. (2017). Effects of technology in mathematics on achievement, motivation, and attitude: A meta-analysis. *Journal of Educational Computing Research, 57*(2), 283–319. Retrieved from http://journals.sagepub.com/doi/abs/10.1177/0735633117748416

Hill, J. (2016). Engaging your beginners. *Educational Leadership, 73*(5), 18–23. Retrieved from http://www.ascd.org/publications/educational-leadership/feb16/vol73/num05/Engaging-Your-Beginners.aspx

Himmel, J. (n.d.). Language objectives: The key to effective content area instruction for English learners [web log comment]. *¡Colorín Colorado!* Retrieved from http://colorincolorado.org/article/language-objectives-key-effective-content-area-instruction-english-learners

Himmele, P., & Himmele, W. (2017). *Total participation techniques: Making every student an active learner* (2nd ed.). Alexandria, VA: ASCD.

Hinds, J. d C. (2015). A curriculum staple: Reading aloud to teens. *School Library Journal.* Retrieved from http://www.slj.com/?detailStory=a-curriculum-staple-reading-aloud-to-teens

Hmelo-Silver, C. E., Duncan, R. G., & Chinn, C. A. (2007). Scaffolding and achievement in problem-based and inquiry learning: A response to Kirschner, Sweller, and Clark (2006). *Educational Psychologist, 42*(2), 99–107. Retrieved from http://www.tandfonline.com/doi/full/10.1080/00461520701263368

Hodge, L. L., & Walther, A. (2017). Building a discourse community: Initial practices. *Mathematics Teaching in the Middle School, 22*(7), 430–437. Retrieved from http://www.nctm.org/Publications/Mathematics-Teaching-in-Middle-School/2017/Vol22/Issue7/Building-a-Discourse-Community_-Initial-Practices

Holstead, C. E. (2015, March 4). The benefits of no-tech note-taking. *Chronicle of Higher Education*. Retrieved from http://www.chronicle.com/article/The-Benefits-of-No-Tech-Note/228089

Hoogerheide, V., Vink, M., Finn, B., Raes, A. K., & Paas, F. (2017). How to bring the news … peak-end effects in children's affective responses to peer assessments of their social behavior. *Cognition and Emotion, 32*(5), 1114–1121. Retrieved from http://www.tandfonline.com/doi/full/10.1080/02699931.2017.1362375

Hord, C., Marita, S., Walsh, J. B., Tomaro, T.-M., & Gordo, K. (2016). Encouraging students with learning disabilities. *Mathematics Teacher, 109*(8), 612–617. Retrieved from http://www.nctm.org/Publications/Mathematics-Teacher/2016/Vol109/Issue8/Encouraging-Students-with-Learning-Disabilities

How to write a unit plan. (n.d.). *wikiHow*. Retrieved from http://www.wikihow.com/Write-a-Unit-Plan

Huebner, T. (2010). Differentiated learning. *Educational Leadership, 67*(5), 79–81. Retrieved from http://www.ascd.org/publications/educational-leadership/feb10/vol67/num05/Differentiated-Learning.aspx

Hulleman, C. S., Godes, O., Hendricks, B. L., & Harackiewicz, J. M. (2010). Enhancing interest and performance with a utility value intervention. *Journal of Educational Psychology, 102*(4), 880–895. Retrieved from http://psycnet.apa.org/record/2010-21220-001

Hunsader, P. D., Zorin, B., & Thompson, D. R. (2015). Enhancing teachers' assessment of mathematical processes through test analysis in university courses. *Mathematics Teacher Educator, 4*(1). Retrieved from http://www.nctm.org/Publications/Mathematics-Teacher-Educator/2015/Vol4/Issue1/Enhancing-Teachers_-Assessment-of-Mathematical-Processes-Through-Test-Analysis-in-University-Courses

Hunt, J. H. (2010). Master geometry while co-teaching. *Mathematics Teaching in the Middle School, 16*(3), 154–161. Retrieved from http://www.nctm.org/Publications/mathematics-teaching-in-middle-school/2010/Vol16/Issue3/Master-Geometry-while-Coteaching

Igo, L. B., Bruning, R. A., & Riccomini, P. R. (2009). Should middle school students with learning problems copy and paste notes from the internet? Mixed-methods evidence of study barriers. *RMLE Online: Research in Middle Level Education, 33*(2). Retrieved from http://files.eric.ed.gov/fulltext/EJ867141.pdf

Igo, L. B., Riccomini, P. R., Bruning, R. A., & Pope, G. (2006). How should middle school students with LD approach online note taking? A mixed-methods study. *Learning Disability Quarterly, 29*(2), 89–100. Retrieved from http://journals.sagepub.com/doi/abs/10.2307/30035537

Jackson, B. (2014). Algebra homework: A sandwich. *Mathematics Teacher*, *107*(7), 528–533. Retrieved from http://www.nctm.org/Publications/Mathematics-Teacher/2014/Vol107/Issue7/Algebra-Homework_-A-Sandwich

Jacobs, G. M. (2014). *Collaborative learning or cooperative learning? The name is not important; flexibility is*. Retrieved from http://www.academia.edu/6997708/Collaborative_Learning_or_Cooperative_Learning_The_Name_Is_Not_Important_Flexibility_Is

Jao, L., & McDougall, D. (2016). Moving beyond the barriers: Supporting meaningful teacher collaboration to improve secondary school mathematics. *Teacher Development*, *20*(4), 557–573. Retrieved from http://www.tandfonline.com/doi/full/10.1080/13664530.2016.1164747

Jasmine, J., & Schiesl, P. (2009). The effects of word walls and word wall activities on the reading fluency of first-grade students. *Reading Horizons*, *49*(4), 301–314. Retrieved from http://scholarworks.wmich.edu/reading_horizons/vol49/iss4/5

Jensen, E. (2005). *Teaching with the brain in mind* (2nd ed.). Alexandria, VA: ASCD. Retrieved from http://www.ascd.org/publications/books/104013/chapters/Movement-and-Learning.aspx

Johnson, B. (2011, August 11). Student learning groups: Homogeneous or heterogeneous? [web log comment]. *Edutopia*. Retrieved from http://www.edutopia.org/blog/student-grouping-homogeneous-heterogeneous-ben-johnson.

Johnson, D. W., & Johnson, R. T. (2002). Learning together and alone: Overview and meta-analysis. *Asia Pacific Journal of Education*, *22*(1), 95–105. Retrieved from http://www.tandfonline.com/doi/abs/10.1080/0218879020220110

Johnson, D. W., & Johnson, R. T. (2009). An educational psychology success story: Social interdependence theory and cooperative learning. *Educational Researcher*, *38*(5), 365–379. Retrieved from http://journals.sagepub.com/doi/10.3102/0013189X09339057

Johnson, E. J. (2014). From the classroom to the living room: Eroding academic inequities through home visits. *Journal of School Leadership*, *24*(2), 357–385. Retrieved from http://www.k12.wa.us/MigrantBilingual/HomeVisitsToolkit/pubdocs/Johnson2014.pdf

Johnson, M. (2018). The struggle is real: How difficult work strengthens student achievement. *ASCD Express*, *14*(11). Retrieved from http://www.ascd.org/ascd-express/vol14/num11/the-struggle-is-real-how-difficult-work-strengthens-student-achievement.aspx

Jones, B., Hopper, P., & Franz, D. (2008). Mathematics: A second language. *Mathematics Teacher*, *102*(4), 307–312. Retrieved from http://www.nctm.org/Publications/mathematics-teacher/2008/Vol102/Issue4/Connecting-Research-to-Teaching_-Mathematics_-A-Second-Language

Joseph, G. G. (2011). *The crest of the peacock: Non-European roots of mathematics* (3rd ed.). Princeton, NJ: Princeton University Press.

Joseph, N. M., Hailu, M. F., & Matthews, J. S. (2019). Normalizing black girls' humanity in mathematics classrooms. *Harvard Educational Review*, 89(1), 132–155. Retrieved from http://www.hepgjournals.org/doi/abs/10.17763/1943-5045-89.1.132

Jung, L. A., & Guskey, T. R. (2010). Grading exceptional learners. *Educational Leadership*, 67(5), 31–35. Retrieved from http://www.ascd.org/publications/educational-leadership/feb10/vol67/num05/Grading-Exceptional-Learners.aspx

Kahneman, D., Fredrickson, B. L., Schreiber, C. A., & Redelmeier, D. A. (1993). When more pain is preferred to less: Adding a better end. *Psychological Science*, 4(6), 401–405. Retrieved from http://journals.sagepub.com/doi/10.1111/j.1467-9280.1993.tb00589.x

Kana'iaupuni, S., Ledward, B., & Jensen, U. (2010). *Culture-based education and its relationship to student outcomes*. Honolulu: Kamehameha Schools, Research & Evaluation. Retrieved from http://www.ksbe.edu/_assets/spi/pdfs/CBE_relationship_to_student_outcomes.pdf

Kapusnick, R. A., & Hauslein, C. M. (2001). The "silver cup" of differentiated instruction. *Kappa Delta Pi Record*, 37(4), 156–159. Retrieved from http://www.tandfonline.com/doi/abs/10.1080/00228958.2001.10518493

Karp, K. S., Bush, S. B., & Dougherty, B. J. (2016). Establishing a mathematics whole-school agreement. *Teaching Children Mathematics*, 23(2), 61–63. Retrieved from http://www.nctm.org/Publications/Teaching-Children-Mathematics/2016/Vol23/Issue2/Establishing-a-mathematics-whole-school-agreement

Keeley, R. G. (2015). Measurements of student and teacher perceptions of co-teaching models. *Journal of Special Education Apprenticeship*, 4(1). Retrieved from http://files.eric.ed.gov/fulltext/EJ1127778.pdf

Kenney, J. M., Hancewicz, E., Heuer, L., Metsisto, D., & Tuttle, C. L. (2005). *Literacy strategies for improving mathematics instruction*. Alexandria, VA: ASCD.

Kidd, R. (1992). Teaching ESL grammar through dictation. *TESL Canada Journal*, 10(1), 49–61. Retrieved from http://www.teslcanadajournal.ca/index.php/tesl/article/view/611

King, R. B., & McInerney, D. M. (2016). Culturalizing motivation research in educational psychology. *British Journal of Educational Psychology*, 86(1), 1–7. Retrieved from http://onlinelibrary.wiley.com/doi/pdf/10.1111/bjep.12106

Kingston, N., & Nash, B. (2011). Formative assessment: A meta-analysis and a call for research. *Educational Measurement: Issues and Practice, 30*(4), 28–37. Retrieved from http://onlinelibrary.wiley.com/doi/full/10.1111/j.1745-3992.2011.00220.x

Kipfer, B. A. (2013, July 3). 9 reasons why print dictionaries are better than online dictionaries. *The Week.* Retrieved from http://theweek.com/articles/462575/9-reasons-why-print-dictionaries-are-better-than-online-dictionaries

Kirschner, P. A., Sweller, J., & Clark, R. E. (2006). Why minimal guidance during instruction does not work: An analysis of the failure of constructivist, discovery, problem-based, experiential, and inquiry-based teaching. *Educational Psychologist, 41*(2), 75–86. Retrieved from http://www.tandfonline.com/doi/abs/10.1207/s15326985ep4102_1

Kohn, A. (1994). *The risks of rewards.* Retrieved from http://www.alfiekohn.org/article/risks-rewards

Kohn, A. (2008). Rethinking homework. http://Teachers.net Gazette, 5(2). Retrieved from http://www.teachers.net/gazette/FEB08/kohn

Kotsopoulos, D. (2007). Mathematics discourse: It's like hearing a foreign language. *Mathematics Teacher, 101*(4), 301–305. Retrieved from http://www.nctm.org/Publications/mathematics-teacher/2007/Vol101/Issue4/Mathematics-Discourse_-It_s-Like-Hearing-a-Foreign-Language

Kraft, M. A., & Papay, J. P. (2014). Can professional environments in schools promote teacher development? Explaining heterogeneity in returns to teaching experience. *Educational Evaluation and Policy Analysis, 36*(4), 476–500. Retrieved from http://www.ncbi.nlm.nih.gov/pmc/articles/PMC4392767

Kumar, R., Zusho, A., & Bondie, R. (2018). Weaving cultural relevance and achievement motivation into inclusive classroom cultures. *Educational Psychologist, 53*(2), 78–96. Retrieved from http://www.tandfonline.com/doi/pdf/10.1080/00461520.2018.1432361

Kunnath, J. (2017). Creating meaningful grades. *Journal of School Administration Research and Development, 2*(1), 53–56. Retrieved from http://files.eric.ed.gov/fulltext/EJ1158167.pdf

La Belle, T. J. (1971). What's deprived about being different? *Elementary School Journal, 72*(1), 13–19. Retrieved from http://www.journals.uchicago.edu/doi/10.1086/460671

Ladson-Billings, G. (1994). *The dreamkeepers: Successful teachers of African American children.* San Francisco, CA: Jossey-Bass.

Ladson-Billings, G. (1995a). But that's just good teaching! The case for culturally relevant pedagogy. *Theory Into Practice, 34*(3), 159–165. Retrieved from http://www.tandfonline.com/doi/abs/10.1080/00405849509543675

Ladson-Billings, G. (1995b). Toward a theory of culturally relevant pedagogy. *American Educational Research Journal, 32*(3), 465–491. Retrieved from http://journals.sagepub.com/doi/abs/10.3102/00028312032003465

Ladson-Billings, G. (2006). Yes, but how do we do it? Practicing culturally relevant pedagogy. In J. Landsman & C. W. Lewis (Eds.). *White teachers, diverse classrooms: A guide to building inclusive schools, promoting high expectations, and eliminating racism* (pp. 29–42). Sterling, VA: Stylus Publishing.

Laitsch, D. (2007). Design-based learning and student achievement. *ResearchBrief, 5*(6). Retrieved from http://www.ascd.org/publications/researchbrief/v5n06/toc.aspx

Laldin, M. (2016, February 11). *The psychology of belonging (and why it matters)* [web log comment]. Retrieved from http://www.learningandthebrain.com/blog/psychology-of-belonging

Lam, U. F., Chen, W.-W., Zhang, J., & Liang, T. (2015). It feels good to learn where I belong: School belonging, academic emotions, and academic achievement in adolescents. *School Psychology International, 36*(4), 393–409. Retrieved from http://journals.sagepub.com/doi/10.1177/0143034315589649

Language. (n.d.-a) *Merriam-Webster Dictionary*. Retrieved from http://www.merriam-webster.com/dictionary/language

Language. (n.d.-b) *Oxford Dictionaries*. Retrieved from http://en.oxforddictionaries.com/definition/language

Larmer, J., & Mergendoller, J. R. (2010). Seven essentials for project-based learning. *Educational Leadership, 68*(1), 34–37. Retrieved from http://www.ascd.org/publications/educational_leadership/sept10/vol68/num01/Seven_Essentials_for_Project-Based_Learning.aspx

Larson, K. E., Pas, E. T., Bradshaw, C. P., Rosenberg, M. S., & Day-Vines, N. L. (2018). Examining how proactive management and culturally responsive teaching relate to student behavior: Implications for measurement and practice. *School Psychology Review, 47*(2), 153–166. Retrieved from http://naspjournals.org/doi/10.17105/SPR-2017-0070.V47-2

Latunde, Y. (2017a). The role of skills-based interventions and settings on the engagement of diverse families. *School Community Journal, 27*(2), 251–273. Retrieved from http://files.eric.ed.gov/fulltext/EJ1165642.pdf

Latunde, Y. (2017b). Welcoming black families: What schools can learn from churches. *Educational Leadership, 75*(1). Retrieved from http://www.ascd.org/publications/educational-leadership/sept17/vol75/num01/Welcoming-Black-Families@-What-Schools-Can-Learn-from-Churches.aspx

Leana, C. R., & Pil, F. K. (2014, October 14). *A new focus on social capital in school reform efforts* [web log comment]. Retrieved from http://www.shankerinstitute.org/blog/new-focus-social-capital-school-reform-efforts

Learning disabilities vs. differences. (n.d.). Learning Disabilities Association of New York. Retrieved from Learning Disabilities vs. Differences

Lee, H. Y., Jamieson, J. P., Miu, A. S., Josephs, R. A., & Yeager, D. S. (2018). An entity theory of intelligence predicts higher cortisol levels when high school grades are declining. *Child Development.* Retrieved from http://onlinelibrary.wiley.com/doi/abs/10.1111/cdev.13116

Leith, C., Rose, E., & King, T. (2016). Teaching mathematics and language to English learners. *Mathematics Teacher, 109*(9), 670–678. Retrieved from http://www.nctm.org/Publications/Mathematics-Teacher/2016/Vol109/Issue9/Teaching-Mathematics-and-Language-to-English-Learners

Lemov, D. (2017, October 6). Using the do now for retrieval practice—An update from Alex Laney [web log comment]. *Teach Like a Champion.* Retrieved from http://teachlikeachampion.com/blog/using-now-retrieval-practice-update-alex-laney

Lenihan, E. (2015, June 11). Using classroom walls to create a learning-rich environment [web log comment]. Retrieved from http://eoinlenihan.weebly.com/blog/using-classroom-walls-to-create-a-thinking-rich-environment

Lepak, J. (2014). Enhancing students' written mathematical arguments. *Mathematics Teaching in the Middle School, 20*(4), 213–219. Retrieved from http://www.nctm.org/Publications/mathematics-teaching-in-middle-school/2014/Vol20/Issue4/Enhancing-Students_-Written-Mathematical-Arguments

Levin, M. (2018). Conceptual and procedural knowledge during strategy construction: A complex knowledge systems perspective. *Cognition and Instruction, 36*(3), 247–278. Retrieved from http://www.tandfonline.com/doi/abs/10.1080/07370008.2018.1464003

Lewis, B. (2017, July 29). Should you offer extrinsic classroom incentives for good behavior? Retrieved from http://www.thoughtco.com/classroom-rewards-for-good-behavior-2080992

Li, Q., & Ma, X. (2010). A meta-analysis of the effects of computer technology on school students' mathematics learning. *Educational Psychology Review, 22*(3), 215–243. Retrieved from http://link.springer.com/article/10.1007/s10648-010-9125-8

Lim, W., Kim, H., Stallings, L., & Son, J.-W. (2015). Celebrating diversity by sharing multiple solution methods. *Mathematics Teacher, 109*(4), 290–297. Retrieved from http://www.nctm.org/Publications/Mathematics-Teacher/2015/Vol109/Issue4/Celebrating-Diversity-by-Sharing-Multiple-Solution-Methods

Linsin, M. (2014). Why you shouldn't let your students decide the class rules. Retrieved from http://www.smartclassroommanagement.com/2014/08/02/why-you-shouldnt-let-your-students-decide-the-class-rules

Little, C., Hauser, S., & Corbishley, J. (2009). Constructing complexity for differentiated learning. *Mathematics Teaching in the Middle School*, 15(1), 34–42. Retrieved from http://www.nctm.org/Publications/mathematics-teaching-in-middle-school/2009/Vol15/Issue1/Constructing-Complexity-for-Differentiated-Learning

Liu, Y., & Hou, S. (2017). Potential reciprocal relationship between motivation and achievement: A longitudinal study. *School Psychology International*, 39(1), 38–55. Retrieved from http://journals.sagepub.com/doi/10.1177/0143034317710574

Lortie, D. (1975). *Schoolteacher: A sociological study*. Chicago, IL: University of Chicago Press.

Loveless, T. (2013). The 2013 Brown Center Report on American education [web log comment]. Retrieved from http://www.brookings.edu/wp-content/uploads/2016/06/2013-brown-center-report-web-3.pdf

Lui, A., & Andrade, H. (2015). Student peer assessment. In R. Gunstone (Ed.). *Encyclopedia of science education*. Dordrecht: Springer. Retrieved from http://link.springer.com/content/pdf/10.1007/978-94-007-6165-0_461-3.pdf

Maloney, E. A., Ramirez, G., Gunderson, E. A., Levine, S. C., & Beilock, S. L. (2015). Intergenerational effects of parents' math anxiety on children's math achievement and anxiety. *Psychological Science*, 26(9), 1480–1488. Retrieved from http://journals.sagepub.com/doi/abs/10.1177/0956797615592630

Margolis, H. (2014). Giving students a reason to try. *Educational Leadership*, 72(1). Retrieved from http://www.ascd.org/publications/educational-leadership/sept14/vol72/num01/Giving-Students-a-Reason-to-Try.aspx

Markant, D. B., Ruggeri, A., Gureckis, T. M., & Xu, F. (2016). Enhanced memory as a common effect of active learning. *Mind, Brain, and Education*, 10(3), 142–152. Retrieved from http://onlinelibrary.wiley.com/doi/abs/10.1111/mbe.12117

Markworth, K. A. (2016). A repeat look at repeating patterns. *Teaching Children Mathematics*, 23(1), 22–29. Retrieved from http://www.nctm.org/Publications/Teaching-Children-Mathematics/2016/Vol23/Issue1/A-Repeat-Look-at-Repeating-Patterns

MARS. (2015). Evaluating statements about radicals. Retrieved from http://map.mathshell.org/download.php?fileid=1714

Marsh, E., & Butler, A. (2013, March). Memory in educational settings. In D. Reisberg (Ed.). *The Oxford handbook of cognitive psychology*. Oxford, UK: Oxford University Press. Retrieved from http://marshlab.psych.duke.edu/publications/Marsh&Butler2013_Chapter.pdf

Martin, K. (2019, May 17). What is the difference between learning centers and stations in class? *The Classroom*. Retrieved from http://www.theclassroom.com/difference-learning-centers-stations-class-7969725.html

Marzano, R. J. (2000). *Designing a new taxonomy of educational objectives*. Thousand Oaks, CA: Corwin Press.

Marzano, R. J. (2010). *Formative assessment and standards-based grading*. Bloomington, MN: Marzano Research Laboratory.

Marzano, R. J. (2011). Art & science of teaching. The perils and promises of discovery learning. *Educational Leadership, 69*(1), 86–87. Retrieved from http://www.ascd.org/publications/educational-leadership/sept11/vol69/num01/The-Perils-and-Promises-of-Discovery-Learning.aspx

Marzano, R. J. (2015). Using formative assessment with SEL skills. In J. A. Durlak, C. E. Domitrovich, R. P. Weissberg, & T. P. Gullotta (Eds.). *Handbook of social and emotional learning: Research and practice*. New York, NY: The Guilford Press.

Marzano, R. J., Pickering, D. J., & Pollock, J. E. (2001). *Classroom instruction that works: Research-based strategies for increasing student achievement*. Alexandria, VA: Association for Supervision and Curriculum Development.

Matteson, S. (2016). The "brick wall" graphic organizer. *Mathematics Teaching in the Middle School, 20*(1), 38–45. Retrieved from http://www.nctm.org/Publications/Mathematics-Teaching-in-Middle-School/2016/Vol22/Issue1/The-%E2%80%9CBrick-Wall%E2%80%9D-Graphic-Organizer

Matthiessen, C. (2018, December 1). The hidden benefits of reading aloud—even to older kids. Retrieved from http://www.greatschools.org/gk/articles/read-aloud-to-children

Mattson, M. P. (2014). Superior pattern processing is the essence of the human brain. *Frontiers in Neuroscience, 8*(265). Retrieved from http://www.frontiersin.org/articles/10.3389/fnins.2014.00265/full

Maxwell, V. L., & Lassak, M. (2008). An experiment in using portfolios in the middle school classroom. *Mathematics Teaching in the Middle School, 13*(7), 404–409. Retrieved from http://www.nctm.org/Publications/mathematics-teaching-in-middle-school/2008/Vol13/Issue7/An-Experiment-in-Using-Portfolios-in-the-Middle-School-Classroom

Mayer, R. E. (2004). Should there be a three-strikes rule against pure discovery learning? *American Psychologist, 59*(1), 14–19. Retrieved from http://psycnet.apa.org/record/2004-10043-002

McCarthy, J (2015, August 28). 3 ways to plan for diverse learners: What teachers do [web log comment]. *Edutopia*. Retrieved from http://www.edutopia.org/blog/differentiated-instruction-ways-to-plan-john-mccarthy

McCarthy, J. (2018, January 10). Extending the silence. *Edutopia*. Retrieved from http://www.edutopia.org/article/extending-silence

McCormick, R. (1997). Conceptual and procedural knowledge. *International Journal of Technology and Design Education, 7*(1–2), 141–159. http://link.springer.com/article/10.1023/A:1008819912213

McKibben, S. (2016). Homing in on family relationships. *Educational Leadership, 58*(5). Retrieved from http://www.ascd.org/publications/newsletters/education-update/may16/vol58/num05/Homing-In-on-Family-Relationships.aspx

McKnight, K., Venkateswaran, N., Laird, J., Robles, J., & Shalev, T. (2017). *Mindset shifts and parent teacher home visits*. Berkeley, CA: RTI International. Retrieved from http://www.pthvp.org/wp-content/uploads/2018/12/171030-MindsetShiftsandPTHVReportFINAL.pdf

McLaughlin, C. (2016, September 1). The lasting impact of mispronouncing students' names. *NEA Today*. Retrieved from http://neatoday.org/2016/09/01/pronouncing-students-names

McMillan, J. H., Venable, J. C., & Varier, D. (2013). Studies of the effect of formative assessment on student achievement: So much more is needed. *Practical Assessment, Research & Evaluation, 18*(2). Retrieved from http://pareonline.net/pdf/v18n2.pdf

McNeal, R. B., Jr. (2014). Parent involvement, academic achievement and the role of student attitudes and behaviors as mediators. *Universal Journal of Educational Research, 2*(8), 564–576. Retrieved from http://files.eric.ed.gov/fulltext/EJ1053945.pdf

McNeal, R. B. (2012). Checking in or checking out? Investigating the parent involvement reactive hypothesis. *Journal of Educational Research, 105*(2), 79–89. Retrieved from http://www.tandfonline.com/doi/abs/10.1080/00220671.2010.519410

McTighe, J., & Lyman, F. T., Jr. (1988). Cueing theory in the classroom: The promise of theory-embedded tools. *Educational Leadership, 45*(7), 18–24. Retrieved from http://www.ascd.org/ASCD/pdf/journals/ed_lead/el_198804_mctighe.pdf

Meador, D. (2017, May 5). Preparing a dynamic lesson plan. Retrieved from http://www.thoughtco.com/preparing-a-dynamic-lesson-plan-3194650

Medina, P. (2001). The intricacies of initiate-response-evaluate in adult literacy education. Presented at Adult Education Research Conference, East Lansing, MI, 2001. East Lansing, MI, Adult Education Research Conference. Retrieved from http://newprairiepress.org/aerc/2001/papers/49

Meyer, D. (2018, November 20). *That isn't a mistake* [web log comment]. Retrieved from http://blog.mrmeyer.com/2018/that-isnt-a-mistake

Miemis, V. (2010, April 10). Essential skills for 21st century survival: Part I: Pattern recognition. Retrieved from http://emergentbydesign.com/2010/04/05/essential-skills-for-21st-century-survival-part-i-pattern-recognition

Miller, R. D. (2016). Contextualizing instruction for English language learners with learning disabilities. *Teaching Exceptional Children*, 49(1), 58–65. Retrieved from http://journals.sagepub.com/doi/full/10.1177/0040059916662248

Min, H.-T. (2006). The effects of trained peer review on EFL students' revision types and writing quality. *Journal of Second Language Writing*, 15(2), 118–141. Retrieved from http://www.sciencedirect.com/science/article/abs/pii/S106037430600004X

Mitchell, C. (2016, May 10). Mispronouncing students' names: A slight that can cut deep. *Edweek*. Retrieved from http://www.edweek.org/ew/articles/2016/05/11/mispronouncing-students-names-a-slight-that-can.html

Mo, J. (2019). How is students' motivation related to their performance and anxiety? In *PISA in focus*, No. 92. Paris, France: OECD Publishing. Retrieved from http://doi.org/10.1787/d7c28431-en

Mohr-Schroeder, M. J., Jackson, C., Cavalcanti, M., Jong, C., Craig Schroeder, D., & Speler, L. G. (2017). Parents' attitudes toward mathematics and the influence on their students' attitudes toward mathematics: A quantitative study. *School Science and Mathematics*, 117(5), 214–222. Retrieved from http://onlinelibrary.wiley.com/doi/abs/10.1111/ssm.12225

Moller, S., Mickelson, R. A., Stearns, E., Banerjee, N., & Bottia, M. C. (2013). Collective pedagogical teacher culture and mathematics achievement: Differences by race, ethnicity, and socioeconomic status. *Sociology of Education*, 86(2), 174–194. Retrieved from http://journals.sagepub.com/doi/10.1177/0038040712472911

Moran, A. S., Swanson, H. L., Gerber, M. M., & Fung, W. (2014). The effects of paraphrasing interventions on problem-solving accuracy for children at risk for math disabilities. *Learning Disabilities Research & Practice*, 29(3), 97–105. Retrieved from http://onlinelibrary.wiley.com/doi/pdf/10.1111/ldrp.12035

Morehead, K., Dunlosky, J., & Rawson, K. A. (2019). How much mightier is the pen than the keyboard for note-taking? A replication and extension of Mueller and Oppenheimer (2014). *Educational Psychology Review*, 1–28. Retrieved from http://link.springer.com/article/10.1007/s10648-019-09468-2

Morgan, H. (2013). Maximizing student success with differentiated learning. *The Clearing House: A Journal of Educational Strategies, Issues and Ideas*, 87(1), 34–38. Retrieved from http://www.tandfonline.com/doi/full/10.1080/00098655.2013.832130

Morin, R. (2015). *Exploring racial bias among biracial and single-race adults: The IAT*. Pew Research Center. Retrieved from http://www.pewsocialtrends.org/2015/08/19/exploring-racial-bias-among-biracial-and-single-race-adults-the-iat

Moschkovich, J. N. (2015). Scaffolding student participation in mathematical practices. *ZDM, 47*(7), 1067–1078. Retrieved from http://link.springer.com/article/10.1007%2Fs11858-015-0730-3

Mueller, J. (n.d.). Constructing good items. Authentic assessment toolbox. Retrieved from http://jfmueller.faculty.noctrl.edu/toolbox/tests/gooditems.htm

Mueller, P. A., & Oppenheimer, D. M. (2014). The pen is mightier than the keyboard: Advantages of longhand over laptop note taking. *Psychological Science, 25*(6), 1159–1168. Retrieved from http://journals.sagepub.com/doi/abs/10.1177/0956797614524581

Mulcahy, C. A., Maccini, P., Wright, K., & Miller, J. (2014). An examination of intervention research with secondary students with EBD in light of Common Core state standards for mathematics. *Behavioral Disorders, 39*(3), 146–164. Retrieved from http://journals.sagepub.com/doi/10.1177/019874291303900304

Mulvahill, E. (2018, February 5). Anchor charts 101: Why and how to use them. Retrieved from http://www.weareteachers.com/anchor-charts-101-why-and-how-to-use-them-plus-100s-of-ideas

Mun, R. U., Langley, S. D., Ware, S., Gubbins, E. J., Siegle, D., Callahan, C. M., . . . Hamilton, R. (2016). *Effective practices for identifying and serving English learners in gifted education: A systematic review of the literature*. Storrs, CT: National Center for Research on Gifted Education. Retrieved from http://ncrge.uconn.edu/wp-content/uploads/sites/982/2016/01/NCRGE_EL_Lit-Review.pdf

Murdock, T. B. (1999). Discouraging cheating in your classroom. *Mathematics Teacher, 92*(7), 587–591. Retrieved from http://www.nctm.org/Publications/mathematics-teacher/1999/Vol92/Issue7/Discouraging-Cheating-in-Your-Classroom

Murphy, T. (1999). Changing assessment practices: Will this be on the test? *Mathematics Teacher, 92*(3), 247–249. Retrieved from http://www.nctm.org/Publications/mathematics-teacher/1999/Vol92/Issue3/Changing-Assessment-Practices-in-an-Algebra-Class,-or-Will-this-Be-on-the-Test_

Namkung, J. M., Peng, P., & Lin, X. (2019). The relation between mathematics anxiety and mathematics performance among school-aged students: A meta-analysis. *Review of Educational Research, 89*(3), 459–496. Retrieved from http://journals.sagepub.com/doi/abs/10.3102/0034654319843494

National Assessment of Educational Progress (2017). *The nation's report card*. Washington, DC: U.S. Department of Education. Retrieved from http://www.nationsreportcard.gov/ndecore/xplore/NDE

National Council of Teachers of Mathematics (2000). *Principles and standards for school mathematics*. Reston, VA: National Council of Teachers of Mathematics. Retrieved from http://www.nctm.org/standards

National Governors Association Center for Best Practices & Council of Chief State School Officers (2010). Common Core state standards for mathematics. Retrieved from http://www.corestandards.org/wp-content/uploads/Math_Standards1.pdf

National Research Council (2000). *How people learn: Brain, mind, experience, and school: Expanded edition*. Washington, DC: The National Academies Press. Retrieved from http://www.nap.edu/catalog/9853/how-people-learn-brain-mind-experience-and-school-expanded-edition

National Research Council (2004). *Engaging schools: Fostering high school students' motivation to learn*. Washington, DC: National Research Council. Retrieved from http://www.nap.edu/download/10421

Nirode, W. (2018). Doing geometry with geometry software. *Mathematics Teacher, 112*(3), 179–184. Retrieved from http://www.nctm.org/Publications/Mathematics-Teacher/2018/Vol112/Issue3/Doing-Geometry-with-Geometry-Software

Orlin, B. (2018, September 18). What does math look like to mathematicians? *Popular Science*. Retrieved from http://www.popsci.com/what-does-math-look-like-to-mathematicians

Osterman, K. F. (2000). Students' need for belonging in the school community. *Review of Educational Research, 70*(3), 323–367. Retrieved from http://journals.sagepub.com/doi/10.3102/00346543070003323

Otten, S., Cirillo, M., & Herbel-Eisenmann, B. A. (2015). Making the most of going over the homework. *Mathematics Teaching in the Middle School, 21*(2), 99–105. Retrieved from http://www.nctm.org/Publications/Mathematics-Teaching-in-Middle-School/2015/Vol21/Issue2/Making-the-Most-of-Going-over-Homework

Palmer, G., Peters, R., & Streetman, R. (2018). Cooperative learning. In P. Lombardi (Ed.). *Instructional methods, strategies, and technologies to meet the needs of all learners*. Creative Commons NonCommercial ShareAlike. Retrieved from http://granite.pressbooks.pub/teachingdiverselearners

Parker, W. C., Mosborg, S., Bransford, J., Vye, N., Wilkerson, J., & Abbott, R. (2011). Rethinking advanced high school coursework: Tackling the depth/breadth tension in the AP US government and politics course. *Journal of Curriculum Studies*,

43(4), 533–559. Retrieved from http://www.tandfonline.com/doi/abs/10.1080/00220272.2011.584561

Parsons, S. A., Vaughn, M., Scales, R. Q., Gallagher, M. A., Parsons, A. W., Davis, S. G., ... Allen, M. (2018). Teachers' instructional adaptations: A research synthesis. *Review of Educational Research, 88*(2), 205–242. Retrieved from http://journals.sagepub.com/doi/10.3102/0034654317743198

Pashler, H., McDaniel, M., Rohrer, D., & Bjork, R. (2008). Learning styles: Concepts and evidence. *Psychological Science in the Public Interest, 9*(3), 105–119. Retrieved from http://journals.sagepub.com/doi/full/10.1111/j.1539-6053.2009.01038.x

Passolunghi, M. C., Cargnelutti, E., & Pellizzoni, S. (2018). The relation between cognitive and emotional factors and arithmetic problem-solving. *Educational Studies in Mathematics, 100*(3), 271–290. Retrieved from http://link.springer.com/article/10.1007/s10649-018-9863-y

Pauk, W. (2001). *How to study in college* (7th ed.). Boston, MA: Houghton Mifflin.

Pavitt, C. (1998). *Small group communication: A theoretical approach* (3rd ed.). New York, NY: Pearson. Retrieved from http://www.uky.edu/%7Edrlane/teams/pavitt/ch2.htm

Paznokas, L. (2003). Teaching math through cultural quilting. *Teaching Children Mathematics, 9*(5), 250–256. Retrieved from http://www.nctm.org/Publications/teaching-children-mathematics/2003/Vol9/Issue5/Teaching-Mathematics-through-Cultural-Quilting

Perkins, I., & Flores, A. (2002). Mathematical notations and procedures of recent immigrant students. *Mathematical Teaching in the Middle School, 7*(6), 346–352. Retrieved from http://www.nctm.org/Publications/mathematics-teaching-in-middle-school/2002/Vol7/Issue6/Mathematical-Notations-and-Procedures-of-Recent-Immigrant-Students

Perry, R. R., & Lewis, C. C. (2009). What is successful adaptation of lesson study in the US? *Journal of Educational Change, 10*(4), 365–391. Retrieved from http://www.researchgate.net/publication/227070986_What_is_successful_adaptation_of_lesson_study_in_the_US

Petty, B. (2018, July 23). 4 tools for a flipped classroom. *Edutopia.* Retrieved from http://www.edutopia.org/article/4-tools-flipped-classroom

Phillips, P. A., & Smith, L. R. (1992). *The effect of teacher dress on student perceptions* [PDF file]. Retrieved from http://files.eric.ed.gov/fulltext/ED347151.pdf

Piaget, J. (1952). *The origins of intelligence in children.* New York, NY: International Universities Press.

Pickhardt, C. E. (2011, March 7). Adolescence and parental favoritism [web log comment]. *Psychology Today*. Retrieved from http://www.psychologytoday.com/us/blog/surviving-your-childs-adolescence/201103/adolescence-and-parental-favoritism

Pierce, R., & Adams, C. (2005). Using tiered lessons in mathematics. *Mathematics Teaching in the Middle School, 11*(3), 144–149. Retrieved from http://www.nctm.org/Publications/Mathematics-Teaching-in-Middle-School/2005/Vol11/Issue3/Using-Tiered-Lessons-in-Mathematics

Pinkin, S. (2016, February 22). Putting standards-based grading into action. *Edweek*. Retrieved from http://www.edweek.org/tm/articles/2016/02/22/putting-standards-based-grading-into-action.html

Pólya, G. (1945). *How to solve it: A new aspect of mathematical method*. Princeton, NJ: Princeton University Press.

Posamentier, A., Smith, B. S., & Stepelman, J. (2010). *Teaching secondary mathematics: Techniques and enrichment units* (8th ed.). Boston, MA: Allyn & Bacon.

Post, G., & Varoz, S. (2008). Supporting teacher learning: Lesson-study groups with prospective and practicing teachers. *Teaching Children Mathematics, 14*(8), 472–478. Retrieved from http://www.nctm.org/Publications/teaching-children-mathematics/2008/Vol14/Issue8/Supporting-Teacher-Learning_-Lesson-Study-Groups-with-Prospective-and-Practicing-Teachers

Preston, J. P., Wiebe, S., Gabriel, M., McAuley, A., Campbell, B., & MacDonald, R. (2015). Benefits and challenges of technology in high schools: A voice from educational leaders with a Freire echo. *Interchange, 46*(2), 169–185. Retrieved from http://link.springer.com/article/10.1007%2Fs10780-015-9240-z

Putwain, D., & Remedios, R. (2014). The scare tactic: Do fear appeals predict motivation and exam scores? *School Psychology Quarterly, 29*(4), 503–516. Retrieved from http://psycnet.apa.org/doiLanding?doi=10.1037%2Fspq0000048

Qin, Z., Johnson, D. W., & Johnson, R. T. (1995). Cooperative versus competitive efforts and problem solving. *Review of Educational Research, 65*(2), 129–143. Retrieved from http://journals.sagepub.com/doi/abs/10.3102/00346543065002129

Raleigh, E., & Kao, G. (2010). Do immigrant minority parents have more consistent college aspirations for their children? *Social Science Quarterly, 91*(4), 1083–1102. Retrieved from http://onlinelibrary.wiley.com/doi/abs/10.1111/j.1540-6237.2010.00750.x

Ramirez, G., Hooper, S. Y., Kersting, N. B., Ferguson, R., & Yeager, D. (2018). Teacher math anxiety relates to adolescent students' math achievement. *AERA Open, 4*(1). Retrieved from http://journals.sagepub.com/doi/full/10.1177/2332858418756052

Ramirez, G., Shaw, S. T., & Maloney, E. A. (2018). Math anxiety: Past research, promising interventions, and a new interpretation framework. *Educational Psychologist, 53*(3), 145–164. Retrieved from http://www.tandfonline.com/doi/full/10.1080/00461520.2018.1447384

Ramos, M. F. (2014). *The strengths of Latina mothers in supporting their children's education: A cultural perspective* [PDF file]. Retrieved from http://www.childtrends.org/wp-content/uploads/2014/06/Strengths-of-Latinas-Mothers-formatted-6-10-14.pdf

Rapke, T. (2017). Involving students in developing math tests. *Mathematics Teacher, 107*(8), 613–616. Retrieved from http://www.nctm.org/Publications/Mathematics-Teacher/2017/Vol110/Issue8/Involving-Students-in-Developing-Math-Tests

Raymond, K., Gunter, M., & Conrady, K. (2018). Developing communication and metacognition through question generating. *Mathematics Teaching in the Middle School, 23*(5), 276–281. Retrieved from http://www.nctm.org/Publications/Mathematics-Teaching-in-Middle-School/2018/Vol23/Issue5/Developing-Communication-and-Metacognition-through-Question-Generating

Raymond, K. M. (2015). Making sense of representations: Card sorting. *Mathematics Teacher, 109*(5), 380–385. Retrieved from http://www.nctm.org/Publications/Mathematics-Teacher/2015/Vol109/Issue5/Making-Sense-of-Representations_-Card-Sorting

Read, S. (2010). A model for scaffolding writing instruction: IMSCI. *The Reading Teacher, 64*(1), 47–52. Retrieved from http://digitalcommons.usu.edu/cgi/viewcontent.cgi?article=1224&context=teal_facpub

Reeve, J., Deci, E. L., & Ryan, R. M. (2004). Self-determination theory: A dialectical framework for understanding the sociocultural influences on student motivation. In D. McInerney & S. Van Etten (Eds.). *Research on sociocultural influences on motivation and learning: Big theories revisited* (Vol. 4, pp. 31–59). Retrieved from http://www.researchgate.net/publication/309563565_Self-determination_theory_A_dialectical_framework_for_understanding_sociocultural_influences_on_student_motivation

Reeve, J., & Halusic, M. (2009). How K-12 teachers can put self-determination theory principles into practice. *School Field, 7*(2), 145–154. Retrieved from http://journals.sagepub.com/doi/abs/10.1177/1477878509104319

Reeves, D. B. (2004). The case against the zero. *Phi Delta Kappan, 86*(4), 324–325. Retrieved from http://journals.sagepub.com/doi/10.1177/003172170408600418

Reeves, R. V., & Halikias, D. (2017). *Race gaps in SAT scores highlight inequality and hinder upward mobility*. Brookings Institution. Retrieved from http://www.brookings.edu/research/race-gaps-in-sat-scores-highlight-inequality-and-hinder-upward-mobility

Reis, H. T., Wilson, I. M., Monestere, C., Bernstein, S., Clark, K., Seidl, E., . . . Radoane, K. (1990). What is smiling is beautiful and good. *European Journal of Social Psychology, 20*, 259–267. Retrieved from http://onlinelibrary.wiley.com/doi/pdf/10.1002/ejsp.2420200307

Reis, S. M., & Renzulli, J. S. (1992). Using curriculum compacting to challenge the above-average. *Educational Leadership, 50*(2), 51–57. Retrieved from http://www.ascd.org/publications/educational_leadership/oct92/vol50/num02/Using_Curriculum_Compacting_To_Challenge_the_Above-Average.aspx

Rentenbach, B., Prislovsky, L., & Gabriel, R. (2017). Valuing differences. *Phi Delta Kappan, 98*(8), 59–63. Retrieved from http://journals.sagepub.com/doi/10.1177/0031721717708297

Reynolds, K. J., Lee, E., Turner, I., Bromhead, D., & Subasic, E. (2017). How does school climate impact academic achievement? An examination of social identity processes. *School Psychology International, 38*(1), 78–97. Retrieved from http://journals.sagepub.com/doi/10.1177/0143034316682295

Rice, L., Barth, J. M., Guadagno, R. E., Smith, G. P. A., & McCallum, D. A. (2013). The role of social support in students' perceived abilities and attitudes toward math and science. *Journal of Youth and Adolescence, 42*(7), 1028–1040. Retrieved from http://link.springer.com/article/10.1007%2Fs10964-012-9801-8

Rimm-Kaufman, S., & Sandilos, L. (n.d.). *Improving students' relationships with teachers to provide essential supports for learning*. American Psychological Association. Retrieved from http://www.apa.org/education/k12/relationships.aspx

Rittle-Johnson, B., Schneider, M., & Star, J. R. (2015). Not a one-way street: Bidirectional relations between procedural and conceptual knowledge of mathematics. *Educational Psychology Review, 27*(4), 587–597. Retrieved from http://link.springer.com/article/10.1007/s10648-015-9302-x

Robertson, K. (n.d.). Supporting ELLs in the mainstream classroom: Language tips. *¡Colorín Colorado!* Retrieved from http://www.colorincolorado.org/article/supporting-ells-mainstream-classroom-language-tips

Robison, J. E. (2013, October 7). What is neurodiversity? [web log comment]. *Psychology Today*. Retrieved from http://www.psychologytoday.com/us/blog/my-life-aspergers/201310/what-is-neurodiversity

Rockwood, H. S., III (1995). Cooperative and collaborative learning. *The National Teaching & Learning Forum, 4*(6), 8–9. Retrieved from https://onlinelibrary.wiley.com/toc/21663327/1995/4/6

Rogers, K. B. (2007). Lessons learned about educating the gifted and talented. *Gifted Child Quarterly, 51*(4), 382–396. Retrieved from http://journals.sagepub.com/doi/10.1177/0016986209334964

Romagnano, L. (2001). Implementing the assessment standards: The myth of objectivity in mathematics assessment. *Mathematics Teacher, 94*(1), 31–37. Retrieved from http://www.nctm.org/Publications/mathematics-teacher/2001/Vol94/Issue1/Implementing-the-Assessment-Standards_-The-Myth-of-Objectivity-in-Mathematics-Assessment

Ronfeldt, M., Farmer, S. O., McQueen, K., & Grissom, J. A. (2015). Teacher collaboration in instructional teams and student achievement. *American Educational Research Journal, 52*(3), 475–514. Retrieved from http://journals.sagepub.com/doi/10.3102/0002831215585562

Routman, R. (2018, January 14). Treating your classroom like "prime real estate." Retrieved from http://www.middleweb.com/36760/treating-your-classroom-like-prime-real-estate

Rowh, M. (2012). First impressions count. *gradPSYCH, 10*(4), 32. Retrieved from http://www.apa.org/gradpsych/2012/11/first-impressions.aspx

Roy, G. J., Fueyo, V., Vahey, P., Knudsen, J., Rafanan, K., & Lara-Meloy, T. (2016). Connecting representations: Using predict, check, explain. *Mathematics Teaching in the Middle School, 21*(8), 492–496. Retrieved from http://www.nctm.org/Publications/Mathematics-Teaching-in-Middle-School/2016/Vol21/Issue8/Connecting-Representations_-Using-Predict,-Check,-Explain

Rubenstein, R. N., & Thompson, D. R. (2002). Understanding and supporting children's mathematical vocabulary development. *Teaching Children Mathematics, 9*(2), 107–112. Retrieved from http://www.nctm.org/Publications/teaching-children-mathematics/2002/Vol9/Issue2/Understanding-and-Supporting-Children_s-Mathematical-Vocabulary-Development

Ruef, J. (2018, November 1). Think you're bad at math? You may suffer from "math trauma." *The Conversation*. Retrieved from http://theconversation.com/think-youre-bad-at-math-you-may-suffer-from-math-trauma-104209

Ruffalo, R. (2018). *Unlocking learning II: Math as a lever for English learner equity*. Oakland, CA: Education Trust-West. Retrieved from http://files.eric.ed.gov/fulltext/ED588756.pdf

Rumack, A. M., & Huinker, D. (2019). Capturing mathematical curiosity with notice and wonder. *Mathematics Teaching in the Middle School, 24*(7), 394–399. Retrieved from http://www.nctm.org/Publications/Mathematics-Teaching-in-Middle-School/2019/Vol24/Issue7/Capturing-Mathematical-Curiosity-with-Notice-and-Wonder

Ryan, R. M., & Deci, E. L. (2000). Intrinsic and extrinsic motivations: Classic definitions and new directions. *Contemporary Educational Psychology, 25*, 54–67. Retrieved from http://mmrg.pbworks.com/f/Ryan,+Deci+00.pdf

Sadker, M., & Sadker, D. (1995). *Failing at fairness: How our schools treat girls*. New York, NY: Touchstone.

Safi, F., & Desai, S. (2017). Promoting mathematical connections using three-dimensional manipulatives. *Mathematics Teaching in the Middle School, 22*(8), 488–492. Retrieved from http://www.nctm.org/Publications/Mathematics-Teaching-in-Middle-School/2017/Vol22/Issue8/Promoting-Mathematical-Connections-Using-Three-Dimensional-Manipulatives

Sampson, E. C. (2016). Teachers' perceptions of the effect of their attire on middle-school students' behavior and learning (Unpublished doctoral dissertation). Walden University. Retrieved from http://search.proquest.com/docview/1767387293

Samson, J. F., & Collins, B. A. (2012). *Preparing all teachers to meet the needs of English language learners: Applying research to policy and practice for teacher effectiveness*. Washington, DC: Center for American Progress. Retrieved from http://eric.ed.gov/?id=ED535608

Scarlett, M. H. (2018). "Why did I get a C?": Communicating student performance using standards-based grading. *InSight: A Journal of Scholarly Teaching, 13*, 59–75. Retrieved from http://files.eric.ed.gov/fulltext/EJ1184948.pdf

Schleifer, D., Rinehart, C., & Yanisch, T. (2017). *Teacher collaboration in perspective: A guide to research* [PDF file]. New York, NY: Public Agenda. Retrieved from http://files.eric.ed.gov/fulltext/ED591332.pdf

Schmeichel, M. (2012). Good teaching? An examination of culturally relevant pedagogy as an equity practice. *Journal of Curriculum Studies, 44*(2), 211–231. Retrieved from http://www.tandfonline.com/doi/abs/10.1080/00220272.2011.591434

Schniedewind, N., & Davidson, E. (2000). Differentiating cooperative learning. *Educational Leadership, 58*(1), 24–27. Retrieved from http://www.ascd.org/publications/educational-leadership/sept00/vol58/num01/Differentiating-Cooperative-Learning.aspx

Schulten, K. (2018, November 15). Writing for an audience beyond the teacher: 10 reasons to send student work out into the world. *New York Times*. Retrieved from http://www.nytimes.com/2018/11/15/learning/writing-for-audience-beyond-teacher.html

Scriffiny, P. L. (2008). Seven reasons for standards-based grading. *Educational Leadership, 66*(2), 70–74. Retrieved from http://www.ascd.org/publications/educational_leadership/oct08/vol66/num02/Seven_Reasons_for_Standards-Based_Grading.aspx

Sebastian, J., Moon, J.-M., & Cunningham, M. (2016). The relationship of school-based parental involvement with student achievement: A comparison of principal and parent survey reports from PISA 2012. *Educational Studies*, *43*(2), 123–146. Retrieved from http://www.tandfonline.com/doi/full/10.1080/03055698.2016.1248900

Segalowitz, N. (2008). Automaticity and second languages. In C. J. Doughty & M. H. Long (Eds.). *The handbook of second language acquisition* (pp. 382–408). Hoboken, NJ: Blackwell Publishing. Retrieved from http://onlinelibrary.wiley.com/doi/10.1002/9780470756492.ch13

Senn, G., McMurtrie, D., & Coleman, B. (2019). Collaboration in the middle: Teachers in interdisciplinary planning. *Current Issues in Middle Level Education*, *24*(1). Retrieved from http://digitalcommons.georgiasouthern.edu/cimle/vol24/iss1/6

Setren, E., Greenberg, K., Moore, O., & Yankovich, M. (2019). *Effects of the flipped classroom* [PDF file]. Cambridge, MA: Massachusetts Institute of Technology, School Effectiveness & Inequality Initiative. Retrieved from http://seii.mit.edu/research/study/effects-of-the-flipped-classroom-evidence-from-a-randomized-trial

Shalaway, L. (n.d.). Creating classroom rules together. Scholastic. Retrieved from http://www.scholastic.com/teachers/articles/teaching-content/creating-classroom-rules-together

Shanahan, T. (2012). *Disciplinary literacy is NOT the same as content area reading*. Metairie, LA: Center for Development and Learning. Retrieved from http://www.cdl.org/articles/disciplinary-literacy-is-not-the-same-as-content-area-reading

Sharpe, W. (2009). Reading aloud—is it worth it? In *Education world*. Retrieved at http://www.educationworld.com/a_curr/curr213.shtml

Shaw, F. (1963). Educating culturally deprived youth in urban centers. *The Phi Delta Kappan*, *45*(2), 91–97. Retrieved from http://www.jstor.org/stable/20343043

Shea, V. (1994). *Netiquette*. San Rafael, CA: Albion. Retrieved from http://www.albion.com/netiquette/book/index.html

Sheldon, S. B., & Jung, S. B. (2015). *The family engagement partnership: Student outcome evaluation* [PDF file]. Baltimore, MD: Center on School, Family, & Community Partnerships, Johns Hopkins University. Retrieved from http://www.pthvp.org/wp-content/uploads/2016/09/JHU-STUDY_FINAL-REPORT.pdf

Sheldon, S. B., & Jung, S. B. (2018). *Student outcomes and parent teacher home visits* [PDF file]. Baltimore, MD: Center on School, Family, & Community Partnerships, Johns Hopkins University. Retrieved from http://www.pthvp.org/wp-content/uploads/2018/12/18-11-30-Student-Outcomes-and-PTHV-Report-FINAL.pdf

Shields, D. J. (2007). Taking math anxiety out of math instruction. *NADE Digest*, *3*(1), 55–64. Retrieved from http://files.eric.ed.gov/fulltext/EJ1097774.pdf

Shifrer, D. (2013). Stigma of a label: Educational expectations for high school students labeled with learning disabilities. *Journal of Health and Social Behavior*, *54*(4), 462–480. Retrieved from http://journals.sagepub.com/doi/10.1177/0022146513503346

Shumaker, H. (2016, March 8). Why parents should not make kids do homework. *Time*. Retrieved at http://time.com/4250968/why-parents-should-not-make-kids-do-homework

Silver, H. F., Strong, R. W., & Perini, M. J. (2009). How the strategic teacher plans for concept attainment. *ASCD Express*, *4*. Retrieved from http://www.ascd.org/ascd-express/vol4/420-silver.aspx

Simmons, K., Carpenter, L., Crenshaw, S., & Hinton, V. M. (2015). Exploration of classroom seating arrangement and student behavior in a second grade classroom. *Georgia Educational Researcher*, *12*(1), 51–68. Retrieved from http://files.eric.ed.gov/fulltext/EJ1194750.pdf

Slavin, R. E. (2014). Making cooperative learning powerful. *Educational Leadership*, *72*(2), 22–26. Retrieved from http://www.ascd.org/publications/educational-leadership/sept98/vol56/num01/Making-Cooperative-Learning-Equitable.aspx

Slavin, R. E., Leavey, M. B., & Madden, N. A. (1984). Combining cooperative learning and individualized instruction: Effects on student mathematics achievement, attitudes, and behaviors. *The Elementary School Journal*, *84*(4), 409–422. Retrieved from http://www.journals.uchicago.edu/doi/10.1086/461373

Small, M., & Lin, A. (2010). *More good questions: Great ways to differentiate secondary mathematics instruction*. New York, NY: Teachers College Press.

Smith, B. (2015). The evolution of my rapport: One professor's journey to building successful instructor/student relationships. *College Teaching*, *63*(2), 35–36. Retrieved from http://www.tandfonline.com/doi/full/10.1080/87567555.2014.999023

Smith, M., & Weinstein, Y. (2016, June 23). *Learning how to study using...retrieval practice* [web log comment]. Retrieved from http://www.learningscientists.org/blog/2016/6/23-1

Smith, M. S., Hillen, A. F., & Catania, C. L. (2007). Using pattern tasks to develop mathematical understandings and set classroom norms. *Mathematics Teaching in the Middle School*, *13*(1), 38–44. Retrieved from http://www.nctm.org/Publications/mathematics-teaching-in-middle-school/2007/Vol13/Issue1/Using-Pattern-Tasks-to-Develop-Mathematical-Understandings-and-Set-Classroom-Norms

Sousa, D. A. (2017). *How the brain learns* (5th ed.). Thousand Oaks, CA: Corwin.

Sparks, D. (2019, August 23). Flipped classrooms may exacerbate student achievement gaps. Here's how [web log comment]. Retrieved from http://blogs.edweek.org/edweek/inside-school-research/2019/08/flipped_classrooms_may_exacerb.html

Spear-Swerling, K. (2006). *The importance of teaching handwriting*. Retrieved from http://www.ldonline.org/spearswerling/The_Importance_of_Teaching_Handwriting

Spooner-Lane, R. (2016). Mentoring beginning teachers in primary schools: Research review. *Professional Development in Education, 43*(2), 253–273. Retrieved from http://www.tandfonline.com/doi/full/10.1080/19415257.2016.1148624

Srougi, M. C., & Miller, H. B. (2018). Peer learning as a tool to strengthen math skills in introductory chemistry laboratories. *Chemistry Education Research and Practice, 19*(1), 319–330. Retrieved from http://pubs.rsc.org/en/content/articlelanding/2018/RP/C7RP00152E

Staples, M., & Colonis, M. (2007). Making the most of mathematical discussions. *Mathematics Teacher, 101*(4), 257–261. Retrieved from http://www.nctm.org/Publications/mathematics-teacher/2007/Vol101/Issue4/Making-the-Most-of-Mathematical-Discussions

Starch, D., & Elliott, E. C. (1912). Reliability of the grading of high-school work in English. *American Journal of Education, 20*(7), 442–457. Retrieved from http://www.journals.uchicago.edu/doi/pdfplus/10.1086/435971

Starch, D., & Elliott, E. C. (1913). Reliability of the grading work in mathematics. *American Journal of Education, 21*(4), 254–259. Retrieved from http://www.journals.uchicago.edu/doi/pdfplus/10.1086/436086

Steele, C. M. (1997). A threat in the air: How stereotypes shape intellectual identity and performance. *American Psychologist, 52*(6), 613–629. Retrieved from https://psycnet.apa.org/doiLanding?doi=10.1037%2F0003-066X.52.6.613

Steenbergen-Hu, S., Makel, M. C., & Olszewski-Kubilius, P. (2016). What one hundred years of research says about the effects of ability grouping and acceleration on K–12 students' academic achievement. *Review of Educational Research, 86*(4), 849–899. Retrieved from http://journals.sagepub.com/doi/abs/10.3102/0034654316675417

Stein, C. C. (2007). Let's talk: Promoting mathematical discourse in the classroom. *Mathematics Teacher, 101*(4), 285–289. Retrieved from http://www.nctm.org/Publications/mathematics-teacher/2007/Vol101/Issue4/Let_s-Talk_-Promoting-Mathematical-Discourse-in-the-Classroom

Stolle, E. P., & Frambaugh-Kritzer, C. (2014). Putting professionalism back into teaching: Secondary preservice and in-service teachers engaging in interdisciplinary unit planning. *Action in Teacher Education, 36*(1), 61–75. Retrieved from http://www.tandfonline.com/doi/abs/10.1080/01626620.2013.850123

Strijbosch, W., Mitas, O., van Gisbergen, M., Doicaru, M., Gelissen, J., & Bastiaansen, M. (2019). From experience to memory: On the robustness of the peak-and-end-rule for complex, heterogeneous experiences. *Frontiers in Psychology, 10*(1705). Retrieved from http://www.frontiersin.org/articles/10.3389/fpsyg.2019.01705/full

Strobel, J., & van Barneveld, A. (2009). When is PBL more effective? A meta-synthesis of meta-analyses comparing PBL to conventional classrooms. *Interdisciplinary Journal of Problem-Based Learning, 3*(1), 44–58. Retrieved from http://docs.lib.purdue.edu/ijpbl/vol3/iss1/4

Strunk, W., Jr., & White, E. B. (1979). *The elements of style* (3rd ed.). New York, NY: Macmillan.

Summers, E. J., & Dickinson, G. (2012). A longitudinal investigation of project-based instruction and student achievement in high school social studies. *Interdisciplinary Journal of Problem-Based Learning, 6*(1), 82–103. Retrieved from http://docs.lib.purdue.edu/ijpbl/vol6/iss1/6

Suprayogi, M. N., Valcke, M., & Godwin, R. (2017). Teachers and their implementation of differentiated instruction in the classroom. *Teaching and Teacher Education, 67*, 291–301. Retrieved from http://www.sciencedirect.com/science/article/pii/S0742051X16303894

Suwantarathip, O., & Wichadee, S. (2010). The impacts of cooperative learning on anxiety and proficiency in an EFL class. *Journal of College Teaching & Learning (TLC), 7*(11), 51–58. Retrieved from http://clutejournals.com/index.php/TLC/article/view/252

Swinford, A. (2016). Constructed-response problems. *Teaching Children Mathematics, 22*(9), 517–519. Retrieved from http://www.nctm.org/Publications/Teaching-Children-Mathematics/2016/Vol22/Issue9/Constructed-response-problems

Szyszka, M. (2011). Foreign language anxiety and self-perceived English pronunciation competence. *Studies in Second Language Learning and Teaching, 1*(2), 283–300. Retrieved from http://files.eric.ed.gov/fulltext/EJ1136571.pdf

Tang, X., Zhang, S., Li, Y., & Zhao, M. (2013). Study on correlation of English pronunciation self-concept to English learning. *English Language Teaching, 6*(4), 74–79. Retrieved from http://files.eric.ed.gov/fulltext/EJ1076977.pdf

Tannenbaum, M. (2013, October 14). "But I didn't mean it!" Why it's so hard to prioritize impacts over intents [web log comment]. *Scientific American.* Retrieved from http://blogs.scientificamerican.com/psysociety/e2809cbut-i-didne28099t-mean-ite2809d-why-ite28099s-so-hard-to-prioritize-impacts-over-intents

Taylor, B. K. (2015). Content, process, and product: Modeling differentiated instruction. *Kappa Delta Pi Record, 51*(1), 13–17. Retrieved from http://www.tandfonline.com/doi/full/10.1080/00228958.2015.988559

Taylor, M. (2016). From effective curricula toward effective curriculum use. *Journal for Research in Mathematics Education, 47*(5), 440–453. Retrieved from http://www.nctm.org/Publications/Journal-for-Research-in-Mathematics-Education/2016/Vol47/Issue5/Research-Commentary_-From-Effective-Curricula-Toward-Effective-Curriculum-Use

Terada, Y. (2018, October 15). Multiple intelligences theory: Widely used, yet misunderstood. *Edutopia*. Retrieved from http://www.edutopia.org/article/multiple-intelligences-theory-widely-used-yet-misunderstood

Terada, Y. (2019, March 14). The science of drawing and memory. *Edutopia*. Retrieved from http://www.edutopia.org/article/science-drawing-and-memory

The Education Alliance (n.d.). *Teaching diverse learners: Are the tests culturally responsive? Brown University*. RI: Providence. Retrieved from https://www.brown.edu/academics/education-alliance/teaching-diverse-learners/question-iv-0

The jigsaw classroom. (n.d.). Retrieved from https://www.jigsaw.org/#overview

Thompson, B. C., Mazer, J. P., & Flood Grady, E. (2015). The changing nature of parent-teacher communication: Mode selection in the smartphone era. *Communication Education, 64*(2), 187–207. Retrieved from http://www.tandfonline.com/doi/abs/10.1080/03634523.2015.1014382

Thompson, D. R., & Rubenstein, R. N. (2000). Learning mathematics vocabulary: Potential pitfalls and instructional strategies. *Mathematics Teacher, 93*(7), 568–574. Retrieved from http://www.nctm.org/Publications/Mathematics-Teacher/2000/Vol93/Issue7/Learning-Mathematics-Vocabulary_-Potential-Pitfalls-and-Instructional-Strategies

Thomson, M. M. (2012). Labelling and self-esteem: Does labelling exceptional students impact their self-esteem? *Support for Learning, 27*(4), 158–165. Retrieved from http://journals.sagepub.com/doi/full/10.1177/0040059915569361

Tomlinson, C. A. (2014). *The differentiated classroom: Responding to the needs of all learners* (2nd ed.). Alexandria, VA: ASCD.

Trocki, A., Taylor, C., Starling, T., Sztajn, P., & Heck, D. (2014). Launching a discourse-rich mathematics lesson. *Teaching Children Mathematics, 21*(5), 277–281. Retrieved from http://www.nctm.org/Publications/teaching-children-mathematics/2014/Vol21/Issue5/Launching-a-Discourse-Rich-Mathematics-Lesson

U.S. Department of Education, National Center for Education Statistics (2015). Table 203.50: Enrollment and percentage distribution of enrollment in public elementary and secondary schools, by race/ethnicity and region: Selected years, fall 1995 through fall 2025. In U.S. Department of Education, National Center for Education Statistics (Ed.) *Digest of education statistics*. (2015 ed.). Retrieved from http://nces.ed.gov/programs/digest/d15/tables/dt15_203.50.asp

U.S. Department of Education, National Center for Education Statistics (2017). Table 204.80: Number of public school students enrolled in gifted and talented programs, by sex, race/ethnicity, and state: Selected years, 2004 through 2013-14. In U.S. Department of Education, National Center for Education Statistics (Ed.) *Digest of education statistics*. (2017 ed.). Retrieved from http://nces.ed.gov/programs/digest/d17/tables/dt17_204.80.asp

U.S. Department of Education, National Center for Education Statistics (2018a). Table 204.20. English language learner (ELL) students enrolled in public elementary and secondary schools, by state: Selected years, fall 2000 through fall 2016. In U.S. Department of Education, National Center for Education Statistics (Ed.) *Digest of education statistics*. (2018 ed.). Retrieved from http://nces.ed.gov/programs/digest/d18/tables/dt18_204.20.asp

U.S. Department of Education, National Center for Education Statistics (2018b). Table 204.30: Children 3 to 21 years old served under individuals with disabilities education act (IDEA), part B, by type of disability: Selected years, 1976-77 through 2017-18. In US Department of Education, National Center for Education Statistics (Ed.) *Digest of education statistics*. (2018 ed.). Retrieved from http://nces.ed.gov/programs/digest/d18/tables/dt18_204.30.asp

Understand the problem (n.d.). Retrieved from http://mathforum.org/pow/support/activityseries/understandtheproblem.html

Usher, A., & Kober, N. (2012a). *Student motivation: An overlooked piece of school reform* [PDF file]. Washington, DC: Center on Education Policy. Retrieved from http://files.eric.ed.gov/fulltext/ED532666.pdf

Usher, A., & Kober, N. (2012b). *What is motivation and why does it matter?* [PDF file]. Washington, DC: Center on Education Policy. Retrieved from http://files.eric.ed.gov/fulltext/ED532670.pdf

Usher, E. L. (2009). Sources of middle school students' self-efficacy in mathematics: A qualitative investigation. *American Educational Research Journal, 46*(1), 275–314. Retrieved from http://journals.sagepub.com/doi/abs/10.3102/0002831208324517

Usher, E. L. (2018). Acknowledging the whiteness of motivation research: Seeking cultural relevance. *Educational Psychologist, 53*(2), 131–144. Retrieved from http://www.tandfonline.com/doi/full/10.1080/00461520.2018.1442220

Utt, J. (2013, July 30). Intent vs. impact: Why your intentions don't really matter. Retrieved from http://everydayfeminism.com/2013/07/intentions-dont-really-matter

Valsiner, J. (1998). *The guided mind*. Cambridge, MA: Harvard University Press.

Valsiner, J., & van der Veer, R. (1993). The encoding of distance: The concept of the zone of proximal development and its interpretations. In R. R. Cocking & K. A. Renninger (Eds.). *The development and meaning of psychological distance* (pp. 35–62). Hillsdale, NJ: Erlbaum.

van der Lans, R. M. (2018). On the association between two things: The case of student surveys and classroom observations of teaching quality. *Educational Assessment, Evaluation and Accountability, 30*, 347–366. Retrieved from http://link.springer.com/content/pdf/10.1007%2Fs11092-018-9285-5.pdf

van der Valk, A., & Malley, A. (2019). What's my complicity? Talking white fragility with Robin DiAngelo. *Teaching Tolerance, 62*. Retrieved from http://www.tolerance.org/magazine/summer-2019/whats-my-complicity-talking-white-fragility-with-robin-diangelo

van Geel, M., Keuning, T., Frèrejean, J., Dolmans, D., van Merriënboer, J., & Visscher, A. J. (2018). Capturing the complexity of differentiated instruction. *School Effectiveness and School Improvement, 30*(1), 51–67. Retrieved from http://www.tandfonline.com/doi/full/10.1080/09243453.2018.1539013

VanAusdal, K. (2019). Collaborative classrooms support social-emotional learning. *ASCD Express, 4*(22). Retrieved from http://www.ascd.org/ascd-express/vol14/num22/collaborative-classrooms-support-social-emotional-learning.aspx

Vangrieken, K., Dochy, F., Raes, E., & Kyndt, E. (2015). Teacher collaboration: A systematic review. *Educational Research Review, 15*, 17–40. Retrieved from http://www.researchgate.net/publication/275723807_Teacher_collaboration_A_systematic_review

Vatterott, C. (2010). Five hallmarks of good homework. *Educational Leadership, 68*(1), 10–15. Retrieved from http://www.ascd.org/publications/educational-leadership/sept10/vol68/num01/Five-Hallmarks-of-Good-Homework.aspx

Venables, D. R. (2015). The case for protocols. *Educational Leadership, 72*(7). Retrieved from http://www.ascd.org/publications/educational-leadership/apr15/vol72/num07/The-Case-for-Protocols.aspx

Verenikina, I. (2010). Vygotsky in twenty-first-century research. In J. Herrington & B. Hunter (Eds.). *Proceedings of world conference on educational multimedia, hypermedia and telecommunications* (pp. 16–25). Chesapeake, VA: AACE. Retrieved from http://ro.uow.edu.au/cgi/viewcontent.cgi?article=2337&context=edupapers

Vescio, V., Ross, D., & Adams, A. (2008). A review of research on the impact of professional learning communities on teaching practice and student learning. *Teaching and Teacher Education, 24*(1), 80–91. Retrieved from http://pdfs.semanticscholar.org/7005/0f51d928cbedba2056a77a8f2c9b225c6821.pdf

Von der Embse, N. P., Schultz, B. K., & Draughn, J. D. (2015). Readying students to test: The influence of fear and efficacy appeals on anxiety and test performance. *School Psychology International, 36*(6), 620–637. Retrieved from http://journals.sagepub.com/doi/10.1177/0143034315609094

von Károlyi, C., Winner, E., Gray, W., & Sherman, G. F. (2003). Dyslexia linked to talent: Global visual-spatial ability. *Brain and Language, 85*(3), 427–431. Retrieved from http://www.sciencedirect.com/science/article/abs/pii/S0093934X0300052X

Vygotsky, L. S. (1978). In M. Cole, V. John-Steiner, S. Scribner, & E. Souberman (Eds.). *Mind in society: The development of higher psychological processes.* Cambridge, MA: Harvard University Press.

Walk, L., & Lassak, M. (2017). Making homework matter to students. *Mathematics Teaching in the Middle School, 22*(9), 546–553. Retrieved from http://www.nctm.org/Publications/Mathematics-Teaching-in-Middle-School/2017/Vol22/Issue9/Making-Homework-Matter-to-Students

Walker, A., & Leary, H. (2009). A problem-based learning meta analysis: Differences across problem types, implementation types, disciplines, and assessment levels. *Interdisciplinary Journal of Problem-ased Learning, 3*(1), 12–43. Retrieved from http://docs.lib.purdue.edu/ijpbl/vol3/iss1/3

Walkington, C., Sherman, M., & Howell, E. (2014). Personalized learning in algebra. *Mathematics Teacher, 108*(4), 272–279. Retrieved from http://www.nctm.org/Publications/mathematics-teacher/2014/Vol108/Issue4/Personalized-Learning-in-Algebra

Walmsley, A., & Hickman, A. (2006). A study of note taking and its impact on student perception of use in a geometry classroom. *Mathematics Teacher, 99*(9), 614–621. Retrieved from http://www.nctm.org/Publications/mathematics-teacher/2006/Vol99/Issue9/Connecting-Research-to-Teaching_-A-Study-of-Note-Taking-and-Its-Impact-on-Student-Perception-of-Use-in-a-Geometry-Classroom

Wammes, J. D., Meade, M. E., & Fernandes, M. A. (2016). The drawing effect: Evidence for reliable and robust memory benefits in free recall. *Quarterly Journal of Experimental Psychology, 69*(9), 1752–1776. Retrieved from http://www.tandfonline.com/doi/full/10.1080/17470218.2015.1094494

Webel, C., & Otten, S. (2015). Teaching in a world with PhotoMath. *Mathematics Teacher, 109*(5), 368–373. Retrieved from http://www.nctm.org/

Publications/Mathematics-Teacher/2015/Vol109/Issue5/Teaching-in-a-World-with-PhotoMath

Weisling, N. F., & Gardiner, W. (2018). Making mentoring work. *Phi Delta Kappan*, *99*(6), 64–69. Retrieved from http://journals.sagepub.com/doi/10.1177/0031721718762426

What is PBL? (n.d.). PBL Works. Retrieved from https://www.pblworks.org/what-is-pbl

White, S. A. (2013). A comparison of teacher evaluation, student surveys and growth scores to identify effective teaching traits (Unpublished doctoral dissertation). Retrieved from http://search.proquest.com/docview/1459764498.

Wieman, R. (2011). Sound off!: Students' beliefs about mathematics: Lessons learned from teaching note taking. *Mathematics Teacher*, *104*(6), 406–407. Retrieved from http://www.nctm.org/Publications/mathematics-teacher/2011/Vol104/Issue6/Sound-Off!_-Students_-Beliefs-about-Mathematics_-Lessons-Learned-from-Teaching-Note-Taking

Wilburne, J. M., & Peterson, W. (2007). Using a before-during-after model to plan effective secondary mathematics lessons. *Mathematics Teacher*, *101*(3), 209–213. Retrieved from http://www.nctm.org/Publications/mathematics-teacher/2007/Vol101/Issue3/Using-a-Before-During-After-(BDA)-Model-to-Plan-Effective-Secondary-Mathematics-Lessons

Wilder, S. (2013). Effects of parental involvement on academic achievement: A meta-synthesis. *Educational Review*, *66*(3), 377–397. Retrieved from http://www.tandfonline.com/doi/abs/10.1080/00131911.2013.780009

Wiliam, D. (2014). The right questions, the right way. *Educational Leadership*, *71*(6), 16–19. Retrieved from http://www.ascd.org/publications/educational-leadership/mar14/vol71/num06/The-Right-Questions,-The-Right-Way.aspx

Williams, N. B., & Wynne, B. D. (2000). Journal writing in the mathematics classroom: A beginner's approach. *Mathematics Teacher*, *93*(2), 132–135. Retrieved from http://www.nctm.org/Publications/mathematics-teacher/2000/Vol93/Issue2/Sharing-Teaching-Ideas_-Journal-Writing-in-the-Mathematics-Classroom_-A-Beginner_s-Approach

Willingham, D. T. (2002). Ask the cognitive scientist. inflexible knowledge: The first step to expertise. *American Educator*, *26*(4), 31–33. Retrieved from http://www.aft.org/periodical/american-educator/winter-2002/ask-cognitive-scientist

Willingham, D. T., Hughes, E. M., & Dobolyi, D. G. (2015). The scientific status of learning styles theories. *Teaching of Psychology*, *42*(3), 266–271. Retrieved from http://journals.sagepub.com/doi/abs/10.1177/0098628315589505

Willis, J. (2007). *Brain-friendly strategies for the inclusion classroom.* Alexandria, VA: ASCD. Retrieved from http://www.ascd.org/publications/books/107040/chapters/Success-for-all-Students-in-Inclusion-Classes.aspx

Willis, J., & Todorov, A. (2006). First impressions: Making up your mind after a 100-ms exposure to a face. *Psychological Science, 17*(7), 592–598. Retrieved from http://journals.sagepub.com/doi/full/10.1111/j.1467-9280.2006.01750.x

Wilson, S. D. (2018). Implementing co-creation and multiple intelligence practices to transform the classroom experience. *Contemporary Issues in Education Research, 11*(4), 127–132. Retrieved from http://files.eric.ed.gov/fulltext/EJ1193191.pdf

Winebrenner, S. (2000). Gifted students need an education, too. *Educational Leadership, 58*(1), 52–56. Retrieved from http://www.ascd.org/publications/educational-leadership/sept00/vol58/num01/Gifted-Students-Need-an-Education,-Too.aspx

Witherspoon, M. (1999). And the answer is...Symbolic literacy. *Teaching Children Mathematics, 5*(7), 396–399. Retrieved from http://www.nctm.org/Publications/teaching-children-mathematics/1999/Vol5/Issue7/And-the-Answer-Is-___-Symbolic-Literacy

Wong, B. (2018, August 12). Think inside the box (Part 1): Using the box with numbers [web log comment]. Retrieved from http://bobsonwong.com/blog/20-think-inside-the-box-1

Wong, B., & Bukalov, L. (2013). Improving student reasoning in geometry. *Mathematics Teacher, 107*(1), 54–60. Retrieved from http://www.nctm.org/Publications/mathematics-teacher/2013/Vol107/Issue1/Improving-Student-Reasoning-in-Geometry

Wong, H. K., & Wong, R. T. (2004). *How to be an effective teacher: The first days of school.* Mountain View, CA: Harry K. Wong Publications.

Wood, D., Bruner, J., & Ross, G. (1976). The role of tutoring in problem solving. *Journal of Child Psychology and Child Psychiatry, 17*(2), 89–100. Retrieved from http://onlinelibrary.wiley.com/doi/abs/10.1111/j.1469-7610.1976.tb00381.x

Woodbury, S. (2000). Teaching toward the big ideas of algebra. *Mathematics Teaching in the Middle School, 6*(4), 226–231. Retrieved from http://www.nctm.org/Publications/mathematics-teaching-in-middle-school/2000/Vol6/Issue4/Teaching-Toward-the-Big-Ideas-of-Algebra

Woodcock, S., & Hitches, E. (2017). Potential or problem? An investigation of secondary school teachers' attributions of the educational outcomes of students with specific learning difficulties. *Annals of Dyslexia, 67*(3), 299–317. Retrieved from http://link.springer.com/article/10.1007%2Fs11881-017-0145-7

Wormeli, R. (2006). Teaching in the middle: Turning zeroes to 60s. *Middle Ground, 9*(3), 21–23. Retrieved from http://www.amle.org/portals/0/pdf/mg/feb2006.pdf

Wormeli, R. (2011). Redos and retakes done right. *Educational Leadership*, *69*(3), 22–26. Retrieved from http://www.ascd.org/publications/educational-leadership/nov11/vol69/num03/Redos-and-Retakes-Done-Right.aspx

Wormeli, R. (2014). Motivating young adolescents. *Educational Leadership*, *72*(1), 26–31. Retrieved from http://www.ascd.org/publications/educational-leadership/sept14/vol72/num01/Motivating-Young-Adolescents.aspx

Wright, T. (2017). Supporting students who have experienced trauma. *NAMTA Journal*, *42*(2), 141–152. Retrieved from http://files.eric.ed.gov/fulltext/EJ1144506.pdf

Xu, X., Kauer, S., & Tupy, S. (2016). Multiple-choice questions: Tips for optimizing assessment in-seat and online. *Scholarship of Teaching and Learning in Psychology*, *2*(2), 147–158. Retrieved from http://www.academia.edu/27248548/TEACHER-READY_RESEARCH_REVIEW_Multiple-Choice_Questions_Tips_for_Optimizing_Assessment_In-Seat_and_Online

Yale Poorvu Center for Teaching and Learning. (n.d.). Public speaking for teachers II: The mechanics of speaking. Retrieved from http://poorvucenter.yale.edu/teaching/ideas-teaching/public-speaking-teachers-ii-mechanics-speaking

Young, J. (2017). Technology-enhanced mathematics instruction: A second-order meta-analysis of 30 years of research. *Educational Research Review*, *22*, 19–33. Retrieved from http://www.sciencedirect.com/science/article/pii/S1747938X1730026X

Yue, C. L., Storm, B. C., Kornell, N., & Bjork, E. L. (2015). Highlighting and its relation to distributed study and students' metacognitive beliefs. *Educational Psychology Review*, *27*(1), 69–78. Retrieved from https://sites.williams.edu/nk2/files/2011/08/Yue.Storm_.Kornell.Bjork_.inpress.pdf

Zhang, X., Clements, M. A., & Ellerton, N. F. (2015). Engaging students with multiple models. *Teaching Children Mathematics*, *22*(3), 139–147. Retrieved from http://www.nctm.org/Publications/Teaching-Children-Mathematics/2015/Vol22/Issue3/Engaging-Students-with-Multiple-Models

Zhao, M., & Lapuk, K. (2019). Supporting English learners in the math classroom: Five useful tools. *Mathematics Teacher*, *112*(4), 288–293. Retrieved from http://www.nctm.org/Publications/Mathematics-Teacher/2019/Vol112/Issue4/Supporting-English-Learners-in-the-Math-Classroom_-Five-Useful-Tools

Zieger, L. B., & Tan, J. (2012). Improving parent involvement in secondary schools through communication technology. *Journal of Literacy and Technology*, *13*(2), 30–54. Retrieved from http://www.literacyandtechnology.org/uploads/1/3/6/8/136889/jlt_vol13_2_zieger_tan.pdf

Index

A

Abbott, Edwin, 15
Ability grouping, for differentiated instruction, 310
Academic achievement: from classroom environment, 246; from culturally responsive teaching, 29; from differentiated instruction, 305; parents and, 272, 276; of students with unique needs, 337; teacher collaboration and, 286
Achieve the Core, 109, 350
Achievement gaps: with ability grouping, 310; culturally responsive teaching and, 29; in flipped classrooms, 422
Active learning, in differentiated instruction, 309
ADHD, 341, 348; cooperative learning for, 386
Adrenaline, 2
Advanced students, 335; cooperative learning for, 386; curriculum compacting for, 341; ELLs as, 336; independent study for, 346; manipulatives for, 346; problems with, 348–349; projects by, 364; strengths and challenges of, 340; technology for, 349–350
Affect, differentiated instruction by, 320
Alternative teaching, 291–292
Always-sometimes-never questions, 192–193
American Mathematics Competition, 350
Analytic rubrics, 318
Anchor charts: for mathematical language, 53, 54; productive struggle and, 342; for two-way tables, 106–107
Annotated work, 256, 268
Anxiety. *See* Math anxiety
Apple Calendar, for unit plans, 145
Application skills, in test questions, 195–196
Area models (box method): for division, 99–100, 114; for equivalence, 99–100; for multiplication, 99–100, 112–113
Artzt, A. F., 41
Assessments, 189–229; alternate forms of, 208; always-sometimes-never questions on, 192–193; analysis of results of, 203–204, 216; application of, 190–208; cheating on, 212–213; as classroom management, 210; Common Core Standards for, 190; constructed-response questions on, 193; for cooperative learning, 383–384, 385; corrections to, 215, 226–228; different versions of, 213–214; for ELLs, 207–208; error analysis for, 202–203; fill-in-the-blank questions on, 193; format for, 193–196, 202; grading of, 202–203, 214–215; grid-in-questions on, 192; handouts for, 208, 217–229; for learning differences, 207–208; matching questions on, 193; mistakes on, 211; multiple-choice questions on, 190–191; peer grading of, 203; poor scheduling

515

Assessments (*continued*) and preparation for, 209–210; practice for, 196–199; problems with, 208–215; question number in, 194; questions on, 215–216; question order in, 194–195; question skills in, 195–196; question types in, 190–193; questions poorly chosen for, 210–211; reflection on, 206, 229; research on, 190; retaking, 206–207, 215; for reteaching assignments, 312; returning to students, 204–206, 214–215; review for, 196–199, 210, 216; scoring guidelines for, 199–201, 202–203, 225; teacher collaboration and, 287; technology for, 215–216; topics in, 194; true-false questions on, 192–193. *See also* Formative assessment
Association for Supervision and Curriculum Development, 240
Autism, 341
Automaticity, 76
Autonomy: classroom rules and, 251; Common Core and, 9; from learning differences motivation, 18; math anxiety and, 10; motivation and, 10; in projects, 359; rewards and, 14; self-determination theory and, 8
Average: in linear regressions, 104; of rate of change, 102

B

"Bad" think-alouds, 73–74
"Bag of tricks," mathematical language as, 54–55
A Beautiful Mind (film), 16

Behavioral problems: assessments and, 210; in classroom environment, 258; grading and, 239; lesson plans and, 151; parents and, 272; self-reflection on, 31–32; technology and, 420
Beliefs: in co-teaching, 293; in culturally responsive teaching, 30–32; about curriculum, 144; in PLCs, 296; in teacher collaboration, 297
Belonging: "color-blind" teaching and, 36, 37; from culturally responsive teaching, 29; stereotypes and, 19
Benchmarks, 199, 200
Bias: cognitive, of curse of knowledge, 144; implicit, in "color-blind" teaching, 36–37
Bilingual glossaries, 55–56, 90; for assessments, 207
Bivariate data: in Common Core, 9; in functions, 102
Box method. *See* Area models
Brain: culturally responsive teaching and, 29, 35; end of school year and, 433
Brown University, Education Alliance at, 39
Brown v. Board of Education, 28
Bruffee, Kenneth, 380
Bruner, Jerome, 43

C

Calculators, 2; for assessments, 212; for learning differences, 17–18; trust in, 427
California Department of Education: on homework, 182; on mathematical language, 55
Card sorting, 396

Cardone, Tina, 110
Cell phones: for cheating, 212–213; problems with, 420
Census at School, 368
Census Bureau, US, 368
Center for Innovation in Engineering and Science Education, 368
Center for Teaching, at Vanderbilt University, 350, 400
Center on Online Learning and Students with Disabilities, 350
Centers, for cooperative learning, 393
Cheating: on assessments, 212–213; reteaching assignments and, 312
Chicago Lesson Study Group, 299
Choral response, 43
Circles: area model for, 100; center and radius of, 115; proportionality of, , 101–102, 120
ClassDojo, 22; for parents, 281
Classmint, 56
Classroom environment: application of, 246–257; behavioral problems in, 258; building productivity in, 245–269; Common Core Standards for, 246; course descriptions and, 252–253, 263–264; for culturally responsive teaching, 34–35, 250; differentiated instruction and, 303, 320; Do Now for, 247; for ELLs, 339; on first day of classes, 246–247; getting to know students, 248–249; greeting students, 246–247; handouts for, 255–256, 259, 261–269; learning names, 248; note-taking

and, 254–257, 259, 260; problems in, 257–259; procedures in, 250, 251–252, 287; for projects, 362; rearrangement of, 80; research on, 245; routines in, 250, 251–252, 287; rules in, 250–251, 287; seating arrangement in, 249–250; student surveys on, 253–254, 260; teacher collaboration and, 287; technology for, 259–260, 418–422; walls in, 250. *See also specific topics*

Classroom response systems, for formative assessments, 410–411

Classwork: mathematical communication of, 71–72; for PBL, 355–357; in point accumulation grading system, 236; tiered tests in, 316

Clickers. *See* Classroom response systems

Closed captioning, 90; in Google Slides, 160

Cognitive bias, of curse of knowledge, 144

Collaboration. *See* Teacher collaboration

Collaborative learning, 380–381

College Board, 215

"Color-blind" teaching, 36–37

¡Colorín Colorado!, 321; for ELLs, 350; for parents, 281

Common Core Standards: for assessments, 190; for classroom environment, 246; for cooperative learning, 381; for culturally responsive teaching, 29–30; for differentiated instruction, 305; for end of school year, 434; for formative assessment, 406; for grading, 232, 238; for homework, 170; for lesson plans, 152; for mathematical communication, 64; for mathematical connections, 98, 109; for mathematical language, 42, 55; motivation and, 9; parents and, 272; for PBL, 355; for students with unique needs, 337; teacher collaboration and, 286; for technology, 418; for unit plans, 138

Communication: in cooperative learning, 383; with parents, 272–275, 282–284. *See also* Mathematical communication

Competence: from learning differences motivation, 18; motivation and, 10; self-determination theory and, 8

Completing the square, 100, 115; in projects, 359

Computational errors: on assessments, 202; in writing, 82

Concept attainment strategy: for mathematical language, 43; for tiered activities, 308

Concept maps: for ELLs, 107, 344; for students with unique needs, 343–344, 352; in Underground Mathematics, 109; in unit plans, 140, 147

Conceptual errors: on assessments, 202; in writing, 82

Conceptual skills, in test questions, 195–196

Conditional probability, 104, 106

Confidence: from classroom rearrangement, 80; from cooperative learning, 380; correspondence journal on, 84; from grading, 232; from homework, 170; from lesson plans, 156; from mathematical communication, 65; from multiple-choice question assessments, 192; from note-taking, 254; nurturing of, 10–11; from scaffolding, 304; from test topics, 194; unit plans for, 142

Connecting Representations, 110

Connections. *See* Mathematical connections

Constructed-response questions (free-response questions): for advanced students, 349; on assessments, 193

Constructivism, 2; from cooperative learning, 380; PBL and, 354

Content: in curriculum compacting, 312; differentiated instruction by, 305–313; technology and, 422–424

Contingency tables. *See* Two-way tables

Cooperative learning: for advanced students, 386; assessments for, 383–384, 385; centers for, 393; collaborative learning and, 380–381; Common Core Standards for, 381; communication in, 383; confidence from, 380; constructivism from, 380; culturally responsive teaching from, 380; elements of, 379; for ELLs, 41, 384; at end of school year, 437; group formation for, 382; handouts for, 399, 401–404; jigsaws for, 390–391, 401–402; for

Cooperative learning: (*continued*)
learning differences, 386; metacognition in, 393; *Notice and Wonder* for, 388–390; peer editing for, 394–395, 404; problems of, 398–399; protocols for, 383; questions in, 383–384; relays for, 395–396; research on, 380–381; role assignment in, 382–383; rubrics for, 385; SEL from, 380; stations for, 392–394, 403; for students with unique needs, 384–386; task cards for, 396–397; techniques for, 381–384; technology for, 400; *Think-Pair-Share* for, 387

Cornell system, for note-taking, 257, 260

Correlation coefficient, in linear regressions, 128

Correspondence journal, for writing, 84

Cosine function, 103, 126–127

Co-teaching, 291–293

Course descriptions, 252–253, 263–264

Cram, 56

Credits for Teachers, 240

Cuban, Larry, 22

Cult of Pedagogy, 38

Cultural deprivation, 28

Culturally Responsive Leadership, 39

Culturally responsive teaching, 27–39; application of, 30–35; beliefs in, 30–32; brain and, 29, 35; classroom environment for, 34–35, 250; "color-blind" teaching in, 36–37; Common Core Standards for, 29–30; from cooperative learning, 380; engagement in, 30; good intentions in, 37; historical context for, 33–34; in home visits, 278; information processing in, 35; in lesson plans, 156, 158; math anxiety and, 35; mindset for, 30, 38; with parents, 278–279; PBL in, 35; problems with, 36–38; research on, 28–29; right time or place for, 38; routines in, 35; scaffolding in, 35; self-determination theory and, 10; self-reflection in, 28, 30–32, 38; technology for, 38–39, 425; trust in, 33; unit plans for, 142–143

Culture, motivation and, 10

Curcio, F. R., 41

Curriculum compacting: for advanced students, 341; in differentiated instruction, 312–313, 328–330

Curse of knowledge, 144

D

Dad's Worksheets, for homework, 181

Danielson, Christopher, 70, 89

Deci, Edward L., 7

Decision making, self-determination theory and, 9

Delay discounting, 14

Deliberate practice, technology for, 424

DeltaMath, 428; for assessments, 216; for differentiated instruction, 322; for end of school year, 443; for grading, 240; for homework, 181; for test review, 216

Desmos, 21, 428; for formative assessments, 415; for homework, 182; for lesson plans, 159

Dewey, John, 354

Diagnostic assessment, 405

Didax, 21, 428

Differentiated instruction: ability grouping for, 310; by affect, 320; application of, 305–320; classroom environment and, 303, 320; Common Core Standards for, 305; by content, 305–313; for cooperative learning, 399; curriculum compacting in, 312–313, 328–330; at end of school year, 440–441; in flipped classrooms, 421; for formative assessments, 413–414; grading for, 318–319; handouts for, 321, 323–334; portfolios in, 317; presentation style for, 309; problems with, 320–321; by process, 313–314; by product, 315–320; for projects, 359, 363–364; research on, 304–305; reteaching assignments in, 310–312; review material in, 317–318; rubrics for, 318–319; scaffolding in, 304; for students with unique needs, 335–352, 384–386; technology for, 321–322, 425; tiered activities in, 306–309, 323–327; tiered tests in, , 315–317, 331

Disciplinary literacy, 42

Discovery learning, 354; jigsaws as, 391, 402; in PBL, 367

Division, area model for, 99–100, 114

Division of fractions by fractions, in Common Core, 9

Do Now (Warm-Up): for classroom environment,

247; for curriculum compacting, 313; for lesson plans, 153–154, 162–166; paired dictation and, 44; routine of, 252; symbols and terms and, 45; tiered activities and, 306
Do2Learn, 350
Double-entry journal, 257, 269
Dray, Barbara, 32
Drop outs, 7
Dropbox, 427; on teacher collaboration, 299

E

Easy Notecards, 56
Edmodo: for parents, 281; for teacher collaboration, 299
Edpuzzle, 428; for formative assessments, 415
Edublogs: for parents, 281; for teacher collaboration, 299
Education Alliance, at Brown University, 39
Edutopia, 38; for ELLs, 350; for end of school year, 443; for formative assessments, 415; for parents, 281; for teacher collaboration, 299
EdWeek blog, 260
Ehlert, Dale, 233, 240
Electronic polling systems. *See* Classroom response systems
The ELL Teacher's Toolbox (Ferlazzo and Sypnieski), 3
Elliott, William, 232
ELLs. *See* English Language Learners
EMathInstruction, 145
End of school year: Common Core Standards for, 434; cooperative learning at, 437; differentiated instruction at, 440–441; ELLs at, 440–441; engagement at, 437; feedback at, 439–440; field trips at, 438; learning differences at, 441; long-term memory and, 433; metacognition at, 438; portfolios at, 438; problems of, 441–443; projects at, 437; reflection at, 438–439; research on, 433; retrieval practice at, 435; review at, 433, 434–438; student surveys at, 438; teacher-student conferences at, 438–439; technology for, 443
Endorphins, 29
Engagement: in culturally responsive teaching, 30; at end of school year, 437; from lesson plans, 151; motivation and, 8; by parents, 271
EngageNY, 102; on assessments, 215; on unit plans, 145
English, symbols and terms in, 46
English Language Learners (ELLs), 2, 3, 335; assessments for, 207–208; bilingual glossaries for, 55–56, 90; classroom environment for, 339; closed captioning for, 160; concept maps for, 107, 344; cooperative learning for, 41, 384; co-teaching for, 291; course descriptions for, 253; differentiated instruction for, 303, 321; double-entry journals for, 257; at end of school year, 440–441; feedback for, 363; formative assessments for, 410, 413; Frayer models for, 107, 343; grading for, 237–238; guided questions for, 154; homework for, 177, 181–182; learning differences of, 336; learning names of, 248; lesson plan for, 155–156, 158–159; lesson summaries for, 82; manipulatives for, 346; matching question assessments for, 193; mathematical communication for, 63, 87; mathematical connections with, 107; motivation for, 16–17; note-taking by, 256–257; pairing with English speakers, 249; parents of, 273; problems with, 348; projects by, 363–364; review by, 357; SEL for, 347; simplifying language for, 344–345; stations for, 392; stereotypes of, 297; strengths and challenges of, 338–339; student surveys for, 254; symbols and, 45; technology for, 425; terms for, 49; think-alouds with, 74; two-way tables for, 106; unit plans for, 139, 143; wait time for, 65; word problems for, 81; word walls for, 52
EquatIO, 90; for unit plans, 145
Error analysis: for assessments, 202–203; for writing, 81–82
Escalante, Jaime, 15
Ethics, of cheating, 213
Even functions, 103–104, 124–125; multiple intelligences and, 315
Evernote, 322
Excel, Microsoft, 240
Exceptional children, 336
Exploding Dots, 109
Expressions, Common Core Standards for, 109

Extra credit: for correcting mistakes, 205; in differentiated instruction, 313; for homework, 174; for mathematical communication, 71; in point accumulating grading system, 236
Extrinsic motivation, 7–8

F

Factual skills, in test questions, 195–196
Fear, motivation from, 18
Feedback: for advanced students, 340; in culturally responsive teaching, 31; for ELLs, 363; at end of school year, 439–440; from Google Classroom, 90; from homework, 171, 176; with peer editing, 394–395; for projects, 359, 362–363; in teacher collaboration, 298
Ferlazzo, Larry, 3; EdWeek blog of, 260; on end of school year, 443; on picture dictation, 48
Field trips, at end of school year, 438
5280 Math, 400
Fill-in-the-blank questions, 193
The First Days of School (Wong, H., and Wong, R.), 259
Flash cards, 51
Flatland (Abbott), 15
Flipgrid, 415
Flipped classrooms, 420–422
Flipped Learning Network, 428
Formative assessments, 405–415; alternative methods for, 412–413; application of, 406–414; classroom response systems for, 410–411; Common Core Standards for, 406; differentiated instruction for, 413–414; for ELLs, 410, 413; initiate-response-evaluate model for, 407; for learning differences, 414; problems of, 414; questions for, 407–409; research on, 406; responding to student answers, 412; scoring guidelines for, 413; student responses in, 409–412; technology for, 415; *Think-Pair-Share* for, 413, 414
Frayer, D., 343
Frayer model, 351, 352; for ELLs, 107, 343; ThoughtCo on, 350
Frederick, W., 343
Free-response questions. *See* Constructed-response questions
Fun, in lesson plans, 158
Functions: characteristics of, 102; Common Core Standards for, 109; even and odd, 103–104, 124–125; graphs for, 120; interdisciplinary collaboration with, 289; mathematical connections with, 100, 102–104; multiple intelligences and, 315; of polynomials, 123–124; probabilities and, 105; tables for, 120, 125; Wikipedia on, 110

G

Gallery walk: of jigsaws, 391; for lesson plan summary, 157; with peer editing, 395
Gardner, Howard, 321
Gay, Geneva, 28–29
Geogebra, 21, 428; for formative assessments, 415; for lesson plans, 159
Gifted students. *See* Advanced students
Given, Find, Solve, and Check, 74–79
Glossaries. *See* Bilingual glossaries
Goals: in curriculum compacting, 312–313; in flipped classrooms, 422; in lesson plan, 152–153; of projects, 359; of teacher collaboration, 286; of tiered activities, 307
Good intentions, in culturally responsive teaching, 37
Google Calendar, 145
Google Classroom, 56, 428; for end of school year, 443; for formative assessments, 415; for homework, 182; for mathematic communication, 89–90; for parents, 281; for test review, 216
Google Docs, 89; for lesson plans, 160; for teacher collaboration, 299; for unit plans, 145
Google Drive, 427; for teacher collaboration, 299
Google Forms, 56; for end of school year, 443; for formative assessments, 415; for student surveys, 260; for test review, 216
Google Sheets: for grading, 240; for test analysis, 216
Google Sites, 322
Google Slides, 90; for lesson plans, 160; for test review, 216
Google Translate, 428; for ELLs, 350; for parents, 281
Grade calculation sheets, 237
Grading: application of, 232–239; of assessments, 202–203, 214–215; Common Core Standards

for, 232, 238; for differentiated instruction, 318–319; for ELLs, 237–238; fixed intelligence and, 141–142; handouts for, 240, 241–242; of homework, 176; for learning differences, 237–238; minimum grading, 234–236; of multiple-choice question assessments, 191; PLCs and, 296; point accumulation system for, 236–237; policy for, 231–242; problems with, 239–240; for projects, 359, 364–367; research on, 232; standards-based, 232–233; teacher collaboration and, 288; technology for, 240; of tiered tests, 316; of writing, 85–86. *See also* Standards-based grading

Graphic organizers, for ELLs, 107, 343

Graphs: in Common Core, 9; for even and odd functions, 125; for functions, 103, 120, 125; for homework, 172; interdisciplinary collaboration with, 289; for linear regressions, 127–128; multiple mathematical methods for, 314; for probability tables, 105; in projects, 359; for rate of change, 121–122; for relative frequency, 129–130; for sines and cosines, 126

Grid-in-questions, 192

Grounding, 12

Growth mindset: from productive struggle, 342; for projects, 363; standards-based grading for, 233; stereotypes and, 19, 297

Guided questions, in lesson plan, 154–155, 173

H

Hammond, Zaretta, 28–29, 30–31, 38

Handouts: for assessments, 208, 217–229; for classroom environment, 255–256, 259, 261–269; for cooperative learning, 399, 401–404; for differentiated instruction, 321, 323–334; for grading, 240, 241–242; for homework, 176, 178, 183–188; for lesson plans, 157, 162–168; for mathematical communication, 89, 91–95; for mathematical connections, 108, 111–134; for mathematical language, 53, 57–61; on motivation, 18, 22–26; for parents, 281, 282–284; for PBL, 368, 369–377; for students with unique needs, 349, 351–352; for technology, 427, 429–432; for unit plans, 143, 145–150

Heinrich, A. K., 22

Higher-order thinking: cognitive routines for, 35; error analysis for, 81; multiple-choice question assessments and, 191; trust for, 33; unit plans for, 138

Historical context, for culturally responsive teaching, 33–34

Holt, Rinehart, and Winston glossaries, 55–56, 90

Home visits, 277–278

Homeschoolmath.net, 181

Homework: application of, 170–177; choosing wrong problems, 180; collection of, 175–176; Common Core Standards for, 170; as discovery, 173, 184–186; Do Now and, 153; for ELLs, 177, 181–182; flexibility with, 179; format for, 171–172; grading of, 176; handouts for, 176, 178, 183–188; for learning differences, 177; low-floor, high-ceiling problems in, 174; mathematical communication on, 71–72, 174–175; mismanagement of class time and, 179; online review of, 175; parents and, 178, 276; for PBL, 357–358; in point accumulation grading system, 236; as practice, 169, 172–173, 183; problems with, 178–180; prompts for, 173–174; quick writes with, 79; quiet time for, 178–179; references for, 172; research on, 169–170; review challenges with, 179–180; review of, 358; sources for, 177; students who don't do, 178–179; technology for, 180–182; terms in, 174; tiered tests in, 316; as transfer, 173–174, 187–188; whole-group review of, 175

How to Solve It (Pólya), 74, 90

I

I Teach Math, 299

Icebreakers, for mathematical communication, 66–68

IEPs. *See* Individualized Education Plans

Illustrative Mathematics, 368

Implicit biases, in "color-blind" teaching, 36–37

Individualized Education Plans (IEPs): grading for,

Individualized Education Plans (*continued*) 238; for students with unique needs, 348; unit plans and, 143
Individuation, 278
Information processing, in culturally responsive teaching, 35
Initiate-response-evaluate model, for formative assessment, 407
Institute for Mathematics and Education, of University of Arizona, 109
Intercepts, 103
Interdisciplinary collaboration, 288–289
International Association for the Study of Cooperation in Education, 321
Intersection, 105
Intrinsic motivation, 7–8; rewards and, 14
Involvement, by parents, 271
IRIS Center, at Vanderbilt University, 321

J

Al-jabr wa'l muqabalah (al-Khwarizmi), 34
Jigsaw Classroom, 321, 400
Jigsaws, 390–391, 401–402
JMAP, 215
Johnson, David, 379
Johnson, Roger, 379
Journal of Urban Math Education, 39

K

Kahneman, Daniel, 434
Kahoot!, 56; for formative assessments, 415; on homework, 182; on lesson plans, 159; on test review, 216
Khan Academy, 90; on end of school year, 443; for homework, 181; on lesson plans, 159
Al-Khwarizmi, Muhammad, 34
Kidblog, 322
Kidspeak, 51–52; open-ended questions and, 66
Klausmeier, H., 343
Knill, Oliver, 22
Kuta Software, 181

L

Ladson-Billings, Gloria, 27, 28–29
Language: in formative assessments, 413; simplified, 344–345, 347–348. *See also* English Language Learners; Mathematical language
Learning centers. *See* Centers
Learning differences, 335; artistic abilities of, 341; assessments for, 207–208; classroom environment and, 252–252; cooperative learning for, 386; co-teaching for, 291; of ELLs, 336; at end of school year, 441; formative assessments for, 414; Frayer models for, 343; grading for, 237–238; homework for, 177; lesson plan for, 155–156; manipulatives for, 346; mathematical communication for, 65, 87; mathematical connections with, 107; motivation for, 17–18; note-taking by, 255, 256, 257; parents of, 279–280; projects by, 364; strengths and challenges of, 339–340; student surveys for, 254; technology for, 350, 425; unit plans for, 139, 143
Learning Disabilities Online, 321, 350
Learning names, 248
Learning stations. *See* Stations
Lesson plans: aims in, 153; application of, 152–157; Common Core Standards for, 152; components of, 151; culturally responsive teaching in, 156, 158; Do Now in, 153–154, 162–166; for ELLs, 155–156, 158–159; engagement from, 151; fun in, 158; goals in, 152–153; guided questions in, 154–155, 173; handouts for, 157, 162–168; interdisciplinary collaboration with, 289; for learning differences, 155–156; practice in, 154; problems with, 157–159; reflection on, 159; research on, 152; scope of, 152–153; SEL in, 156; for students with unique needs, 340–341; summary in, 156–157, 168; teacher collaboration of, 287, 288; technology for, 159–161; *Think-Pair-Share* for, 156; writing of, 157
Lesson study, 294–295
Lesson Study Group, 299
Lesson summaries, 82–83, 85
Linear equations: in Common Core, 9, 30; homework on, 173; mathematical communication of, 65; open-ended questions on, 356
Linear functions: rate of change of, 102; variability in, 104
Linear regression, 127–128; variability in, 104
LINKS Learning, 350
Living Proof (Heinrich, Lawrence, Pons, & Taylor), 22

Long-term memory: culturally responsive teaching and, 29; end of school year and, 433; mathematical connections in, 108

Lortie, Dan, 285

Low-floor, high-ceiling problems: for homework, 174; for math anxiety, 247; for mathematical communication, 69

M

Major themes, in unit plans, 139

Manipulatives, 17; for mathematical communication, 69-70; for mathematical connections, 107; for students with unique needs, 345-346

Matching questions, 193

Matchwords, 55

Math and Stats, 110

Math anxiety, 1-2; autonomy and, 10; correspondence journal on, 84; culturally responsive teaching and, 35; low-floor, high-ceiling problems for, 247; mathematical language and, 41; motivation and, 8, 20; of parents, 273-276; pronunciation and, 43; self-efficacy and, 10; unit plans and, 142

Math Assessment Project, 368

Math Bits Notebook, 159

Math for America, 296

Math Forum, 400

The Math Lab, 21

Math League, 350

Math Twitter Blogosphere, 299

Mathalicious, 21

MathBitsNotebook, 110

MathBootCamps, 110

MathCounts, 109; for advanced students, 350

Mathematical communication, 63-95; application of, 64-70; of classwork, 71-72; Common Core Standards for, 64; discourse problems with, 88-89; for ELLs, 63, 87; Given, Find, Solve, and Check for, 74-79; guiding students in, 71-74; handouts for, 89, 91-95; on homework, 71-72, 174-175; icebreakers for, 66-68; for learning differences, 65, 87; low-floor, high-ceiling problems for, 69; manipulatives for, 69-70; of mathematical language, 64; open-ended questions for, 64-67, 71-72; problems with, 87-89; Problem-Solving Chart for, 74-79; prompts for, 72; research on, 64; sentence starters for, 72; with student mistakes, 87-88; in teacher collaboration, 287; with teacher mistakes, 88; technology for, 89-90; think-alouds for, 72-74; *Think-Pair-Share* for, 66, 71, 72; writing in, 79-87. *See also* Writing

Mathematical connections, 97-134; by advanced students, 340; application of, 98-107; Common Core Standards for, 98, 109; with ELLs, 107; with equivalence, 98, 99-100; with functions, 100, 102-104; handouts for, 108, 111-134; with learning differences, 107; motivation for, 108-109; problems with, 108-109; with proportionality, 100, 101-102; research on, 98; shortcuts for, 108-109; with variability, 100, 104-107

Mathematical language: anchor charts for, 53, 54; application of, 42; as "bag of tricks," 54-55; Common Core Standards for, 42, 55; concept attainment strategy for, 43; confusing examples of, 46-48; flash cards for, 49-51; handouts for, 53, 57-61; math anxiety and, 41; mathematical communication of, 64; need for, 42-433; precision with, 48; problems with, 53-55; real-world applications for, 42; research on, 42; technology for, 55-56; in unit plans, 143; verbal aids for, 51-52; visual aids for, 51-52; vocabulary charts for, 49-51; word walls for, 52-53, 54, 56. *See also* Symbols; Terms

Mathematicians Project, 259

Mathematics for English Language Learners, 350

MathPapa, 212

Mathslang, 51-52; with open-ended questions, 66

MathType, 90

Mathway, for cheating, 212

Mean Girls (film), 15

Mean Proportional Theorem, 173, 185

Meaningfulness, motivation and, 10

Memory. *See* Long-term memory

Mentoring, 294

Merriam-Webster Dictionary, 55

Metacognition: in cooperative learning, 393; at end of

Metacognition: (continued) school year, 438; from Problem-Solving Chart, 76; from reflection on mistakes, 206
Micromanaging, 258
Microsoft Excel, 240
Microsoft Office: Templates, on grading, 240; for unit plans, 145
Microsoft Outlook, 145
Microsoft PowerPoint, 90; for lesson plans, 160; on test review, 216
Microsoft Translate, 281
Microsoft Windows 10, 160
Microsoft Word: for lesson plans, 160; for unit plans, 145
Milestones, for projects, 362
Mills College, Lesson Study Group at, 299
Mindset: for culturally responsive teaching, 30, 38. *See also* Growth mindset
Minimum grading policy, 234–236
Motivation: classroom rules and, 251; Common Core Standards for, 9; for ELLs, 16–17; extrinsic, 7–8; from fear, 18; in flipped classrooms, 421; handouts on, 18, 22–26; intrinsic, 7–8; knowledge void recognition for, 13–14; for learning differences, 17–18; limitations to, 21; with math, 11–14; math anxiety and, 8; for mathematical connections, 108–109; from minimum grading, 234; misreading students and, 20–21; for non-real-world applications, 19–20; from note-taking, 254; parents and, 272; pattern recognition for, 12–13; with popular culture, 15–16; with real-world applications, 11–12; research on, 8–9; retrieval practice for, 13; rewards for, 14–15; SEL and, 8; stereotypes and, 19; from teacher collaboration, 286, 287; technology for, 21–22; value internalization for, 8
MTBoS, 110
Multiple intelligences, 304, 315
Multiple mathematical methods, 314
Multiple-choice questions, 156; for advanced students, 349; on assessments, 190–191; at end of school year, 436–437
Multiplication, area model for, 99–100, 112–113

N

Names, learning, 248
NASA CONNECT, 428
Nash, John, 16
National Assessment of Educational Progress, 28
National Association of Special Education Teachers, 321, 350
National Center for Research on Gifted Education, 321
National Council of Teachers of Mathematics (NCTM): on lesson plans, 159; on *Notice and Wonder*, 400; PLCs and, 296; on teacher collaboration, 299
National Library of Virtual Manipulatives, 21, 428
National School Reform Faculty, 400
NCTM. *See* National Council of Teachers of Mathematics
Nearpod, 415
Neurodiversity, 339
New Visions for Public Schools, 110
New York City Interscholastic Mathematics League, 350
New York State Education Department, 39; on assessments, 215; ELLs and, 350; on homework, 182; on mathematical language, 55; on teacher collaboration, 299
Nix the Tricks, 110
Non-real-world applications, motivation for, 19–20
Nonroutine questions, in PBL, 354
Notebooks, 255–256; parents and, 276
Note-taking, 254–255; Cornell system for, 257, 260; by ELLs, 256–257; by learning differences, 255, 256, 257, 339; parents and, 276; problems with, 259
Notice and Wonder: for cooperative learning, 388–390; National Council of Teachers of Mathematics on, 400

O

Odd functions, 103–104, 124–125; multiple intelligences and, 315
Office, Microsoft: Templates, on grading, 240; for unit plans, 145
One teach/one assist, for co-teaching, 291–292
Open-ended questions: constructed-response questions as, 193; for mathematical communication, 64–67, 71–72; in PBL, 354, 355–358; wait time for, 88
Outlook, Microsoft, 145
Owens, Kate, 240
Oxford Dictionary, 55

P

Pacing calendar: of curriculum, 139; of unit plans, 144
Paired dictation, 44
Parabolas, homework on, 173, 186
Parallel tasks (tiered activities), 306–309, 323–327
Parallel teaching, 291–202
PARCC, 215–216
Parents: building relationships with, 271–299; "color-blind" teaching and, 36; Common Core Standards and, 272; communication with, 272–275, 282–284; course descriptions and, 252–253; culturally responsive teaching with, 278–279; of ELLs, 273; at end of school year, 440; engagement by, 271; grading and, 240; handouts for, 281, 282–284; home visits with, 277–278; homework and, 178, 276; involvement by, 271; of learning differences, 279–280; math anxiety of, 273–276; motivation from, 8; PLCs and, 296; problems with, 280–281; research on, 272; teacher collaboration and, 288; technology for, 281
Parent-teacher conferences, 277
Parent-Teacher Home Visit Project, 278, 281
Pattern recognition: for functions, 102; for motivation, 12–13; technology for, 423–424
Pauk, Walter, 257
PBL. *See* Project-based learning
PBLWorks, 368
Peak-end effect, 434
Peer coaching, PLCs and, 296
Peer editing, for cooperative learning, 394–395, 404
Peer grading, 203
Perfectionism, motivation and, 20
Performance-based learning, 354
PhotoMath: for cheating, 212; for quick writes, 79
Piaget, Jean, 354
Picture dictation, 48
"The Pit and the Pendulum" (Poe), 16
Planning. *See* Lesson plans; Unit plans
PLCs. *See* Professional learning communities
Plickers: for formative assessments, 415; on lesson plans, 159; on test review, 216
Poe, Edgar Allan, 16
Point accumulation grading system, 236–237
Pólya, George, 74, 90
Polygons, similar areas of, 117
Polynomials: addition and subtraction of, 111; equivalence of, 99–100; Exploding Dots for, 109; functions of, 103, 123–124; unit plans for, 140
Poorvu Center for Teaching and Learning, of Yale University, 260
Popular culture, motivation with, 15–16
Portfolios: in differentiated instruction, 317; at end of school year, 438
Powerpoint, Microsoft, 90; for lesson plans, 160; on test review, 216
Practice: for assessments, 196–199; homework as, 169, 172–173, 183; jigsaws as, 391, 401; in lesson plan, 154; stations for, 393; technology for, 424. *See also* Retrieval practice
Praise: for advanced students, 340; in culturally responsive teaching, 31
Precision, with mathematical language, 48
Presentation style, for differential instruction, 309
Preview-view-review, for ELLs, 177
Probability: Common Core Standards for, 109; of relative frequency, 129–130; tables for, 104–107, 110; two-way tables for, 131–134
Problem solving, self-determination theory and, 9
Problem-Attic, 216
Problem-Solving Chart, 74–79
Procedures: in classroom environment, 250, 251–252, 287; in co-teaching, 293; skills in, in test questions, 195–196; technology for, 422–423
Process, differentiated instruction by, 313–314
Product, differentiated instruction by, 315–320
Productive struggle, 342–343
Professional learning communities (PLCs), 295–296
Project Management Institute Educational Foundation, 368
Project-based learning (PBL): classwork for, 355–357; Common Core Standards for, 355; in culturally responsive teaching, 35; in curriculum compacting, 313; handouts for, 368,

Project-based learning (PBL): (*continued*)
369–377; homework for, 357–358; open-ended questions in, 354, 355–358; problems with, 367; projects in, 236, 358–367; research on, 354–355; technology for, 368

Projects: by advanced students, 364; for assessment, 208; differentiated instruction for, 359, 363–364; by ELLs, 363–364; at end of school year, 437; feedback for, 359, 362–363; grading for, 359, 364–367; individual and group, 360; by learning differences, 364; milestones for, 362; in PBL, 236, 358–367; questions for, 358, 359–361; rubrics for, 364–367; scope of, 358, 359

Prompts: for homework, 173–174; for mathematical communication, 72; for writing, 80–81

Pronunciation: math anxiety and, 43; of symbols, 43–44, 47, 49, 55; of terms, 43–44

Proof Blocks, 21

Proportionality: in circles, 101–102, 120; MathCounts for, 109; mathematical connections with, 100, 101–102; rate of change of, 102; YouCubed for, 109

Protocols, for cooperative learning, 383

Pythagoras, 142

Q

Quadratic equations: area model for, 100; multiple intelligences and, 315; in projects, 359

Questions: always-sometimes-never, 192–193; on assessments, 196, 215–216; in cooperative learning, 383–384; in curriculum compacting, 313; fill-in-the-blank, 193; in flipped classrooms, 422; for formative assessments, 407–409; grid-in, 192; guided, in lesson plan, 154–155, 173; matching, 193; nonroutine, in PBL, 354–357; poorly chosen, for assessments, 210–211; for projects, 358, 359–361; for reteaching assignments, 311; skills in, in assessments, 195–196; in tiered activities, 306, 308; in tiered tests, 315–317; true-false, 192–193; types of, in assessments, 190–193; "what-if," for projects, 360. *See also* Constructed-response questions; Open-ended questions

Question order: in assessments, 194–195; in presentation style, 309

Quick Rubric, 322

Quick writes, 79–80, 86

Quiet time, for homework, 178–179

Quizlet, 56; for formative assessments, 415; for homework, 182; for test review, 216

Quizzes: Nearpod for, 415; in point accumulation grading system, 236; for students with unique needs, 344. *See also* Assessments

R

Radical Math, 39
Rate of change, 102, 121–122

Ratios: concept map for, 344; homework on, 173, 184; multiple mathematical methods for, 314; PBL for, 357; probabilities and, 105; similarity and, 116; YouCubed for, 109

Rcampus, 322

Real-world applications: for functions, 102; for mathematical language, 42; motivation with, 11–12; open-ended questions on, 356–357; for projects, 358, 359; for writing, 84–85

Reflection: on assessments, 206, 229; in cooperative learning, 383; at end of school year, 438–439; on lesson plans, 159; in lesson study, 294. *See also* Metacognition; Self-reflection

Region X Equity Assistance Center, 39

Reis, Sally, 312–313

Relatedness: motivation and, 10; self-determination theory and, 8

Relative frequency, 129–130

Relays, for cooperative learning, 395–396

Remind, for parents, 281

Rent (musical), 15

Renzulli, Joseph, 312–313

Reteaching assignments, 310–312

Retrieval practice, 80; active learning for, 309; in Do Now, 153; at end of school year, 435; with homework, 173; for motivation, 13

Review: by advanced students, 340; for assessments, 196–199, 210, 216; centers for, 394; by ELLs, 357; at end of school year, 433, 434–438; in flipped classrooms, 421; of

homework, 358; material, in differentiated instruction, 317–318; stations for, 393; in unit plans, 144
Review booklets, 318, 333–334
Review sheets, 318, 331–332
Rewards: for motivation, 14–15. *See also* Extra credit
Routines: in classroom environment, 250, 251–252, 287; in culturally responsive teaching, 35; for SEL, 347
Rubrics, 199; for cooperative learning, 385; for differentiated instruction, 318–319; for projects, 364–367; for tiered tests, 316, 319
Rules: in classroom environment, 250–251, 287; in co-teaching, 293
Rules of Netiquette, 428
Ryan, Richard M., 7

S

Scaffolding: in culturally responsive teaching, 35; in differentiated instruction, 304; in lesson summaries, 82–83; in reteaching assignments, 310; for students with unique needs, 341
Scholastic, 260
Schrock, Kathy, 368, 400
Scientific notation, in Common Core, 9
Scoring guidelines: for assessments, 199–201, 202–203, 225; for formative assessments, 413; for standards-based grading, 233
"Seasons of Love" (song), 15
Sectors, 119

SEL. *See* Social-emotional learning
Self-determination theory, 7, 8; culturally responsive teaching and, 10; decision making and, 9; problem solving and, 9
Self-efficacy, 8; differentiated instruction and, 305; math anxiety and, 10; stereotypes and, 19
Self-esteem: ability grouping and, 310; from classroom environment, 245; culturally responsible teaching and, 27; motivation and, 7
Self-reflection: on behavioral problems, 31–32; in culturally responsive teaching, 28, 30–32, 38; strategy for, 32; in teacher collaboration, 287, 298
Sentence starters, 72
Shake Up Learning, 415
Shea, Virginia, 428
Shortcuts: of advanced students, 340; for mathematical connections, 108–109; MTBoS and, 110
Similarity: homework on, 187; MathCounts for, 109; of polygon areas, 117; proportionality and, 101; ratios and, 116; of solid volume, 118
Simplifying language: for ELLs, 344–345; for SEL, 347–348; for students with unique needs, 344–345
Sine function, 103, 126–127
Slope: in linear regressions, 128; rate of change of, 102
SMART Notebook, 90
Snipping Tool, on Windows 10, 160
Social media: guidelines for, 428; for PLCs, 296

Social Science Statistics, 216
Social-emotional learning (SEL), 1; from cooperative learning, 380; for ELLs, 347; in flipped classrooms, 421; formative assessment for, 406; in lesson plans, 156; motivation and, 8; for students with unique needs, 337, 347–348; tiered activities and, 307; unit plans for, 141–142
Special education. *See* Learning differences
Stand and Deliver (film), 15
Standards-based grading, 232–233; for tiered activities, 308
Stanford Graduate School of Education, 350
Starch, Daniel, 232
Station teaching, 291
Stations, for cooperative learning, 392–394, 403
Statistics: Common Core Standards for, 30, 109; interdisciplinary collaboration with, 289; PBL for, 358; two-way tables for, 107
STatistics Education Web, 368
Stereotypes: motivation and, 19; in teacher collaboration, 297
Story of Mathematics, 39
Student Information Sheet, 248, 261–262
Student response systems. *See* Classroom response systems
Student safety, 2
Student surveys: on classroom environment, 253–254, 260; at end of school year, 438
"Students will be able to..." (SWBAT), 153

Students with unique needs: academic achievement of, 337; Common Core Standards for, 337; concept maps for, 343–344, 352; cooperative learning for, 384–386; differentiated instruction for, 335–352, 384–386; Frayer models for, 343; handouts for, 349, 351–352; IEPs for, 348; lesson plans for, 340–341; manipulatives for, 345–346; problems with, 348–349; productive struggle for, 342–343; research on, 336–337; scaffolding for, 341; SEL for, 337, 347–348; simplifying language for, 344–345; strengths and challenges of, 337–340; techniques for, 340–348; technology for, 349–350; unit plans for, 341. *See also* Advanced students; English Language Learners; Learning differences

SugarSync, 428; on teacher collaboration, 299

Summary, in lesson plans, 156–157, 168

Summative assessment, 405

SurveyMonkey, 260; on end of school year, 443

SWBAT. *See* "Students will be able to . . ."

Symbols: ELLs and, 45; in English, 46; EquatIO for, 90; introduction of, 43–45; in lesson plans, 156; MathType for, 90; paraphrasing, 45; pronunciation of, 43–44, 47, 49, 55; translating to terms, 45

Sypnieski, Katie Hull, 3; on picture dictation, 48

T

Tables: for even and odd functions, 125; for functions, 120, 125; interdisciplinary collaboration with, 289; multiple mathematical methods for, 314; for probability, 104–107, 110; for rate of change, 121–122; for sines and cosines, 126–127. *See also* Two-way tables

Talented students. *See* Advanced students

TalkingPoints, 281

Tanton, James, 109

Task cards, 396–397

Taylor, Megan Westwood, 143

"Teach to the test," 189

Teacher collaboration, 285–299; application of, 286–296; beliefs in, 297; Common Core Standards and, 286; co-teaching, 291–293; feedback in, 298; goals of, 286; interdisciplinary, 288–289; of lesson plans, 287, 288; lesson study, 294–295; mentoring, 294; motivation from, 286, 287; observing other teachers, 289–291; PLCs, 295–296; problems with, 297–298; research on, 286; self-reflection in, 287, 298; stereotypes in, 297; technology for, 288, 299; trust in, 286, 297; of unit plans, 287, 288; values for, 287–288

Teacher's clothing, 247

Teacher-student conferences, 438–439

Teacher-student relationship. *See* Classroom environment

TeacherVision: for parents, 281; on students with unique needs, 350

Teaching Diverse Learners, 38–39

TeachThought, 299

Team teaching, 291

Technology, 417–432; for advanced students, 349–350; for assessments, 215–216; behavioral problems and, 420; for classroom environment, 259–260, 418–422; Common Core Standards for, 418; content and, 422–424; for cooperative learning, 400; for culturally responsive teaching, 38–39, 425; for deliberate practice, 424; for differentiated instruction, 321–322, 425; for ELLs, 425; for end of school year, 443; in flipped classrooms, 420–422; for formative assessments, 415; for grading, 240; handouts for, 427, 429–432; for homework, 180–182; for learning differences, 350, 425; for lesson plans, 159–161; for mathematical communication, 89–90; for mathematical connections, 109–110; for mathematical language, 55–56; for motivation, 21–22; for parents, 281; for pattern recognition, 423–424; for PBL, 368; for practice, 154; problems of, 425–427; for procedures, 422–423; research on, 418; for students with unique needs, 349–350; for teacher collaboration, 288, 299; for unit plans, 145

Terms: for ELLs, 49; in English, 46; in homework, 174; introduction of, 43–45; in lesson plans, 156;

pronunciation of, 43–44; translating to symbols, 45
Test reflection form, 206, 229
Tests: in point accumulation grading system, 236; tiered, 315–317, 331. *See also* Assessments
Texas Education Agency: on homework, 182; on mathematical language, 55
Think-alouds: with ELLs, 74; for mathematical communication, 72–74
Think-Pair-Share: for cooperative learning, 387; for formative assessment, 413, 414; for lesson plans, 156; for mathematical communication, 66, 71, 72; with open-ended questions, 66
ThoughtCo, 260; on Frayer model, 350; for parents, 281
Tiered activities (parallel tasks), 306–309, 323–327
Tiered tests, in differentiated instruction, 315–317, 331
Tomlinson, Carol Ann, 305
Tracking, 310
Transfer learning, 97
Translate, Microsoft, 281
True-false questions, 192–193
Trust: in calculators, 427; in co-teaching, 293; in culturally responsive teaching, 33; with differentiated instruction, 303; in teacher collaboration, 286, 297
Turn-and-talk, 71
Twice-exceptional children, 336
Twitter, 296, 299
Two-way tables (contingency tables), 105–107; for probability, 131–134

U

Underground Mathematics, 109
Union, probability of, 105
Unit plans, 137–150; application of, 138–140; Common Core Standards for, 138; concept maps in, 140, 147; for culturally responsive teaching, 142–143; for ELLs, 139, 143; flexibility of, 141; handouts for, 143, 145–150; for learning differences, 139, 143; logical sequence for, 140, 148; major themes in, 139; mathematical language in, 143; organization of topics and problems in, 141; pacing calendar of, 144; problems with, 143–144; re-evaluations of, 144; research on, 138; sample of, 149–150; for SEL, 141–142; for students with unique needs, 341; summary for, 141; teacher collaboration of, 287, 288; technology for, 145
University of Arizona, Institute for Mathematics and Education of, 109
US and Latin American Prime Factorization Methods, 338
Using English for Academic Purposes, 55

V

Values: internalization, for motivation, 8; for teacher collaboration, 287–288
Vanderbilt University: Center for Teaching at, 350, 400; IRIS Center at, 321
Variability: mathematical connections with, 100, 104–107; probability tables for, 104–107
Venn diagrams, 131
Verbal aids: for functions, 102; for mathematical language, 51–52
Virginia Department of Education, 56
Visual aids, for mathematical language, 51–52
Vocabulary: Frayer model for, 343. *See also* Terms
Vocabulary charts: for differentiated instruction, 318; for mathematical language, 51
Vygotsky, Lev, 304

W

Wait time: for guided questions, 154; for open-ended questions, , 65–66, 88
Warm-Up. *See* Do Now
"We the People" National Alliance, 39
Weebly for Educators, 281
Weinstein, Yelena, 240
Well-being: motivation and, 7; unit plans and, 138
What Do You Notice? What Do You Wonder?. *See* Notice and Wonder
"What-if" questions, for projects, 360
Which One Doesn't Belong (Danielson), 70, 89
Whole numbers, area model for, 100
Wikipedia, 110
Windows 10, Snipping Tool on, 160
Wisneski, Debora, 32
Women You Should Know, 259
Wong, Harry K., 252, 259
Wong, Rosemary T., 252, 259

Word, Microsoft: for lesson plans, 160; for unit plans, 145
Word bank, 193
Word problems: for ELLs, 81; table for, 105–106; in unit plans, 140; writing of, 81
Word walls, for mathematical language, 52–53, 54, 56
WordPress, 281
Writing: by advanced students, 340; correspondence journal for, 84; by ELLs, 339; on equivalence, 99; error analysis for, 81–82; finding time for, 89; grading of, 85–86; of guided questions, 154; of lesson plans, 157; lesson summaries for, 82–83, 85; of linear regressions, 104; in mathematical communication, 79–87; of multiple-choice question assessments, 191; note-taking, 254–257; quick writes for, 79–80, 86; real-world applications for, 84–85; of word problems, 81

Y

Yale University, Poorvu Center for Teaching and Learning of, 260
YouCubed, 109

Z

Zapier, 260
Zeigarnik Effect, 195
ZipGrade: for assessments, 216; for formative assessments, 415
Zoho Notebook, 260
Zone of proximal development, 304
ZooWhiz, 181